Basic Concepts in Enhanced Oil Recovery Processes

Critical Reports on Applied Chemistry Volume 33

Basic Concepts in Enhanced Oil Recovery Processes

Edited by M. Bavière

SCI

Published for SCI
by
ELSEVIER APPLIED SCIENCE
LONDON and NEW YORK

ELSEVIER SCIENCE PUBLISHERS LTD
Crown House, Linton Road, Barking, Essex IG11 8JU, England

Sole distributor in the USA and Canada
ELSEVIER SCIENCE PUBLISHING CO., INC.
655 Avenue of the Americas, New York, NY 10010, USA

WITH 14 TABLES AND 169 ILLUSTRATIONS

British Library Cataloguing in Publication Data

Basic concepts in enhanced oil recovery processes.
1. Petroleum. Production
I. Bavière, M. II. Series
622.3382
ISBN 1-85166-617-6

Library of Congress Cataloging-in-Publication Data applied for

Printed by The Universities Press (Belfast) Ltd.

Preface

Oil is accumulated, together with water and gas, in reservoirs consisting of porous and permeable rocks. When an oil reservoir is brought into communication with the surface through a drilling operation, part of the oil is spontaneously produced because of the natural pressure that exists in the reservoir. The amount of oil produced by the energy of the reservoir itself, as well as by pumping of individual wells to assist the natural drive, is known as *primary oil recovery*. At the end of this depletion period, the recovery factor is low, usually lower than 15% of the original oil in place (OOIP). For improving oil recovery, water or gas can be injected in order to maintain or to restore the pressure and to flood the reservoir. These are *secondary oil recovery* processes.

However, these now conventional techniques still leave in the ground about two-thirds of OOIP. This is a result of both unfavorable reservoir characteristics, such as high oil viscosity or rock heterogeneity, which are responsible for areas of the reservoir that have not been flooded by the injected fluid, and the poor microscopic displacement efficiency of water and gas due to capillary forces which are highly active in such porous media. Therefore, methodologies have been developed to increase both the oil production rate and the recovery factor. Due to the great diversity of oil reservoirs, there is no single method to enhance oil recovery, but a group of methods specifically addressed to the various reservoir types. They are called *enhanced oil recovery* (EOR) processes.

A definition of EOR was given in the report *Enhanced oil recovery* published by the US National Petroleum Council in December 1976: 'Enhanced oil recovery is the additional recovery of oil from a petroleum reservoir over that which can be economically recovered by conventional primary and secondary methods.' This is a rather broad concept which could include any new technologies helping to produce more oil under economical conditions.

The definition adopted in this book is technically more restrictive: EOR consists of methods aimed at increasing ultimate oil recovery by injecting appropriate agents not normally present in the reservoir, such as chemicals, solvents, oxidizers and heat carriers, in order to induce new mechanisms for displacing oil. This definition excludes the pressure maintenance by water or gas injection, which uses mechanical energy alone. It also excludes well-stimulation techniques, that affect the vicinity of the wellbore only, as well as infill drilling and horizontal drilling, both considered here as being complementary to EOR. Nevertheless, the borderlines may sometimes be imprecise, for instance when gas injection is concerned.

Oil remaining trapped after primary and secondary recovery is the target of EOR. Therefore EOR methods are often referred to as tertiary recovery processes. However, their use is not restricted to a particular phase in the production life of a reservoir. For field operations at the pilot scale, it is often recommended that EOR methods be tested under tertiary conditions. This is the best way to evaluate the amount of incremental oil displaceable by EOR. However, for field-scale operations, it may be more advisable to start the implementation of EOR at an early stage of production.

Considerable amounts of oil are concerned by EOR: probably more than 100 milliard tonnes (100×10^9 tonnes). As a matter of fact, the challenge of EOR is to raise the oil recovery factor from about 30% to 45% of OOIP. In the 1990s, 1% of the total amount of the oil discovered, estimated to be 700 milliard tonnes, roughly represents world consumption over 2 years.

The development of EOR is strongly dependent on both technical and economic considerations. Indeed, EOR is an ongoing technology open to great improvements, but the economic conjuncture may postpone the development of the most expensive methods. Thus, speculations are extremely hazardous. However, it is generally recognized that oil will remain a major and difficult-to-replace source of energy, especially for transportation, and of raw materials for petrochemicals. 'Oil will remain the leading energy source during the first half of the 21st century' was a conclusion of the World Energy Conference held in Montreal in September 1989.

As a consequence, the development of EOR technology appears to be quite probable, in so far as significant and technical advances continue to improve its efficiency. This development should particularly concern those countries that have relatively small oil reserves recoverable by conventional techniques and which depend on foreign supplies for their energy.

Most principles applied in EOR methods have been known for a long time. Numerous laboratory studies and field pilots have been carried out, starting in the 1960s. However, the increasing demand for energy, and especially the oil crises of 1973 and 1979, have given rise to intense activity in EOR research and development during the last two decades. This has resulted in a great increase in knowledge and the understanding of mechanisms involved in EOR processes. This book aims at summarizing these achievements, which provide the process design with a fundamental basis leading to better mastery of the methods.

Chapter 1 is an introduction devoted to considerations on the entrapment and mobilization of oil as well as on the basic principles of EOR methods. The principal methods are classically divided into three groups, namely chemical, miscible and thermal methods, with each being characterized by the search for particular physical effects.

Processes that call upon chemical additives are presented in Chapter 2 (polymers), Chapter 3 (surfactants) and Chapter 4 (alkaline agents). These methods can be seen as a single method: the polymer–surfactant–alkali process, displaying multiple facets resulting from various combinations and implementation designs of chemicals. When applied to appropriate field cases, this is the most efficient method, but also the one that still requires the greatest effort of research and field testing to reduce the cost–efficiency ratio.

The next three chapters are then assigned to miscible processes. As a preliminary, Chapter 5 describes the thermodynamics considered to be the common background to any type of miscible process. Chapter 6 deals essentially with hydrocarbon-gas flooding, but also includes nitrogen and flue gas injection, and considerations about diffusion–dispersion phenomena. Chapter 7 is devoted to carbon dioxide flooding.

Lastly, thermal processes are presented. Chapter 8 examines the chemistry of crude oil alterations due to the increase of temperature. Chapter 9 concerns the injection of hot fluids, including their effects on fluid and rock–matrix physical properties. General considerations on foams, used for mobility control, are also given in this chapter. Chapter 10 is devoted to in-situ combustion.

All the above-mentioned EOR methods have been technically proved on a field scale. About 75 million tonnes of oil were produced by EOR in 1989, mainly by using miscible gas flooding and steam injection, which are the most advanced processes. This represents approximately 2% of the world production.

Other techniques have been put forward for enhancing oil recovery. Some are based on the same mechanisms as those involved in chemical or solvent flooding (this is the case for the microbial enhanced oil recovery technique) or thermal recovery, such as electric and electromagnetic heating. Some others are related more to extraction techniques (oil mining) or well stimulation (nuclear stimulation). These techniques are described in Chapter 11.

Finally, Chapter 12 gives a general overview concerning the methodology of an EOR project. This is illustrated by a synthetic presentation of the screening criteria of the main EOR methods to help in reaching a decision to implement a process.

The book emphasizes the basic phenomena involved in EOR. Therefore, details concerning mathematics, numerical modeling, experimental procedures and pilot tests will mostly be accessible through the references that are estimated to be particularly significant, dealing with basic findings, new design concepts, recent reviews and illustrative field operations. Also, the impacts of EOR research and development in the field of applied chemistry are indicated where appropriate.

In addition to their own contribution, the authors have greatly contributed to improving and harmonizing the book as a whole by their comments and critics. All are here sincerely acknowledged, as well as G. Brace for his thorough revision and editing of the text.

MARC BAVIÈRE

Contents

Preface v

List of Contributors xi

List of Abbreviations xiii

INTRODUCTION TO ENHANCED OIL RECOVERY

1 **Oil entrapment and mobilization** 3
C. M. Marle

IMPROVED WATERFLOOD: THE CHEMICAL PROCESSES

2 **Mobility control by polymers** 43
G. Chauveteau and K. S. Sorbie

3 **Reduction of capillary forces by surfactants** 89
G. A. Pope and M. Bavière

4 **The use of alkaline agents in enhanced oil recovery processes** 123
J. Labrid

MISCIBLE GAS DISPLACEMENT

5 **Thermodynamics** 159
F. M. Orr, Jr

6 **Miscible gas flooding** 185
D. Morel

7 **Carbon dioxide flooding** 215
M. A. Klins and C. P. Bardon

THERMAL RECOVERY

8 **Thermo-chemical alteration of crude oils** 243
A. Audibert and S. Vossoughi

9 **Hot fluid injection** 271
C. Gadelle and L. M. Castanier

10 **In-situ combustion** 313
J. Burger

UNCONVENTIONAL EOR TECHNOLOGIES

11 **Unconventional EOR concepts** 341
G. J. Stosur

CONCLUSIONS ON THE MAIN EOR PROCESSES

12 **Methodology and screening criteria—implementation of EOR processes** 377
P. Chaumet and R. H. Cottin

Index 403

List of contributors

Annie Audibert Institut Français du Pétrole, BP 311, 92506 Rueil-Malmaison Cedex, France

Charles P. Bardon Institut Français du Pétrole, BP 311, 92506 Rueil-Malmaison Cedex, France

Marc Bavière Institut Français du Pétrole, BP 311, 92506 Rueil-Malmaison Cedex, France

Jacques Burger Comité d'Etudes Pétrolières et Marines, Tour Franklin, 92081 Paris La Défense Cedex 11, France and Institut Français du Pétrole, BP 311, 92506 Rueil-Malmaison Cedex, France

Louis M. Castanier Petroleum Engineering Department, Stanford University, Stanford, California 94305-2224, USA

Pierre Chaumet Institut Français du Pétrole, BP 311, 92506 Rueil-Malmaison Cedex, France

Guy Chauveteau Institut Français du Pétrole, BP 311, 92506 Rueil-Malmaison Cedex, France

René H. Cottin Société Nationale Elf Aquitaine, Chemin Vignancourt, 64000 Pau, France

Claude Gadelle Institut Français du Pétrole, BP 311, 92506 Rueil-Malmaison Cedex, France

Mark A. Klins Chevron USA Inc., 1923, North Dal Paso Street, Hobbes, New Mexico 88240, USA

Jean Labrid Institut Français du Pétrole, BP 311, 92506 Rueil-Malmaison Cedex, France

Charles M. Marle Département de Mathématiques, Université Pierre et Marie Curie, 4 Place Jussieu, 75282 Paris Cedex 05, France and Institut Français du Pétrole, BP 311, 92506 Rueil-Malmaison Cedex, France

Danielle Morel Institut Français du Pétrole, BP 311, 92506 Rueil-Malmaison Cedex, France

Franklin M. Orr, Jr Petroleum Engineering Department, School of Earth Sciences, Stanford University, Stanford, California 94305-2224, USA

Gary A. Pope Department of Petroleum Engineering, The University of Texas at Austin, Austin, Texas 78712, USA

Kenneth S. Sorbie Department of Petroleum Engineering, Heriot-Watt University, Riccarton, Edinburgh EH14 4AS, UK

George J. Stosur US Department of Energy, Washington, District of Columbia 20545, USA

Shapour Vossoughi Chemical and Petroleum Engineering Department, University of Kansas, 4006 Learned Hall, Lawrence, Kansas 66045, USA

List of abbreviations

CEC	Cation exchange capacity (Chapter 4, page 135)
CMC	Critical micelle concentration (Chapter 3, page 93)
COSTALD	Corresponding states liquid density (Correlation) (Chapter 5, page 173)
DSC	Differential scanning calorimetry (Chapter 8, page 257)
DTA	Differential thermal analysis (Chapter 8, page 257)
DTG	Derivative of thermogravimetric analysis curve (Chapter 8, page 257)
EGA	Effluent gas analysis (Chapter 8, page 257)
EOR	Enhanced oil recovery (Preface)
EOS	Equation of state (Chapter 5, page 162)
GOR	Gas/oil ratio (Chapter 7, page 220)
HCPV	Hydrocarbon pore volume (Chapter 7, page 236)
HEC	Hydrogen exchange capacity (Chapter 4, page 139)
HPAM	Hydrolyzed polyacrylamide (Chapter 2, page 45)
IFT	Interfacial tension (Chapter 4, page 123)
LPG	Liquefied petroleum gas (Chapter 6, page 200)
LTO	Low temperature oxidation (Chapter 8, page 259)
MEOR	Microbial enhanced oil recovery (Chapter 11, page 342)
MMC	Minimum miscibility composition (Chapter 6, page 200)
MMP	Minimum miscibility pressure (Chapter 6, page 190)
OOIP	Original oil in place (Preface)
PAM	Polyacrylamide (Chapter 2, page 45)
PR	Peng–Robinson's equation of state (Chapter 5, page 169)
PV	Pore volume (Chapter 6, page 195)
RK	Redlich–Kwong's equation of state (Chapter 5, page 169)

SBR	Simonnet–Behar–Rauzy's equation of state (Chapter 5, page 169)
SRK	Soave–Redlich–Kwong's equation of state (Chapter 5, page 169)
TGA	Thermogravimetric analysis (Chapter 8, page 257)
VDW	van der Waals's equation of state (Chapter 5, page 169)
WAG	Water alternate gas (Process) (Chapter 7, page 222)
WGSR	Water gas shift reaction (Chapter 8, page 251)
WOR	Water/oil ratio (Chapter 3, page 101)

INTRODUCTION TO ENHANCED OIL RECOVERY

1 Oil entrapment and mobilization

Charles M. Marle
Université Pierre et Marie Curie, Paris, France

1.1	**What is enhanced oil recovery?**	3
1.1.1	General features of an oil field	3
1.1.2	The three classical steps in oil recovery	5
1.1.3	Enhanced oil recovery	9
1.2	**Oil displacement by another fluid: the causes of non-total recovery**	10
1.2.1	The scale of an elementary block	10
1.2.2	The scale of a flow tube	21
1.2.3	The scale of the reservoir	24
1.3	**How to improve the factors contributing to oil recovery**	34
1.3.1	Oil displacement in an elementary block	34
1.3.2	Transition zones in a flow tube	35
1.3.3	Sweep efficiency at the scale of the reservoir	35
1.3.4	Basic principles of enhanced oil recovery methods	36
1.4	**References**	37

1.1 What is enhanced oil recovery?

1.1.1 *General features of an oil field*

An oil field is a natural hydrocarbon accumulation in the pores of underground porous rocks, which are called *reservoir rocks,* or simply reservoirs. An oil field may contain several separate reservoirs. The phenomena which lead to the creation of an oil field are very complicated and outside the scope of this book.[1] However, for a proper understanding of the principles involved in oil recovery, it may be useful to have some rough ideas about these phenomena.

In sedimentary basins, layers of sedimentary materials are successively deposited. Under the weight of the overlying layers, they sink to increasing depths. The resulting increase in pressure and temperature induces various chemical reactions; under favorable conditions, organic materials contained in the sediments are transformed into hydrocarbons. These hydrocarbons are, at first, widely disseminated in the rock in which they were created. Under some circumstances, they are expelled from that rock; this stage is called *primary migration*. After traveling through the neighboring rocks, called *secondary migration*, they accumulate in a reservoir rock and form an oil field.

Several mechanisms are involved in primary and secondary migration: compaction of the rock, diffusion, hydrodynamic flow. It seems that, at least in the last stages of secondary migration, the movement of hydrocarbons is mainly due to hydrodynamic flow, driven by buoyancy forces. This hypothesis is supported by the following observation: almost all known natural accumulations of hydrocarbons are encountered in more or less permeable rocks covered by non-permeable rocks, near the upper part of these permeable rocks. It might be supposed that hydrocarbons, which are lighter than the water that initially filled all the pore space, were driven into the reservoir by buoyancy. They accumulated against the covering impermeable rock, which prevented them from flowing any farther upwards.

In a typical oil field (Fig. 1.1), gaseous hydrocarbons, which are the lightest, occupy the highest part of the reservoir; this part of the field is called the *gas cap*. The part just underneath is occupied by liquid hydrocarbons; it is called the *oil zone*. Still deeper, the pores of the reservoir rock contain only water; this area is called the *aquifer*.

The processes that lead to the formation and accumulation of hydrocarbons in sedimentary basins take place in water-saturated

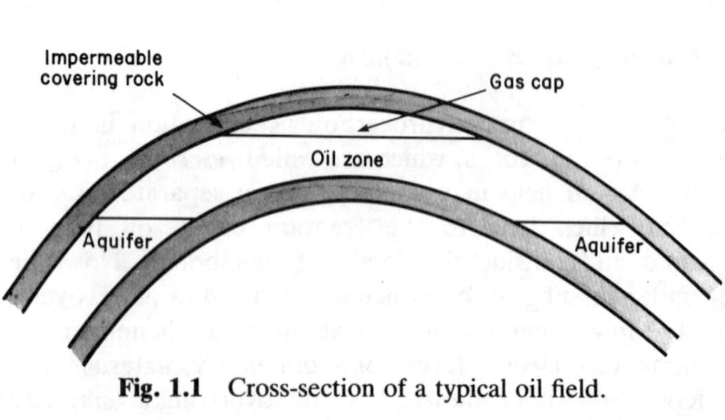

Fig. 1.1 Cross-section of a typical oil field.

sediments. The water phase is, in most cases, not completely displaced by hydrocarbons during their accumulation. Therefore, the gas cap and the oil zone in a reservoir contain some water, in addition to the hydrocarbon phases.

Of course, hydrocarbon fields may contain only gaseous hydrocarbons, or only liquid hydrocarbons.

1.1.2 *The three classical steps in oil recovery*

(a) *Primary recovery.* Hydrocarbons are recovered from a field by means of wells drilled from the surface into the reservoir. The initial pressure in the fluids occupying the pore space in a reservoir is very often higher than the hydrostatic pressure; therefore, by opening a well, the pressure at its bottom is reduced, thus driving reservoir fluids towards the well.

If the natural pressure drop obtained by opening the well is too small, it is possible to operate a pumping system, which further decreases the pressure at the bottom of the well. In this way, a pressure gradient is created in the reservoir. Because of that pressure gradient, the fluids contained in the pore space begin to flow towards the well and are then evacuated to the surface (Fig. 1.2).

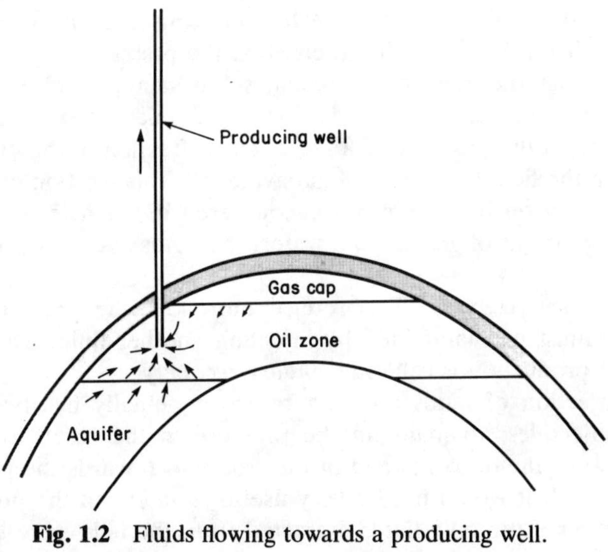

Fig. 1.2 Fluids flowing towards a producing well.

The withdrawal of fluids from the reservoir causes the decompression of the rock and of the remaining fluids. The pressure in the whole reservoir drops progressively.

Figure 1.2 shows a producing well drilled in the oil zone. The withdrawn fluid is mainly oil. But the pressure drop does not occur only in the oil zone; it extends to adjacent zones, i.e. the gas cap and the aquifer, when they exist. Therefore, water from the aquifer and gas from the gas cap may progressively flow into the zone formerly occupied by oil.

Besides, when the pressure in the oil zone drops below a certain value, called the bubble pressure, which depends on the oil composition, some light hydrocarbon components of the oil are expelled from the liquid phase and form a new gas phase. Since the compressibility of this gas phase is higher than that of the oil phase, its appearance slows down the decrease in pressure accompanying oil withdrawal. In some cases this gas phase may remain disseminated in the oil zone and thus flows with the oil towards the producing well. In other cases, it may flow upwards under buoyancy forces and form a gas cap at the top of the reservoir (when no gas cap existed initially) or join the existing gas cap. Similarly, the water contained in the oil zone may flow, together with the oil, towards the producing well and may constitute a part of the fluids evacuated by the well.

The period of production described above is called *primary recovery*. It is obtained merely by reducing the pressure in the reservoir. It reaches its limit either when the pressure in the reservoir is so low that the production becomes too small, or else when the proportion of gas or water in the extracted fluids becomes too high. This results in the recovery of a rather small fraction of the oil initially present in the field (around 10% on average). This fraction depends to a great extent on the conditions encountered in the field, such as the initial proportions of gas, oil and water, the fluids and rock properties.

(b) *Secondary recovery*. To produce more oil, the pressure in the reservoir must be maintained by injecting another fluid. This second period of production is called *secondary recovery*.

The injection of fluids into the reservoir actually has two closely linked objectives: to maintain the pressure in the reservoir, and to push forward the oil contained in the reservoir towards the producing wells. Indeed, it would not be very useful to maintain the pressure in the reservoir if the only fluids extracted from the field were the fluids injected into that field some time before!

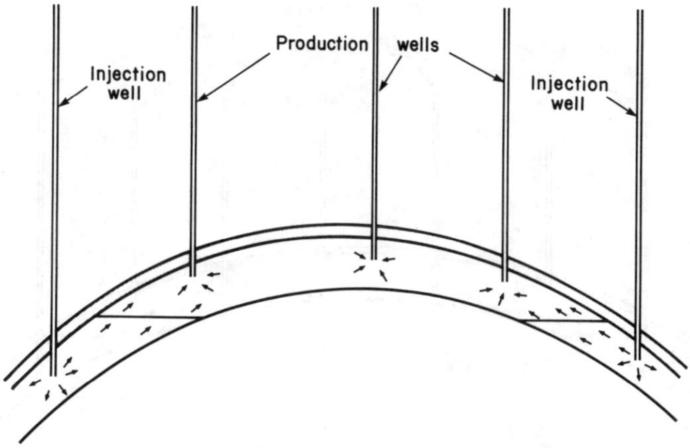

Fig. 1.3 Water injection into the aquifer.

Several types of fluid injection are used according to the local conditions. For a rather small oil field, the pressure may effectively be maintained by injecting water into the aquifer at the outer limit of the reservoir (Fig. 1.3), or gas into the gas cap at the center of the reservoir (Fig. 1.4). For a large field, fluid injection only at the outer limit or only at the center is not an effective way of maintaining the

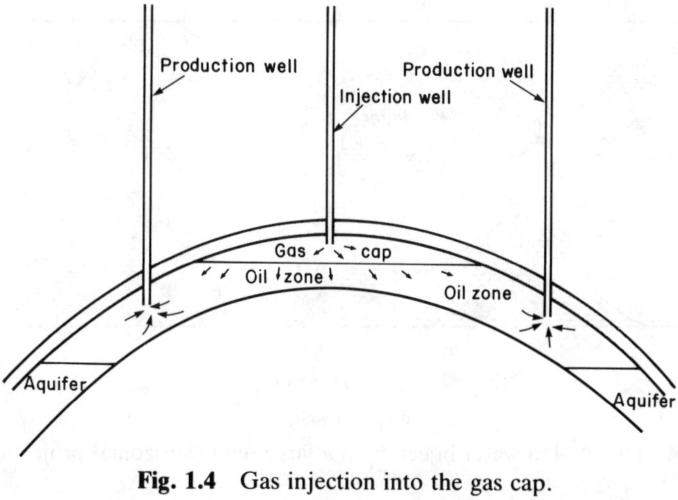

Fig. 1.4 Gas injection into the gas cap.

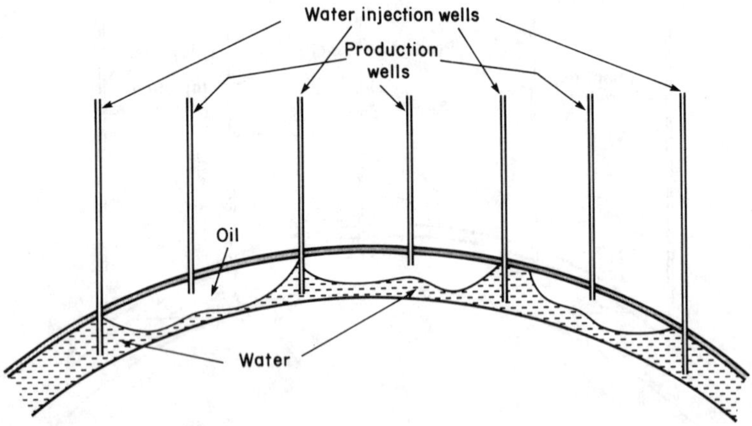

Fig. 1.5 Distributed water injection in a large field (cross-section).

pressure. Very often, fluid injection must be distributed throughout the field. Injection then creates a water zone, if the injected fluid is water (Figs 1.5 and 1.6) or a gas zone, if the injected fluid is gas, around each injection well. These zones extend progressively, and push the oil towards the production wells.

After several years of secondary recovery, the injected fluid reaches the production well and makes up an increasing proportion of the extracted fluids. Finally, production is no longer cost effective;

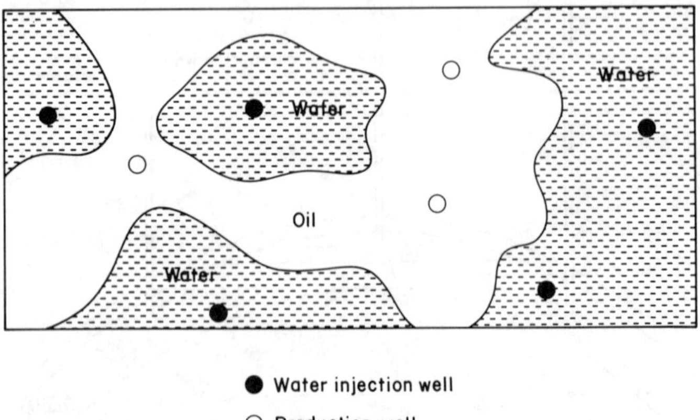

Fig. 1.6 Distributed water injection in a large field (horizontal projection).

secondary recovery has reached its limit. At this time, there are still some parts of the reservoir that have not been swept by the displacing fluid, where the pore space may still contain a large proportion of oil. Even in the part of the reservoir that has been swept by the injected fluid, a fraction of the oil initially in place may still remain, together with the injected fluid. Therefore, by the successive use of primary recovery and secondary recovery, only a fraction (from 15 to 60%) of the oil initially contained in the field can be recovered.

(c) *Tertiary recovery.* When a field has been exploited by primary and secondary recovery techniques, in some cases an attempt is made to recover part of the remaining oil by using more sophisticated techniques. This period of production is called *tertiary recovery*. Some of the techniques involved aim mainly at producing the oil that remains in the part of the reservoir already swept by the displacing fluid; we say that they aim at increasing the *displacement efficiency*. Other techniques aim mainly at producing the oil that remains in the part of the reservoir not swept by the displacing fluid; we say that they aim at increasing the *sweep efficiency*. Of course, some techniques can increase both displacement and sweep efficiency.

1.1.3 *Enhanced oil recovery*

In the previous section primary, secondary and tertiary recovery were described as three successive periods in the exploitation of an oil field, which indeed used to be common practice in the petroleum industry. In newly discovered fields, the trend now is to plan exploitation as a whole. Injection of a displacing fluid may begin very early, long before the complete depletion of the field by primary recovery. Techniques for improving displacement efficiency or sweep efficiency may be used at the very beginning of the first injection of a displacing fluid, rather than under tertiary recovery conditions. The use of such techniques is called *enhanced oil recovery*.

For a proper understanding of the general principles underlying enhanced oil recovery techniques, we must take a closer look at the reasons for non-total recovery when oil is displaced by another fluid. This is the subject of the next section.

1.2 Oil displacement by another fluid: the causes of non-total recovery

In typical oil field exploitation, the oil initially present in the field is partly displaced towards the production well by another fluid. To analyze the causes of non-total recovery, we must look at displacement at three distinct scales:

(i) the scale of a small block of reservoir rock effectively swept by the displacing fluid;
(ii) the scale of a long, narrow flow tube going from an injection well to a production well;
(iii) the scale of the whole field, where the effects of multidimensionality and heterogeneity must be taken into account.

1.2.1 *The scale of an elementary block*

In the displacement of oil by another fluid in an elementary block of porous medium, two cases must be distinguished: the case when the displacing fluid is immiscible with the oil, and the case when it is (at least partially) miscible with the oil.

(a) *Oil displacement by an immiscible fluid*

FORCES ACTING UPON THE FLUID DROPLETS. Any droplet of fluid in the pore space of the rock is acted upon by several forces: the force of gravity, which is distributed throughout the droplet, and contact forces, which are exerted along the surface bounding the droplet.

Let us look more closely at the contact forces exerted along the droplet's boundary (Fig. 1.7). At point A, where the droplet is bounded by the solid part of the porous rock, contact forces are such that the flow of the fluid satisfies a no-slip condition: the velocity of the fluid phase is equal to that of the solid (it vanishes if the solid is at rest).

At point B, where the same fluid phase is present on both sides of the boundary, the contact forces come from the constraints existing inside that fluid phase. These constraints have their origin in the pressure difference between injection and production wells, and also in the nonuniformity of the fluid velocity. When the fluid flows into the pores, it undergoes great deformations since, by the no-slip condition,

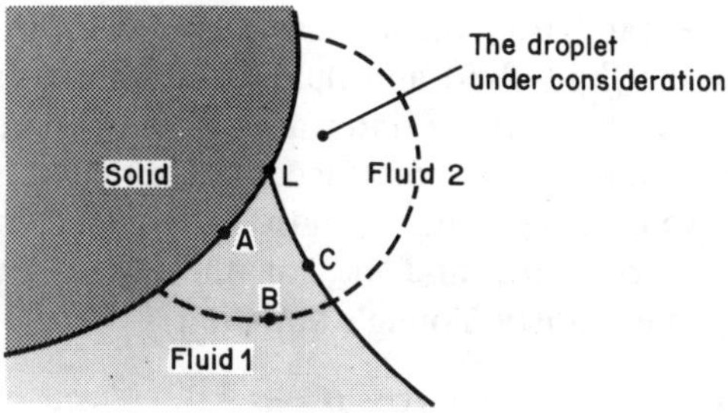

Fig. 1.7 A fluid droplet at the pore scale.

the part of its boundary that is along the solid remains at rest. These deformations cause the appearance of viscosity constraints inside the fluid phase.

At point C, where the droplet is bounded by another fluid phase, contact forces are exerted by the fluid–fluid interface. Such forces are called *capillary forces.* They depend on the physical properties of the interface. Roughly speaking, their tangential component is such that the flow of the fluids on both sides of the interface satisfies a no-slip condition. Moreover, their normal component depends on the curvature of the interface and on the interfacial tension between the two fluid phases. It may also depend on the surface deformations, when the interface exhibits interfacial viscosity.

Near contact line L (Figs 1.7 and 1.8), along which the fluid–fluid interface meets the solid surface, the fluid–fluid interface's position is mainly controlled by the wettability properties of the surface.

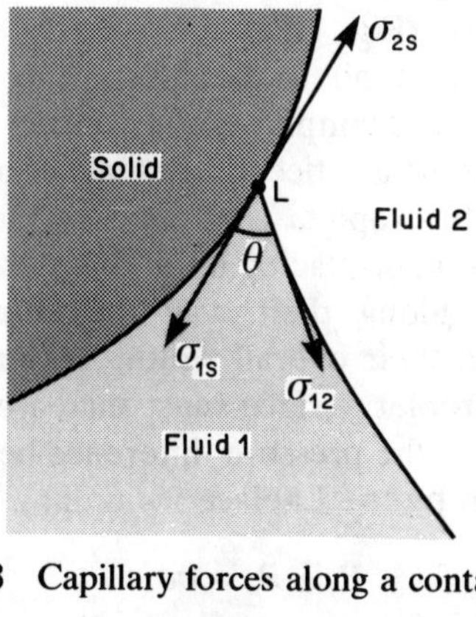

Fig. 1.8 Capillary forces along a contact line.

According to the capillarity theory, capillary forces are exerted on contact line L by the fluid–fluid and fluid–solid interfaces. The force exerted by each of these three interfaces is perpendicular to the contact line, tangential to the interface, and directed towards this interface. Their values, per unit length of contact line, are the interfacial tensions σ_{12}, σ_{1s} and σ_{2s} of the three interfaces. At equilibrium, they must satisfy Young's equation

$$\sigma_{1s} + \sigma_{12} \cos \theta = \sigma_{2s} \tag{1.1}$$

where θ is the equilibrium contact angle (counted in fluid phase 1). The solid is said to be *preferentially wet* by one of the two fluid phases when the equilibrium contact angle, counted in that fluid phase, is smaller than $\Pi/2$. That fluid phase is then called the *wetting fluid phase,* and the other one is called the *nonwetting fluid phase.*

When $|\sigma_{2s} - \sigma_{1s}| > \sigma_{12}$, no equilibrium contact angle can satisfy Young's equation. Therefore at equilibrium, no contact line can exist along which the two fluid phases meet the solid: one of the two fluid phases completely covers the solid surface. The solid is said to be *totally wet* by that fluid phase, which is called, as above, the *wetting fluid phase* (the other one being the *nonwetting fluid phase*).

In all cases, the wetting fluid phase is the phase that has the lowest interfacial tension with the solid.

At pore scale, capillary forces are in general much larger than gravity forces or forces arising from the constraints inside a single fluid phase. Therefore fluid distribution at pore scale is, to a large extent, mainly controlled by capillary forces. The wetting fluid preferentially fills the narrowest parts of the pore space, while the nonwetting fluid occupies the largest parts (Fig. 1.9).

Let us first assume that oil is the nonwetting fluid and that the displacing fluid, water for example, is the wetting fluid. When the oil proportion in the pore space becomes small enough, the oil phase breaks up into separate droplets contained in the largest pores and surrounded by the wetting displacing fluid. Such droplets may still flow inside the pores, but along their way they may encounter some channels narrower than their overall width. To enter such a channel, the curvature of the droplet surface must increase (Fig. 1.10). When the two fluids are at rest, the pressure difference between the two sides of the interface is given by the Laplace formula:

$$P_o - P_w = \sigma c \tag{1.2}$$

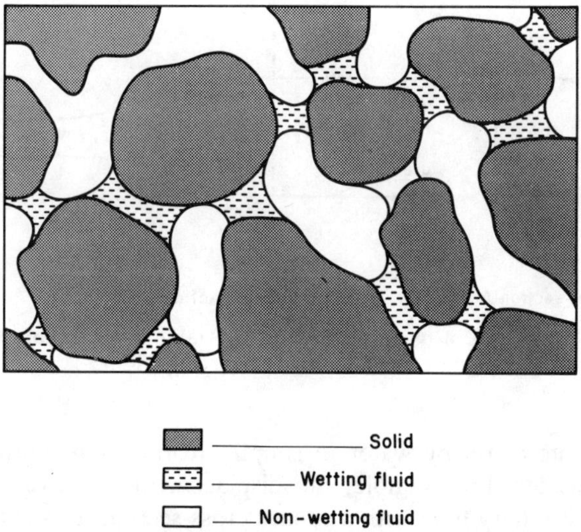

Fig. 1.9 The distribution of two immiscible fluids at the pore scale.

where σ is the interfacial tension of the oil–water interface, and c its mean curvature. This equation remains approximately valid when the fluids are flowing, since the capillary forces are much greater than the viscosity forces. This shows that, to enable the droplet to flow across the narrow channel, the pressure in the oil phase must be higher than the pressure in the surrounding water phase, by a quantity that is proportional to the water–oil interfacial tension and inversely proportional to the channel's width at its narrowest part.

Such an increase in the oil phase pressure can arise only from the pressure drop in the surrounding water phase, over the length of the oil droplet. This fact is most easily understood by looking at

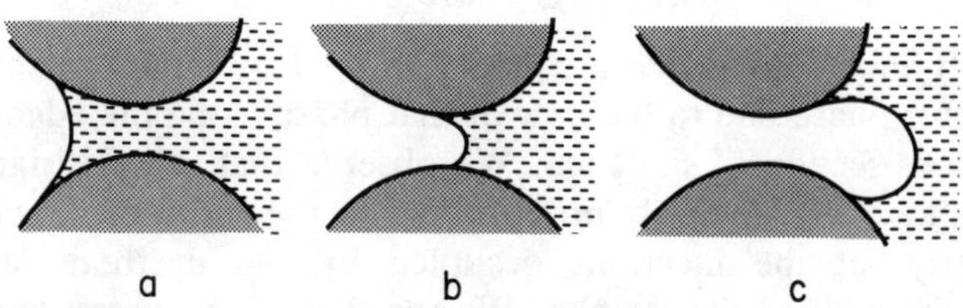

Fig. 1.10 Three successive positions of a droplet of a nonwetting fluid entering a narrow channel.

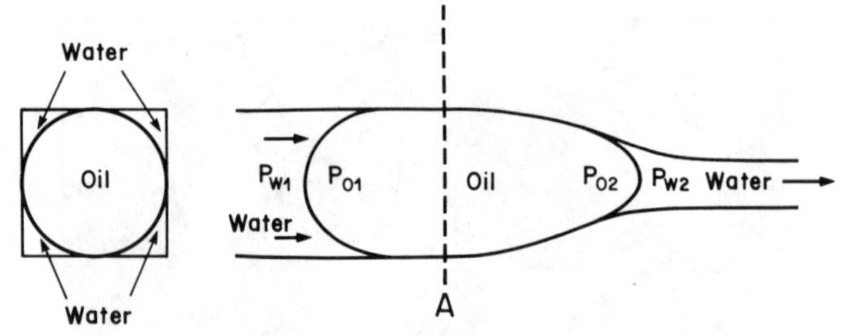

Fig. 1.11 Displacement of an oil droplet by water in a capillary tube with a square cross-section of varying width.

displacements of oil by water in simplified models of porous media. Let us consider, for example, the displacement of an oil droplet by water in a capillary tube with a square cross-section. It is assumed that at some location along the tube the cross-section becomes progressively narrower (Fig. 1.11). Such systems have been studied by Kalaydjian and Legait, for investigating the two-phase flow equations in porous media.[2]

The wetting fluid (water) occupies the part of the tube along the edges of the square cross-section, while the nonwetting fluid (oil) occupies the center of the channel. The pressure difference between oil and water at the upstream and downstream ends of the droplet are:

$$P_{o1} - P_{w1} = \sigma c_1 \tag{1.3}$$

$$P_{o2} - P_{w2} = \sigma c_2 \tag{1.4}$$

where c_1 and c_2 are the mean curvatures of the oil–water interface at the upstream and downstream ends of the droplet, respectively. The oil droplet cannot flow unless $P_{o1} > P_{o2}$, in other words, unless

$$P_{w1} - P_{w2} > \sigma(c_2 - c_1) \tag{1.5}$$

The left-hand side of that inequality, $P_{w1} - P_{w2}$, is the pressure drop in the water phase due to the flow of that phase along the edges of the square cross-section of the tube. We observe that, on the right-hand side, c_2 and c_1 are inversely proportional to the width of the channel, respectively, at the locations occupied by the upstream and the downstream ends of the droplet. We see that, if the pressure drop in the flowing water phase is too small, the oil droplet will be unable to enter the narrower part of the tube and will remain trapped.

Let us now assume that oil is the wetting fluid and that the displacing fluid, gas for example, is the nonwetting fluid. When the proportion of oil becomes small enough, the oil phase breaks into droplets, which are now located in the narrowest parts of the pore space. These droplets may be completely separated from each other, or (when a part of the rock surface is totally oil wet) they may be connected via very thin films covering the oil-wet part of the rock surface. The flow in these films is very slow, and so the remaining oil finally becomes trapped.

EQUATIONS GOVERNING IMMISCIBLE FLUID FLOW. Let us briefly review the classical equations, generally considered to be valid at the scale of an elementary block of porous medium, which describe immiscible fluid flow in a porous medium. For simplicity, it will be assumed that the solid part of the medium is perfectly rigid and at rest. We shall see that many qualitative features of the displacement of oil by another fluid discussed in the previous section also appear in these equations.

In the case when a single fluid phase fills all the pore space, at flow velocities generally encountered in oil reservoirs, inertia forces are negligible, and the flow is governed by Darcy's equation (introduced by Darcy in 1856):[3]

$$\vec{u} = -\frac{k}{\mu}(\overrightarrow{\text{grad}}\, P - \rho\vec{g}) \tag{1.6}$$

together with the mass balance equation for the fluid phase:

$$\text{div}(\rho\vec{u}) + \Phi\frac{\partial\rho}{\partial t} = 0 \tag{1.7}$$

In these equations, $\vec{g}$ is the gravity acceleration, ρ the density, μ the viscosity, P the pressure and $\vec{u}$ the filtration velocity of the fluid phase. The quantities Φ and k are, respectively, the porosity and the permeability of the medium. It is assumed, for simplicity, that the medium is isotropic, so the permeability is a scalar (otherwise the permeability would be a rank 2 tensor).

Let us now consider the case when two immiscible fluids are simultaneously flowing in the pores. By analogy with eqn (1.6), Wyckoff *et al.*[4] and, a little later, Leverett,[5] introduced the relative permeability equations:

$$\vec{u}_1 = -\frac{kK_1}{\mu_1}(\overrightarrow{\text{grad}}\, P_1 - \rho_1\vec{g}) \tag{1.8}$$

$$\vec{u}_2 = -\frac{kK_2}{\mu_2}(\overrightarrow{\text{grad}}\ P_2 - \rho_2\vec{g}) \tag{1.9}$$

In these equations, ρ_i ($i = 1$ or 2) is the density, μ_i the viscosity, P_i the pressure and $\vec{u}_i$ the filtration velocity of fluid i. The quantities Φ and k have the same meaning as in eqns (1.6) and (1.7). The dimensionless coefficients K_i ($i = 1$ or 2) are the so-called relative permeabilities. Wyckoff *et al.*, Leverett, and subsequently many other authors, have investigated the effects of various flow parameters on the relative permeabilities. For a given porous medium and a given pair of fluids at a given temperature and pressure, it is generally assumed that the relative permeabilities K_i depend only on the local values of the fluid saturations S_i ($i = 1$ or 2). Since we assume that the pores of the medium contain no more than two fluids, the saturations S_i must satisfy:

$$S_1 + S_2 = 1 \tag{1.10}$$

We must also take into account the mass balance equation for each fluid phase:

$$\text{div}(\rho_1\vec{u}_1) + \Phi\frac{\partial(\rho_1 S_1)}{\partial t} = 0 \tag{1.11}$$

$$\text{div}(\rho_2\vec{u}_2) + \Phi\frac{\partial(\rho_2 S_2)}{\partial t} = 0 \tag{1.12}$$

Moreover, after Leverett,[6] it is generally assumed that for a given porous medium and a given pair of fluids, at a given temperature and mean pressure, the pressure difference between the two fluid phases $P_2 - P_1$ depends only on the local values of the fluid saturations S_i ($i = 1$ or 2). Since the S_i must satisfy eqn (1.10) we can assume that $P_2 - P_1$ depends only on S_1, and we can write the capillary pressure equation:

$$P_2 - P_1 = P_c(S_1) \tag{1.13}$$

Together with suitably chosen boundary conditions, eqns (1.8)–(1.13) are at the foundations of the mathematical modeling of two-phase fluid flow in porous media. Figures 1.12 and 1.13 show the relative permeabilities K_1 and K_2, and the capillary pressure $P_c = P_2 - P_1$, as a function of the fluid saturation S_1, in a typical case.

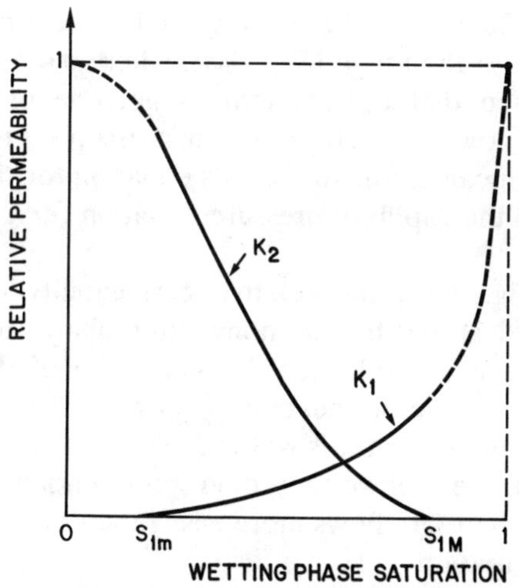

Fig. 1.12 Relative permeabilities as functions of the wetting fluid saturation.

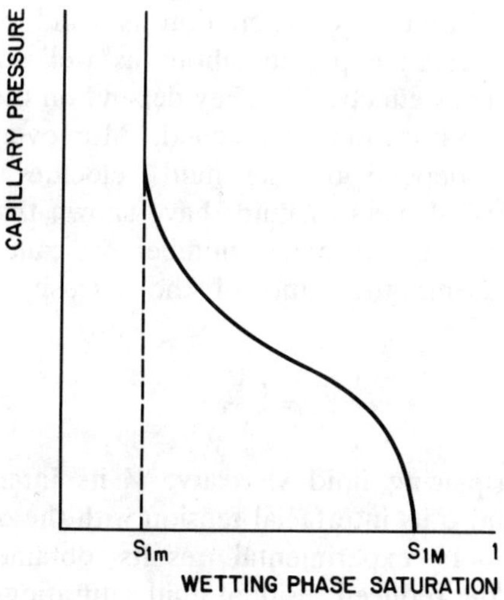

Fig. 1.13 Capillary pressure as a function of the wetting fluid saturation.

Assuming that fluid 1 is the wetting fluid, we observe that its saturation S_1 varies in the range (S_{1m}, S_{1M}), where $0 < S_{1m} < S_{1M} < 1$. It is generally assumed that outside that range, one of the two fluid phases is disconnected. Therefore the macroscopic pressure in that phase is no longer meaningful, so Darcy's equation for that phase (eqn (1.8) or (1.9)) and the capillary pressure equation (eqn (1.13)) are no longer valid.

As shown in Fig. 1.12, the relative permeability curves for the wetting phase (fluid 1) and for the nonwetting phase (fluid 2) are not symmetrical. For a given value of the saturation of the fluid under consideration, the relative permeability is in general higher when that fluid is nonwetting than when it is wetting. This fact can be understood by observing that the nonwetting fluid preferentially occupies the largest pores, and therefore flows more easily than the wetting fluid for equal values of saturation.[7]

More generally, the wettability properties of a reservoir rock have very important effects on all its macroscopic properties, and therefore on the oil recovery.[8] The reader is referred to the very comprehensive survey by Anderson[9] for more information and further references.

Let us now discuss the assumption indicated above that, for a given porous medium and a given pair of fluids at a given temperature and a given mean pressure, the relative permeabilities depend only on the local values of the fluid saturations. A great deal of experimental evidence has shown that this assumption is valid only as a first approximation. The relative permeabilities as well as the capillary pressure show hysteresis effects.[10–12] They depend on the way in which the actual state of saturation is reached. Moreover, the relative permeability curves depend on the fluid velocities. Experimental studies involving several pairs of fluids have shown that this dependency occurs through a dimensionless number N_c, called the *capillary number,* that represents the ratio of the viscosity forces to the interfacial forces,

$$N_c = \frac{\mu V}{\sigma} \tag{1.14}$$

where μ is the displacing fluid viscosity, V its interstitial velocity (Darcy velocity), and σ its interfacial tension with the oil.

Figure 1.14 gathers experimental results obtained by several authors and shows the reduced residual fluid saturations (i.e. the ratio of the actual residual fluid saturation to the residual fluid saturation for

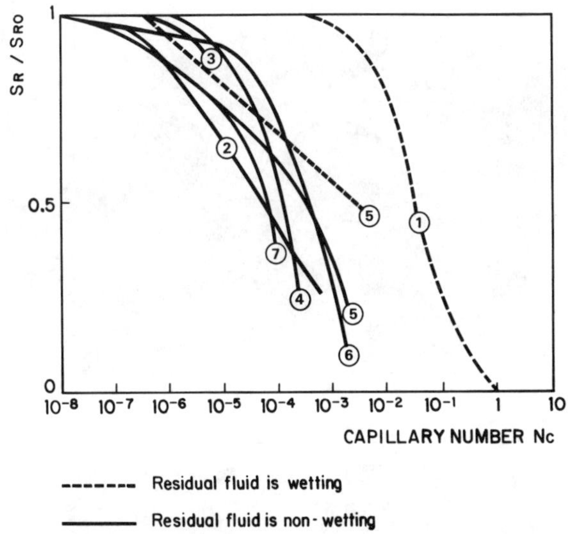

Fig. 1.14 Reduced residual fluid saturations as functions of the flowing fluid's capillary number. (From Ref. 13, reproduced by permission of Academic Press, Inc.) Experimental results gathered from several authors—Curve 1: Dombrowski & Brownell;[14] Curve 2: Moore & Slobod;[15] Curve 3: Wagner & Leach;[16] Curve 4: Taber;[17] Curve 5: Lefebvre du Prey;[18] Curve 6: Foster;[19] Curve 7: Abrams.[20]

small values of the capillary number) as functions of the capillary number. There is some scattering of the data, due to differences in the porous media and experimental procedures used by the authors. However, it appears that residual fluid saturation can be effectively reduced by increasing the flowing fluid's capillary number, but only when that capillary number reaches relatively high values (around 10^{-4} or 10^{-3} if the trapped fluid is nonwetting, even higher if it is wetting). Such values can be reached in practice only by reducing the interfacial tension σ by means of surfactants (see Chapter 3).

(b) *Oil displacement by a totally miscible fluid.* Let us now assume that the displacing fluid is totally miscible with the oil. In the pore space, the oil and the displacing fluid make up a single fluid phase of varying composition. At the pore scale, its flow velocity varies from place to place. It vanishes along the solid surface and reaches its greatest values at the center of the largest pores. Therefore the

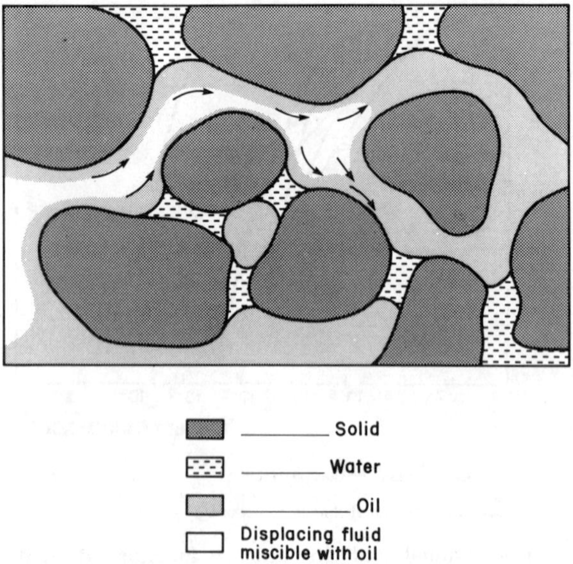

Fig. 1.15 Oil displacement by a miscible fluid.

displacing fluid travels faster in the central part of the largest pores, and bypassed oil may remain along the walls of the pores and in the smaller pores. Contrary to the case when oil is immiscible with the displacing fluid, the bypassed oil is not completely trapped (Fig. 1.15). It is still flowing, even if its flow velocity is much less than that of the displacing fluid in the center of the largest pores. Moreover, since the composition of the flowing fluid phase is nonuniform, diffusion phenomena come into play and help transfer the oil from the bypassed zones towards the zones where flow velocity is faster. If displacement is carried on long enough, all the oil can be recovered.[21]

(c) *Oil displacement by a partially miscible fluid.* In actual practice, displacing fluids are very often only partially miscible with the oil. Such a case features displacement properties that are intermediate between the ones described in the last two sections. As in the case when the displacing fluid is totally immiscible, oil droplets may become hydrodynamically trapped by capillary forces. However, exchanges of components may occur across the fluid–fluid interface, between the trapped oil droplets and the flowing displacing fluid. The exchanges may take place in both directions. Some components of the displacing fluid may become dissolved by the oil phase, and conversely some components of the oil phase may cross the interface and rejoin the flowing displacing fluid phase. Both of these phenomena can, in some cases, increase oil recovery. The oil components that leave the trapped oil droplets to join the flowing displacing fluid are readily recovered, while on the other hand, the dissolution of some light components in the trapped oil droplets may increase the volume of these droplets, reduce their viscosity and their interfacial tension with the displacing fluid, and help their remobilization.

Moreover, these exchanges of components between the displacing fluid and the oil in place may, in some cases, create a mixing zone in which the composition varies continuously between the composition of the displacing fluid and the composition of the oil, with no sharp interface.[22,23] Then the displacement may proceed as when the displacing fluid is totally miscible with the oil.

1.2.2 *The scale of a flow tube*

In a field where oil is displaced by another fluid, let us now consider a flow tube, i.e. a long, narrow part of the reservoir extending from an injection well to a production well, with its boundary being chosen in such a way that at each point it is tangential to the mean flow velocity (Fig. 1.16).

Since the boundary of the flow tube is made up of flow lines, we can study the displacement in the flow tube as if it were isolated from the rest of the reservoir.

Of course, the concept of a flow tube is an idealization and cannot be rigorously defined in actual field exploitation, since the mean velocities of the various flowing phases at a given point in the reservoir are not, in general, colinear and vary with time. However, we will use it merely to give some qualitative explanations that remain valid even when flow tubes cannot be rigorously defined.

By taking mean values of all the physical quantities that are involved over the cross-section, we can consider the displacement in the flow tube as one-dimensional.

Let us first assume that the displacing fluid is immiscible with oil. Figure 1.17 shows, at several successive times, the oil saturation S along the flow tube, as a function of the arc length along its central flow line.

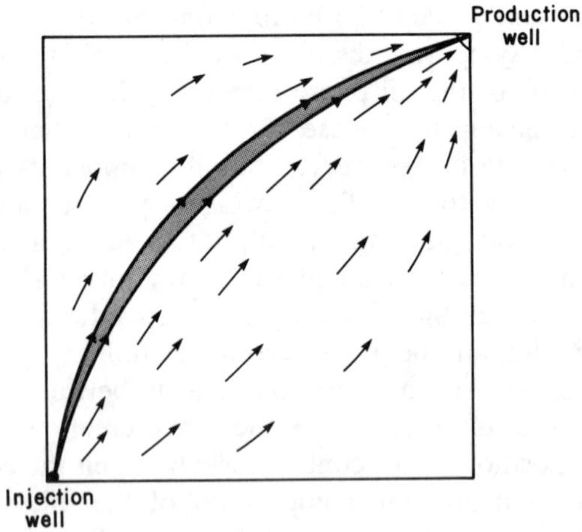

Fig. 1.16 A flow tube in a five-spot pattern.

We observe that the saturation profile spreads over a length that increases with time. When the displacing fluid first reaches the production well (in the example of Fig. 1.17, this occurs approximately at time t_5), a relatively large amount of oil remains in the flow tube. If the displacement is carried out, the proportion of oil in the produced fluid decreases with time, while that of the displacing fluid

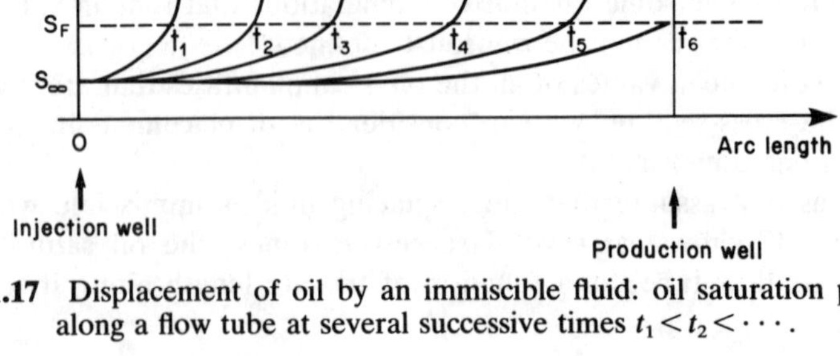

Fig. 1.17 Displacement of oil by an immiscible fluid: oil saturation profiles along a flow tube at several successive times $t_1 < t_2 < \cdots$.

increases. Likewise, when the oil saturation along the flow tube is everywhere equal to the residual oil saturation S_∞ (corresponding to the unrecoverable trapped oil), no more oil will be produced.

The zone of the flow tube where the oil and the displacing fluid are both simultaneously flowing may be divided into two parts. In the upstream part (for oil saturations lower than the value S_F in Fig. 1.17), the saturation profile, as a function of the arc length, is not very steep. When time increases, that part spreads over a length of the flow tube that becomes larger and larger. According to the standard equations governing immiscible fluid displacements, for a one-dimensional displacement at a constant flow rate, the thickness of that part would increase roughly linearly with time. In the downstream part (for saturations greater than S_F) the saturation profile is much steeper. When the time increases, the length of that part remains relatively short, and increases more slowly than linearly in time.[24] That part of the saturation profile corresponds to the abrupt change in saturation that is predicted by the flow equations in the Buckley and Leverett theory,[25] when the capillary pressure is neglected.

We see, therefore, that oil saturation in the whole flow tube may reach values lower than S_F not too long after the first arrival of the displacing fluid at the production well (through that flow tube). However, to reduce the oil saturation still further, until values substantially lower than S_F, near S_∞, are reached, a much longer time may be necessary. Moreover, the additional oil will be produced together with large amounts of displacing fluid, so that such production conditions are in general economically of no interest.

By choosing the displacing fluid properties and the operating conditions we can act on the oil saturation values S_∞ and S_F. Of course, reducing final oil saturation S_∞ is important; but reducing saturation S_F just upstream from the abrupt saturation change in saturation profile may be economically even more important.

Let us now assume that the displacing fluid is totally miscible with oil. Figure 1.18 shows the oil concentration in the flowing fluid phase as a function of the arc length along the flow tube for several successive times. We observe that there is a mixing zone between pure oil and pure displacing fluid. The thickness of that mixing zone increases more slowly than linearly in time. For a one-dimensional displacement at a constant flow rate, the standard equations governing miscible fluid displacements lead to a mixing zone thickness that is proportional to the square root of the time.[21] In most cases, the

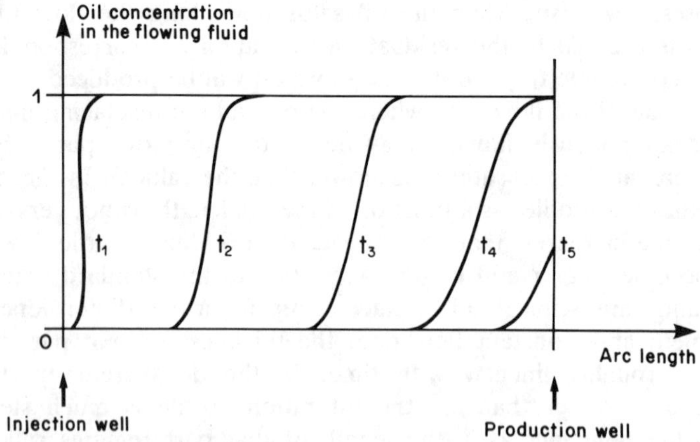

Fig. 1.18 Displacement of oil by a miscible fluid: oil concentration profiles along a flow tube at several successive times $t_1 < t_2 < \cdots$.

thickness of the mixing zone remains small compared to the total length of the flow tube. Therefore, when the displacing fluid in the flow tube reaches the producing well, the proportion of oil already produced may be fairly high. Most of the recoverable oil may be obtained not too long after the first arrival of the displacing fluid at the production well.

When the displacing fluid is only partially miscible with the oil, the behavior of the mixing zone between the oil and the displacing fluid is somewhat intermediate between the two cases described above. According to some numerical simulations,[26] its thickness increases almost linearly with time and may reach values much greater than when the displacing fluid is totally miscible with the oil. It seems that such is the case even when thermodynamic miscibility is obtained (i.e. when, by exchange of components between the oil and the displacing fluid, a mixing zone is created in which the composition varies continuously between the composition of the displacing fluid and the composition of the oil).

1.2.3 *The scale of the reservoir*

As seen in Section 1.2.1, at the pore scale, the distribution of two immiscible fluids in a porous rock is mainly controlled by capillary forces which are much greater than gravity forces and than forces arising from the constraints inside a single fluid phase. This is no longer true at larger scales, such as the scale of the reservoir. The capillary forces, distributed all along the interface inside the pores, are oriented in all directions and largely cancel each other out on the whole. They control fluid distribution at the pore scale (the wetting fluid preferentially occupies the narrowest pores, and the nonwetting fluid the largest), but they do not have an important overall effect on fluid distribution at larger scales. On the contrary, gravity forces and pressure distribution (which greatly depends on the location of the injection and production wells and on the local properties of the reservoir) have very important effects on the fluid distribution as a whole. In Section 1.2.2, we could neglect these effects because, in a flow tube, the flow may be considered one-dimensional. But at the scale of the whole field, we must now take them into account.

At such a scale, we observe that when the displacing fluid first reaches the production wells, the reservoir has not been wholly swept. Some parts of the reservoir have been bypassed and still hold most of their initial oil content. This oil is not completely trapped; if the displacement is continued, the oil flows slowly toward the production wells, and part of it may be recovered. However, the displacing fluid is simultaneously produced by the same wells, and its proportion in the output increases rapidly with time. Finally, reservoir exploitation ceases to be cost effective.

Let us look at a few aspects of this phenomenon.

(a) *Horizontal* (*or areal*) *sweep efficiency*. Injection and production wells in an oil field are often located according to a more or less regular repetitive pattern. Figure 1.19 shows one of the most frequently used patterns—the five-spot pattern. Calculations show that the proportion of the unswept area depends greatly on the mobility ratio:

$$M = \frac{\lambda_d}{\lambda_o} = \frac{k_d \mu_o}{k_o \mu_d} \tag{1.15}$$

In this equation μ_o and μ_d are the oil and displacing fluid viscosities, k_o and k_d the effective oil and displacing fluid permeabilities of the reservoir rock, in the zones occupied by these fluids. The ratios $\lambda_o = k_o/\mu_o$ and $\lambda_d = k_d/\mu_d$ are called the mobilities of the oil and of the displacing fluid, respectively.

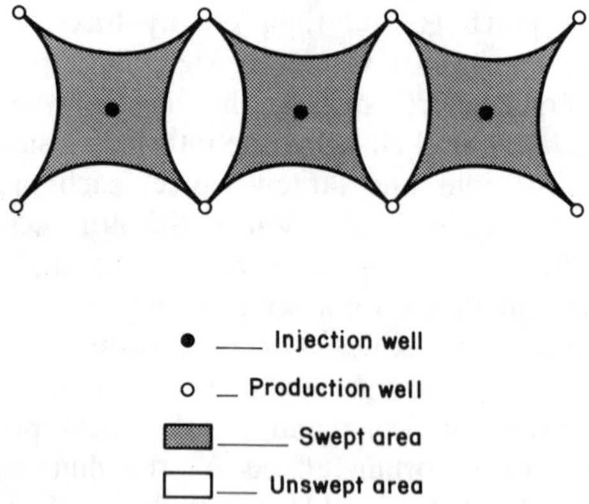

Fig. 1.19 The swept and the unswept areas in a five-spot pattern.

(b) *Vertical sweep efficiency*. In a thick reservoir, when the displacing fluid density and the oil density are different, gravity forces may cause a vertical separation of the flowing fluids. Figure 1.20 illustrates the case when oil is displaced by a lighter fluid, such as gas. The gas overruns the oil, displacing it mainly along the upper boundary of the reservoir. The lower part of the reservoir remains unswept when the displacing gas reaches the producing well. The importance of this effect depends on several parameters, especially the density difference between the displacing and displaced fluids, the mobility ratio, the reservoir

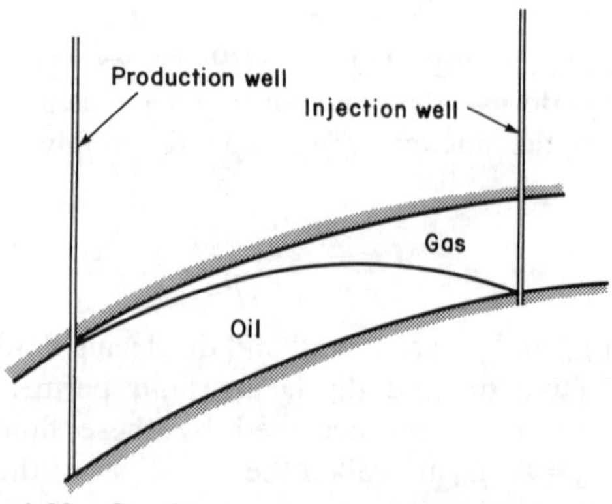

Fig. 1.20 Gravity overrunning in a thick reservoir.

thickness and its dip-angle, the flow rate and the reservoir permeability.

(c) *Effects of reservoir heterogeneity*. Reservoir rocks are very often greatly heterogeneous. Several types of heterogeneities are encountered. Some reservoirs feature several layers of different permeabilities. Their thickness may vary from a few centimeters to a few meters, and their horizontal extension may be several hundred meters. In other reservoirs, heterogeneities are distributed in a less ordered manner. For example, in some cases, low permeability parts are more or less randomly distributed in a more permeable medium. Other reservoirs may contain a network of high permeability fractures surrounding matrix blocks of low permeability.

Reservoir heterogeneities can have very important effects on the displacement of oil by another fluid. Let us look, for example, at a typical reservoir model made up of two layers with very different permeabilities (Fig. 1.21). The flow velocity is greater in the more permeable layer, and therefore displacement occurs earlier in that layer. In some cases the displacing fluid may reach the production wells, through the high permeability layer, while a large quantity of oil has not been swept in the lower permeability layer.

Other mechanisms may help in reducing such heterogeneity effects, or in other cases they may increase these effects. Let us look at the most important ones.

Because of gravity forces, the higher density fluid flows preferentially in the lower layer. For example, when the displacing fluid is heavier than the displaced oil, gravity forces tend to reduce the effects of heterogeneity if the high permeability layer is at the top of the reservoir, and to increase these effects if the high permeability layer is at the bottom of the reservoir.

Heterogeneity effects on displacement are increased when the

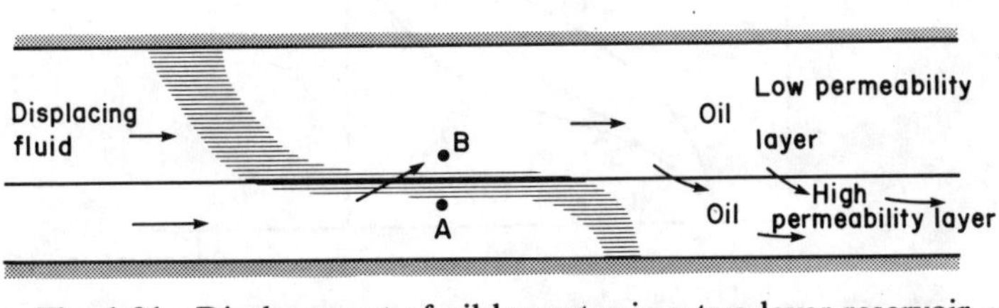

Fig. 1.21 Displacement of oil by water in a two-layer reservoir.

displacing fluid mobility is higher than the oil mobility (mobility ratio $M > 1$), and reduced when the displacing fluid mobility is lower than the oil mobility ($M < 1$). This phenomenon can easily be understood by the typical reservoir model in Fig. 1.21. The transition zone has advanced farther downstream in the high permeability layer than in the low permeability layer. Therefore, in the high permeability layer, oil has been replaced by the displacing fluid in a larger portion of the layer. If the displacing fluid mobility is higher than the oil mobility, this tends to make the flow of fluids in that layer even easier, and therefore, further increases the heterogeneity effect. On the contrary, if the displacing fluid mobility is smaller than the oil mobility, the replacement of oil by the displacing fluid in the high permeability layer tends to reduce the mean flow velocity in that layer, and therefore reduces the heterogeneity effect.

When the displacing fluid is immiscible with oil, capillary forces may have important effects in heterogeneous reservoirs. In some cases, these effects are favorable. Let us consider, for example, the typical reservoir model in Fig. 1.21, and assume that the displacing fluid (water) is the wetting fluid. Capillary forces act mainly in the part of the transition zone between oil and water, which is along the boundary between the two layers. These forces tend to create fluid exchanges between the two layers in such a direction that the capillary pressures on both sides of the boundary between the two layers become equal.

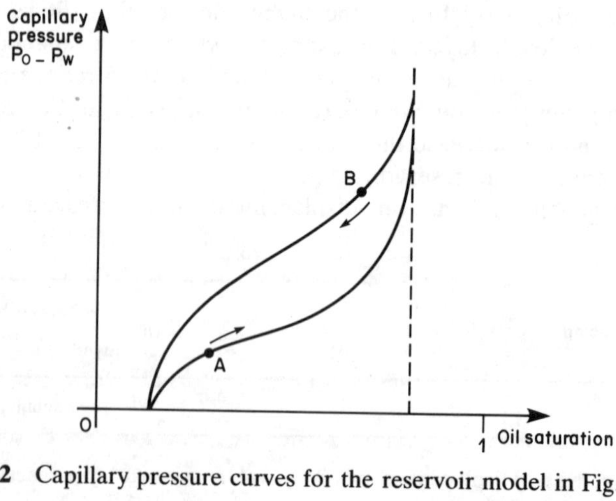

Fig. 1.22 Capillary pressure curves for the reservoir model in Fig. 1.21.

Points A and B in Fig. 1.22 represent the capillary state (oil saturation and capillary pressure) in the high permeability layer and in the low permeability layer, at two nearby points in the transition zone, on each side of the boundary between the two layers (points A and B in Fig. 1.21). The capillary forces, which tend to make the capillary pressures equal, will tend to displace point A towards higher oil saturations and point B towards lower oil saturations. Therefore these forces tend to drive oil from the low permeability layer towards the high permeability layer, and water in the opposite direction.

In other cases, capillary forces in heterogeneous reservoirs may have unfavorable effects. Let us consider, for example, a reservoir containing a low permeability zone surrounded by a higher permeability rock (Fig. 1.23). We now assume that oil is the wetting fluid. Once the displacing fluid (gas) has swept the high permeability rock past the low permeability inclusion, the oil saturation in the low permeability zone may still be somewhat higher than the residual oil saturation, and the corresponding oil remains trapped. This phenomenon (similar to the well-known end effect in laboratory experiments) can be understood by looking at the capillary curves for the two media (Fig. 1.24). We consider two nearby points A and B, one in the high permeability zone and the other in the low permeability zone (Fig. 1.23). Once the oil has been displaced in the high permeability medium, and assuming that capillary equilibrium is established, the capillary pressures at these two points are equal. It can be seen in Fig. 1.24 that the oil saturation at point B remains larger than the residual saturation, which would be reached if the reservoir were homogeneous.

When the displacing fluid is (at least partially) miscible with oil, diffusion and dispersion phenomena can reduce the heterogeneity effects. For example, in the typical reservoir model in Fig. 1.21, assuming that the displacing fluid is miscible with oil, transverse

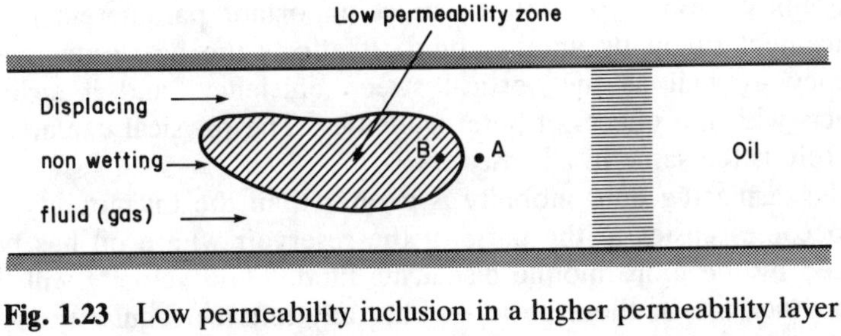

Fig. 1.23 Low permeability inclusion in a higher permeability layer.

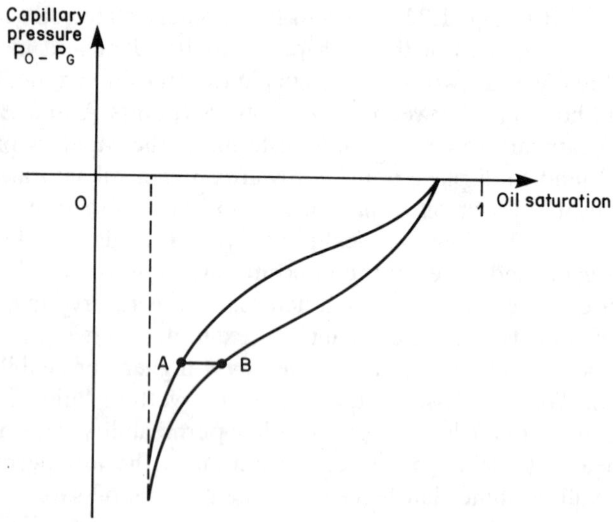

Fig. 1.24 Capillary pressure curves for the reservoir model in Fig. 1.23.

diffusion–dispersion may help the recovery of oil in the less permeable layer. In fractured reservoirs produced by gas injection, an exchange of components between the oil and the displacing gas may account for a fairly large portion of the oil recovery in the low permeability matrix blocks. However, diffusion–dispersion is in general less efficient than capillary forces. Its role can be considered to be important only when the distance over which this phenomenon will have to transport matter is rather small, e.g. in layered reservoirs where the low permeability layers are not very thick, or in fractured reservoirs where the matrix blocks are small.

(d) *Dominant role of the mobility ratio.* At the scale of a whole field, the mobility ratio is one of the most important parameters in the displacement of oil by another fluid. It affects the horizontal sweep efficiency as well as the vertical sweep efficiency, and it strongly interacts with the effects of heterogeneities. The physical explanation of its role is the same in all cases.

If the displacing fluid mobility is greater than the oil mobility, the flow becomes easier in the parts of the reservoir where oil has been replaced by the more mobile displacing fluid. Fluid velocity will thus tend to increase in these parts, so that any initial irregularity in the

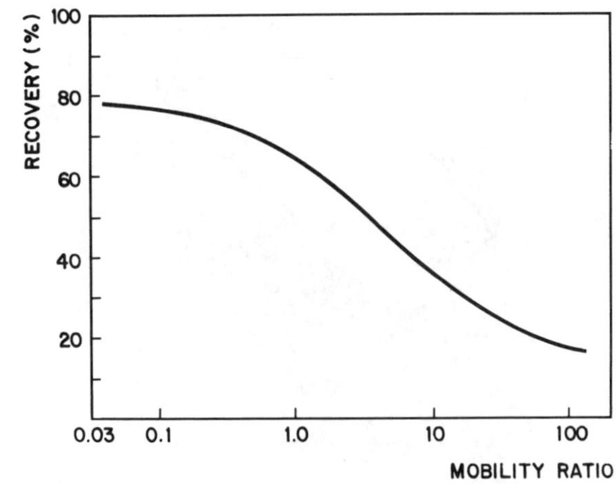

Fig. 1.25 Recovery at breakthrough in a five-spot pattern as a function of the mobility ratio. (From Ref. 27, reproduced by permission of The Society of Petroleum Engineers.)

displacement pattern will tend to increase. On the contrary, if the displacing fluid mobility is less than oil mobility, the same argument shows that the initial irregularities in the displacement pattern will tend to be reduced.

Figure 1.25 shows recovery at breakthrough in a five-spot pattern, as a function of the mobility ratio.[27] These results were obtained in laboratory experiments which may not always be representative of actual field behavior, but the qualitative trend that they indicate can be considered as firmly proven.

These adverse effects of a high mobility ratio may occur even in a homogeneous reservoir, far away from injection or production wells; it is the well-known phenomenon of viscous fingering. Figure 1.26 shows the results of laboratory experiments, concerning miscible fluid displacement, in which a highly viscous fluid is displaced by a less viscous one.[27] The figure shows that 'fingers' of displacing fluid penetrate into the oil zone, while a large part of the reservoir remains unswept. The reader is referred to the recent review by Homsy[28] for a discussion of all aspects of viscous fingering in homogeneous porous media.

Viscous fingering can be a major cause of poor sweep efficiency in recovery methods in which the displacing fluid mobility is much larger than the oil mobility. Such is frequently the case in miscible drive by

Fig. 1.26 Viscous fingering during miscible displacement. Mobility ratio: 71·5. (From Ref. 27, reproduced by permission of The Society of Petroleum Engineers.)

an hydrocarbon gas: the displacing fluid viscosity is often much smaller than the oil viscosity. Moreover, contrary to what happens in immiscible fluid displacements, the effective permeability to the displacing fluid is not reduced in the swept zone since there remains no residual oil in that zone.

Fortunately, other factors can reduce, or even eliminate, the tendency to viscous fingering due to an adverse mobility ratio. The most important ones are briefly discussed below.

When the displacing fluid density is different enough from the oil density, gravity forces can be used to prevent viscous fingering. Let us assume for example that the displacing fluid is lighter than the oil. If the locations of the injection and production wells can be arranged in such a way that the displacement occurs downwards, gravity forces tend to reduce the growth rate of the fingers of displacing fluid. It can be shown[29] that viscous fingering is completely eliminated when the flow rate is smaller than some critical value. For a linear displacement in a homogeneous layer of constant thickness, with the simplifying assumption of a piston-like displacement, this critical value (expressed as a Darcy velocity) is given by the formula

$$u_c = \frac{k(\rho_o - \rho_d)g \sin \alpha}{(\mu_o/K_o) - (\mu_d/K_d)} \tag{1.16}$$

where k is the permeability, g the gravity acceleration, α the dip-angle of the layer, ρ_0, ρ_d, μ_o, μ_d the densities and viscosities of the oil and of the displacing fluid, K_o and K_d the relative permeabilities to the oil in the oil zone and to the displacing fluid in the swept zone, respectively.

Similar considerations apply when the displacing fluid is heavier than the oil: gravity forces tend to prevent the growth of fingers when the displacement occurs upwards.

In immiscible fluid displacements, capillary forces give rise to fluid exchanges between adjacent zones in the reservoir, in such a way that capillary pressure differences tend to decrease. This mechanism has already been discussed in Section 1.2.3(c) for heterogeneous reservoirs (Figs 1.21–1.24). In a homogeneous reservoir in which viscous fingering occurs, the same mechanism tends to eliminate the fingers of displacing fluid by transverse fluid exchanges with the adjacent unswept zones. Such a mechanism is efficient when the width of the fingers is small enough, but does not prevent the appearance and growth of fingers of large width. Theoretical studies (see for example Ref. 29 and references therein, or for a more elaborate and recent treatment, Refs 30 and 31) lead to the concept of a critical wavelength: fingers of width smaller than that critical wavelength cannot grow and are eliminated by the effects of capillary forces.

Similarly, in miscible displacements, the mixing of fluids due to diffusion–dispersion phenomena tends to reduce viscous fingering. However, it seems that for the same value of the mobility ratio, viscous fingering is often more important in miscible displacements than in immiscible displacements: the stabilizing effects of diffusion–dispersion in miscible displacements are, in general, less efficient than the stabilizing effects of capillary forces in immiscible displacements.[28,30,31]

Other effects such as heat transfer in thermal methods and swelling of oil by dissolution of carbon dioxide can also reduce the tendency to viscous fingering.

Special care is required in numerical calculations of displacement patterns when viscous fingering occurs.[32,33]

1.3 How to improve the factors contributing to oil recovery

In Section 1.2 we analyzed the causes of non-total recovery in the displacement of oil by another fluid. We can now understand the basic ideas underlying most enhanced oil recovery techniques. The way in which to improve each factor that contributes to oil recovery is discussed below.

1.3.1 *Oil displacement in an elementary block*

As seen in Section 1.2.1, residual oil, after displacement by an immiscible fluid, is mainly trapped by capillary forces. The factor controlling entrapment or unentrapment is the capillary number (eqn 1.14), i.e. the ratio of the forces exerted by the pressure gradient in the flowing fluid to the capillary forces at the pore level. Figure 1.14, already discussed in Section 1.2.1(b), shows that the reduction of residual oil saturation becomes significant only for relatively high values of the capillary number, which cannot be reached in practice by increasing the velocity of the displacing fluid, nor by increasing its viscosity: this would lead to prohibitively high pressures in the injection wells. However, such high values can be reached by strongly decreasing the interfacial tension, by the use of surfactants. Figure 1.14 also shows that a change of wettability of the rock, in which that rock becomes more strongly wet by the displacing fluid, together with a reduction of the interfacial tension, can help the mobilization of trapped oil. Alkaline agents and some surfactants can have such a favorable effect on rock wettability.

Another way of preventing oil entrapment by capillary forces is to use a displacing fluid that is miscible with oil. Some hydrocarbon gases, with a high proportion of intermediate molecular weight components (C_3, C_4, C_5) are miscible with oil under pressure and temperature conditions encountered in some oil fields. Moreover, under much wider conditions, the displacement of oil by hydrocarbon gases may lead, through component exchanges between the oil and the gas, to the creation of a transition zone in which the composition varies continuously between the composition of the displacing fluid and the composition of the oil. Similarly, oil displacement by carbon dioxide may lead to such a miscible transition zone. In such types of displacements, the exchanges of components between the residual oil and the displacing gas have a favorable effect for reducing oil entrapment, even when total miscibility is not reached. Some components of the residual oil are vaporized by the displacing gas and thus recovered. The dissolution of some components of the gas in the residual oil may increase its volume, decrease its viscosity and its interfacial tension with the gas, and therefore facilitate oil mobilization.

1.3.2 *Transition zones in a flow tube*

We have seen in Section 1.2.2 that, when oil is displaced by another fluid, a transition zone develops between the oil and the displacing fluid. The thickness of this transition zone increases with time, even if we consider only displacement in a flow tube where the effects of multidimensionality can be ignored. To increase the oil recovery at the time of first arrival of the displacing fluid in production wells, the thickness of this transition zone should be reduced as much as possible.

When the displacing fluid is immiscible with oil, an effort should be made to lower, as much as possible, the value of the oil saturation S_F immediately upstream from the displacement front (Fig. 1.17). According to the well-known Buckley and Leverett theory,[25] the main factor that acts on S_F is the mobility ratio. Oil saturation S_F decreases as the mobility ratio decreases.

When the displacing fluid is miscible with oil, the thickness of the transition zone increases more slowly than linearly in time. In most cases, its effects on the oil recovery are far less than those of heterogeneity and multidimensionality.

1.3.3 *Sweep efficiency at the scale of the reservoir*

As seen in Section 1.2.3, the main factor acting on the different aspects of sweep efficiency is the mobility ratio.

The main aim of several enhanced oil recovery methods is to decrease the mobility of the displacing fluid. The most successful method is polymer flooding. Other methods based on the same idea are foam drive and alternate gas and water flooding (WAG process, discussed in Chapter 7).

Other methods aim to increase oil mobility. For example, in thermal methods, an increase in temperature reduces oil viscosity and therefore increases its mobility. In oil drive by carbon dioxide, oil viscosity is reduced by solubilization of carbon dioxide.

Controlling the mobility ratio is of utmost importance in complex enhanced oil recovery techniques such as those described in Chapters 3 and 4, in which several slugs containing various chemicals are successively injected into the reservoir. Otherwise these slugs would be rapidly destroyed or mixed together by viscous fingering or by the effects of heterogeneities.

In some particular cases, gravity forces can be used to obtain good sweep efficiency. Such is the case, for example, in thick, high permeability reservoirs in which oil is displaced by a miscible gas. The low density displacing fluid is injected at the top of the reservoir. If horizontal permeability is great enough and if the flow rate is not too high, a very regular displacement stabilized by gravity forces can be obtained.

Sweep efficiency at the scale of the reservoir can be improved by new technologies such as production through horizontal wells which, in some cases, allow much improved flow patterns. This subject is briefly dealt with in Chapter 12.

1.3.4 *Basic principles of enhanced oil recovery methods*

For analyzing the way in which enhanced recovery methods operate, two separate aspects of oil recovery efficiency are generally distinguished: the displacement efficiency of oil at the pore scale, and the sweep efficiency at the macroscopic scale. These aspects are discussed in Sections 1.3.1 and 1.3.3, respectively. The time evolution of transition zones in a flow tube (discussed in Section 1.3.2) should also be considered, although it is less important for the final oil recovery than the other two aspects.

Some enhanced oil recovery methods act mainly on one of these aspects of oil recovery efficiency; other methods act on both aspects. For example, in-situ combustion acts on the sweep efficiency at the macroscopic scale by decreasing the oil viscosity, and also acts on the displacement efficiency at the pore scale, since essentially no residual oil remains behind the combustion front. Polymer flooding acts mainly on sweep efficiency at a macroscopic scale by increasing the displacing

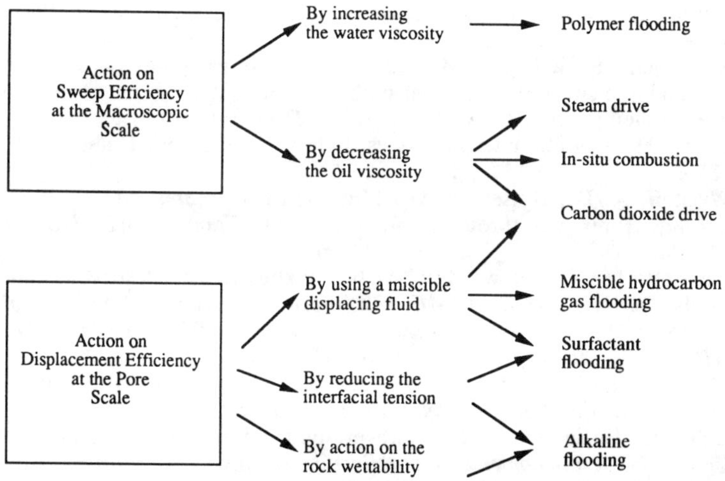

Fig. 1.27 Factors upon which the main types of enhanced oil recovery methods act.

fluid viscosity. Miscible hydrocarbon gas flooding acts mainly on the displacement efficiency at the pore scale, by annihilating the capillary forces. Figure 1.27 summarizes the factors upon which the three great families of methods (chemical, miscible and thermal methods) act for improving oil recovery.

Some enhanced oil recovery methods, such as steam drive, miscible hydrocarbon gas flooding, carbon dioxide drive, are already extensively used, since they meet the economic requirements for applicability. Other methods, such as surfactant flooding, are still under development; additional laboratory studies, research in applied chemistry and field experiments are needed for mastering such complex processes, improving the products used and reducing operating costs.

The basic principles described in this chapter can also be used for understanding the way in which other unconventional methods act (such as those described in Chapter 11). Readers are referred to Ref. 34 for further information and discussion from another viewpoint.

1.4 References

1. Tissot, B. P., Migration of hydrocarbons in sedimentary basins: a geological, geochemical and historical perspective. In *Migration of Hydro-*

carbons in Sedimentary Basins, ed. B. Doligez, Technip, Paris, 1987, p. 1–19.
2. Kalaydjian, F. & Legait, B., Effets de la géométrie des pores et de la mouillabilité sur le déplacement diphasique à contre-courant en capillaire et en milieu poreux. *Revue Phys. Appl.*, **23** (1988) 1071–81.
3. Darcy, H., Les fontaines publiques de la ville de Dijon, Dalmont, Paris, 1856.
4. Wyckoff, R. D., Botset, H. G., Muskat, M. & Meres, M. V., Flow of gas–liquid mixtures through sands. *Petroleum Trans. AIME,* **123** (1937) 247–74.
5. Leverett, M. C., Flow of oil–water mixtures through unconsolidated sands. *Petroleum Trans. AIME,* **132** (reprinted edition) (1938) 381–403.
6. Leverett, M. C., Capillary behavior in porous solids. *Petroleum Trans. AIME,* **142** (second reprinted edition) (1941) 152–69.
7. Mungan, N., Interfacial effects in immiscible liquid–liquid displacement in porous media. *Soc. Petroleum Engineers J.*, **6**(3) (1966) 247–53.
8. Cuiec, L., Evaluation of reservoir wettability and its effect on oil recovery. In *Interfacial Phenomena in Oil Recovery,* ed. N. R. Morrow, Marcel Dekker, New York, 1990, pp. 319–75.
9. Anderson, W. G., Wettability literature survey. *J. Petroleum Technol.*, **38** (1986) 1125–44, 1246–62, 1371–8; **39** (1987) 1283–1300, 1453–68, 1605–22.
10. Melrose, J. C., Wettability as related to capillary action in porous media. *Soc. Petroleum Engineers J.*, **5**(3) (1965) 259–71.
11. Morrow, N. R., Physics and thermodynamics of capillary action. *Ind. Engineering Chem.*, **62** (1970) 32–56.
12. Morrow, N. R. & Harris, C. C., Capillary equilibrium in porous materials. *Soc. Petroleum Engineers J.*, **5**(1) (1965) 15–24.
13. Stegemeier, G. L., Mechanisms of entrapment and mobilization of oil in porous media. In *Improved Oil Recovery by Surfactant and Polymer Flooding,* ed. D. O. Shah & Schechter, R. S., Academic Press, New York, 1977, pp. 55–91.
14. Dombrowski, H. S. & Brownell, L. E., Residual equilibrium saturation of porous media. *Ind. Engineering Chem.*, **46** (1954) 1207–19.
15. Moore, T. F. & Slobod, R. L., The effect of viscosity and capillarity on the displacement of oil by water. *Producers Monthly,* **20**(10) (1956) 20–30.
16. Wagner, O. R. & Leach, R. O., Effect of interfacial tension on displacement efficiency. *Soc. Petroleum Engineers J.*, **6**(4) (1966) 335–44.
17. Taber, J. J., Static and dynamic forces required to remove a discontinuous oil phase from porous media containing both oil and water. *Soc. Petroleum Engineers J.*, **9**(1) (1969) 3–12.
18. Lefebvre du Preÿ, E. G., Factors affecting liquid–liquid relative permeabilities of a consolidated porous medium. *Soc. Petroleum Engineers J.*, **13**(1) (1973) 39–47.
19. Foster, W. R., A low tension water flooding process. *J. Petroleum Technol.*, **25** (1973) 205–10.
20. Abrams, A., The influence of fluid viscosity, interfacial tension and flow velocity on residual saturation left by waterflood. *Soc. Petroleum Engineers J.*, **15**(5) (1975) 437–47.

21. Blackwell, R. J., Rayne, J. R. & Terry, W. M., Factors influencing the efficiency of miscible displacement. *Petroleum Trans. AIME,* **216** (1959) 1–8.
22. Slobod, R. L. & Koch, H. A., High pressure gas injection, mechanism of recovery increase. *Drilling and Production Practice,* **20** (1953) 82–96.
23. Benham, A. L., Dowden, W. E. & Kunzman, W. J., Miscible fluid displacement—prediction of miscibility. *Petroleum Trans. AIME,* **219** (1960) 229–37.
24. Fayers, F. J. & Sheldon, J. W., The effect of capillary pressure and gravity on two-phase fluid flow in a porous medium. *Petroleum Trans. AIME,* **216** (1959) 147–55.
25. Buckley, S. E. & Leverett, M. C., Mechanism of fluid displacement in sands. *Petroleum Trans. AIME,* **146** (1942) 107–16.
26. Nguyen Van Quy, Étude numérique des écoulements diphasiques multicomposants en milieu poreux avec miscibilité partielle et échanges entre phases. Thèse de doctorat d'État, Université Pierre et Marie Curie, Paris, 1976.
27. Habermann, B., The efficiency of miscible displacement as a function of mobility ratio. *J. Petroleum Technol.,* **219** (1960) 264–72.
28. Homsy, G. M., Viscous fingering in porous media. *Ann. Rev. Fluid Mech.,* **19** (1987) 271–311.
29. Marle, C.-M., *Multiphase Flow in Porous Media,* Technip, Paris, 1981, Chapter 7.
30. Yortsos, Y. C. & Huang, A. B., Linear stability analysis of immiscible displacements: Part 1—simple basic flow profiles. *SPE Reservoir Engineering,* **1**(4) (July 1986) 378–90.
31. Chikhlivala, E. D. & Yortsos, Y. C., Investigations on viscous fingering by linear and weakly nonlinear stability analysis. *SPE Reservoir Engineering,* **3**(4) (November 1988) 1268–78.
32. Chorin, A. J., The instability of fronts in a porous medium. *Commun. Math. Phys.,* **91** (1983) 103–16.
33. Christie, M. A., High-resolution simulation of unstable flows in porous media. *SPE Reservoir Engineering,* **4**(3) (August 1989) 297–303.
34. Lake, L. W., *Enhanced Oil Recovery,* Prentice Hall, Englewood Cliffs, 1989.

IMPROVED WATERFLOOD: THE CHEMICAL PROCESSES

2 Mobility control by polymers

Guy Chauveteau
Institut Français du Pétrole, Rueil-Malmaison, France

and

Kenneth S. Sorbie
Heriot-Watt University, Edinburgh, UK

2.1	**Introduction**	44
2.2	**Structure of polymers**	46
2.3	**Macromolecules in solution**	48
2.3.1	Static conformations	48
2.3.2	Dynamic properties	49
2.3.3	Concentration regimes	51
2.4	**Polymer solution rheology**	51
2.4.1	Shear flows	52
2.4.2	Elongational flows	53
2.4.3	Complex flows	54
2.5	**Retention of polymers**	56
2.5.1	Specific features of polymer adsorption	57
2.5.2	Adsorption on mineral surfaces	59
2.5.3	Mechanical entrapment and hydrodynamic retention	60
2.5.4	Retention kinetics and polymer propagation	61
2.6	**Polymer stability**	63
2.6.1	Chemical stability of polymers	63
2.6.2	Mechanical degradation	65
2.6.3	Biological degradation	66
2.7	**Rheology in porous media**	67
2.7.1	Effects of pore geometry	67
2.7.2	Influence of pore size	68
2.7.3	Rheology of adsorbed polymer layers	72
2.8	**Transport in porous media**	74
2.8.1	Transport parameters	74
2.8.2	Viscous slug propagation	75

2.9 Oil displacement mechanisms . . . 76
2.9.1 One-dimensional polymer flooding . . . 76
2.9.2 Improvement in areal sweep efficiency . . . 77
2.9.3 Flow mechanisms in flooding layered systems . . . 78

2.10 Numerical simulation of polymer flooding . . . 79
2.10.1 Polymer properties to be included in simulators . . . 79
2.10.2 Simulations including temperature-dependent degradation . . . 81
2.10.3 Simulation of adsorption effects . . . 81

2.11 Status and perspectives of polymer flooding . . . 82

2.12 References . . . 84

2.1 Introduction

This chapter describes the role of water-soluble polymers in decreasing the mobility ratio in waterflooding and hence improving linear areal and vertical sweep efficiency. Oil is left behind in the reservoir during waterflooding either because it is trapped by capillary forces—residual oil—or because it is in some way bypassed. The mechanisms of oil trapping have been discussed in detail in Chapter 1 and will not be pursued here since it is not the 'target oil' that polymer flooding is aiming at. Central to all polymer oil recovery mechanisms, however, is the idea of mobility ratio, M (see Chapter 1, Section 1.2.3, eqn (1.15)), which is reduced not only by increasing water viscosity but also by reducing water phase effective permeability.

For one-dimensional floods, we need only consider the effect of the unfavorable mobility ratio on the microscopic or local displacement efficiency. This effect is usually taken into account by using Buckley–Leverett frontal displacement theory (see Chapter 1, Ref. 25). In two-dimensional linear or areal floods, poor sweep occurs if there is an unfavorable mobility ratio because of large-scale fingering or channeling, as shown schematically in Fig. 2.1(a). A more stable displacement may be achieved using polymers as shown in Fig. 2.1(b). Poor vertical sweep efficiency arising from reservoir heterogeneity and the possible remedy using polymers are shown in Figs 2.2(a) and (b). It is the oil that has been bypassed, either in the areal or vertical plane along with the 'slowly' recovered oil behind low saturation water fronts, which forms the target oil for the polymer recovery process. Thus, a reduced mobility ratio caused by introducing poly-

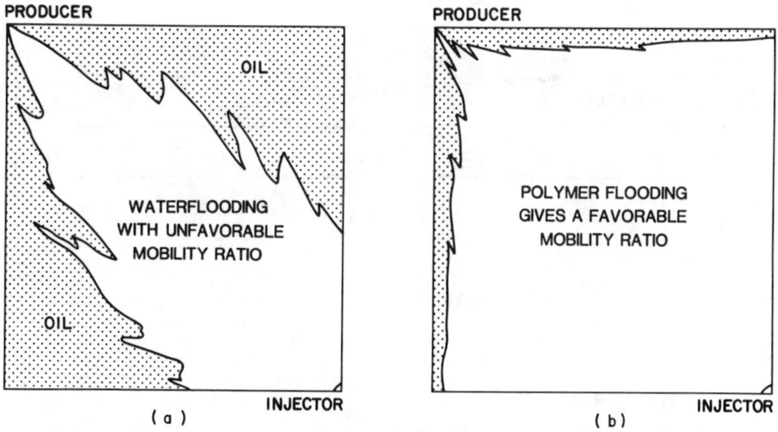

Fig. 2.1 Improvement in areal sweep of an homogeneous five-spot by use of polymers.

mers into the injected brine may improve the microscopic (local), areal and vertical oil displacement efficiencies.

After an intensive period of activity in the 1950s in studying waterflooding as an oil recovery process,[1] the strengths and weaknesses of this technique were fairly well established. In particular, the inefficiency of the waterflood oil displacement mechanism due to an unfavorable mobility ratio and heterogeneity were largely known. In order to remedy this situation, polymers were first suggested in the early 1960s as a means of increasing the drive water viscosity and, also of reducing formation permeability.[2–5] However, the poor filterability and sometimes the formation plugging characteristics of most of these early polymers were mainly due to supermolecular aggregates or microgels resulting from the manufacturing process.[6] One of the attractions of polymers is that they can significantly increase the viscosity of the injected brine by factors of between 3 and 20 at very low concentrations of a few hundreds of ppm. Another factor influencing the emergence of polymers for enhanced oil recovery (EOR) is that they were readily available from other industrial applications. At present, there are two main polymers which are commercially available for oil recovery application—the synthetic polymer polyacrylamide (PAM) which is usually supplied in its ionic form (HPAM), and the biopolymer, xanthan. However, not all industrial products have satisfactory filterability and stability pro-

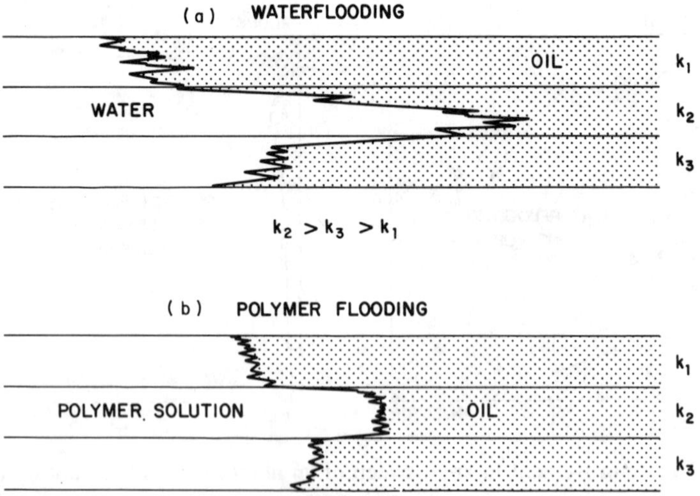

Fig. 2.2 Improvement in vertical sweep of a layered reservoir by use of polymers.

perties. Throughout this early period and through the 1970s a large number of polymer flooding field applications were carried out with varying degrees of success. In this sense, polymer flooding is an improved oil recovery process in which field application very much preceded knowledge, since today there are still several issues concerning the polymer flooding process which are poorly understood.

2.2 Structure of polymers

Polymer molecules, or macromolecules, are long chains of repeating units (monomers) linked by covalent bonds. Their solubility in water is primarily due to the chemical nature of monomers, and it is increased when there are charged groups on the chain (polyelectrolytes). Among commercially-available water-soluble polymers proposed as mobility buffers for EOR, only polyacrylamides and xanthans have been both extensively studied and used in field applications.[7]

Anionic polyacrylamides, which have been used in 95% of field applications, are obtained either by polymerization of acrylamide and subsequent hydrolysis or by direct copolymerization of acrylamide with sodium acrylate, leading to the following chemical structure:

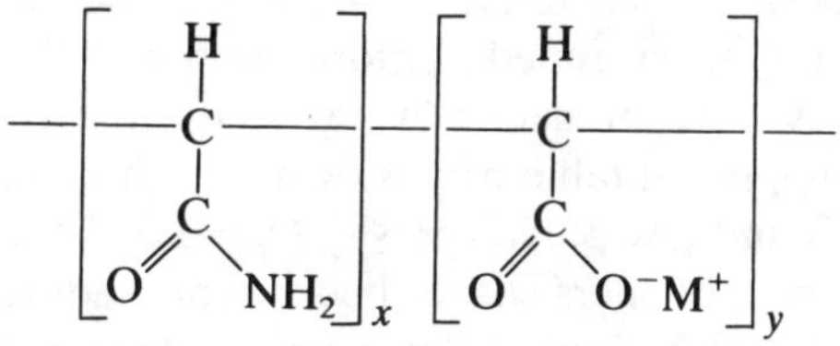

Their degree of hydrolysis or fraction of ionizable groups $\tau_h = y/(x + y)$ varies from 15 to 35% for currently used polyacrylamides. These synthetic polymers are intrinsically flexible, have very high molecular weights and are strongly polydisperse. Their weight average molecular weights $\bar{M}_w$ are around 10^7 and their polydispersity index I_p (weight-to-number molecular weight ratio) ranges from 2 to 3.

Xanthan is an extracellular polysaccharide produced by fermentation of the bacterium *Xanthomonas campestris*. Its chemical structure (Fig. 2.3) consists of a linear 1-4-linked D-glucan chain substituted on every glucose residue by a trisaccharide side chain. In the native form,

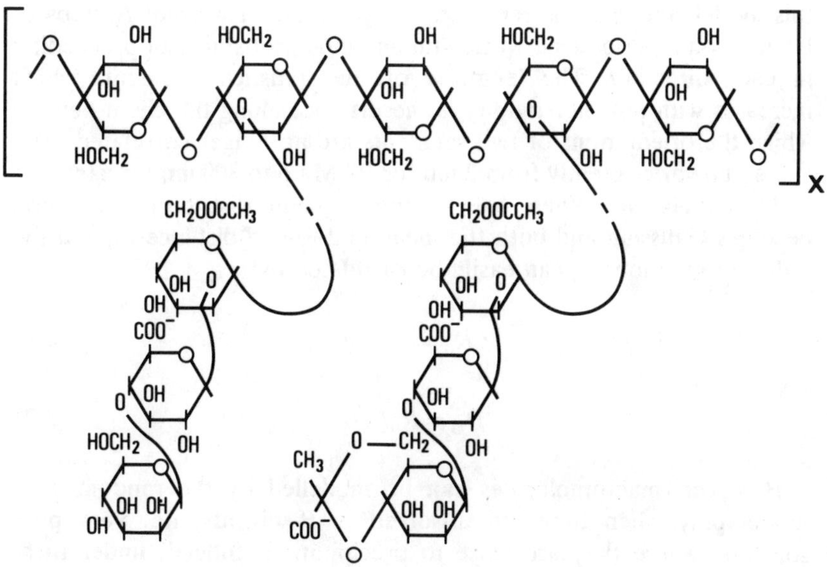

Fig. 2.3 Chemical structure of xanthan.

every internal mannose on the side chain is acetylated, whereas the terminal mannose has a variable degree of pyruvate substitution, depending on both the bacterial strain and fermentation conditions. Xanthan has a double-stranded conformation stabilized by hydrogen bonds when dissolved in saline water and thus is strongly rigid. Microgel-free samples suitable for EOR have a high molecular weight ($\bar{M}_w \approx 4\text{–}5 \times 10^6$) and low polydispersity ($I_p \approx 1{\cdot}3\text{–}1{\cdot}5$).

Other types of polymers have been proposed for use in high temperature and high salinity environments. Among these polymers, scleroglucan, a nonionic polysaccharide with a triple-stranded conformation,[8] and a synthetic terpolymer based on acrylamide, vinylamide and vinylsulfonate[9] seem to be the most promising.

2.3 Macromolecules in solutions

2.3.1 *Static conformations*

Among the various models proposed to describe the conformation of isolated polymer chains,[10,11] the *random walk model* is the simplest. In this model, the chain is represented by a random walk of N steps of length a on a periodic network with an equal probability of proceeding in each direction. The length a of the statistical segment, which increases with polymer rigidity is the distance along the chain beyond which the orientations of two segments are no longer correlated. The value of a varies greatly from 2 nm for PAM up to 300 nm for xanthans and scleroglucan. When $N \rightarrow \infty$, the segment distribution function becomes Gaussian and both the mean end-to-end distance R_0 and the radius of gyration R_G can easily be calculated by:

$$R_0 = aN^{1/2} \tag{2.1}$$

and

$$R_G = R_0/\sqrt{6} \tag{2.2}$$

However, macromolecules can be modelled by the random walk model only when they are dissolved in θ-solvents, i.e. very poor solvents, where they are close to precipitation. Indeed, under these conditions, the volume excluded for random walk is reduced to the volume of the polymer chain itself, i.e. is practically negligible. On the contrary, when polymer is dissolved in a good solvent or is a polyelectrolyte, which is almost always the case in EOR, the polymer chain is surrounded by strongly-bound solvent molecules and/or by a repulsive electrostatic field creating a substantial excluded volume. In this case, the *self avoiding walk model,* which takes into account the effects of excluded volume, must be used. The excluded volume which increases with solvent quality and as salinity decreases for polyelectrolytes causes an expansion of polymer coils, the dimension of which varies for large N as:

$$R_0 = aN^{3/5} \tag{2.3}$$

$$R_G = R_0/\sqrt{7} \tag{2.4}$$

For strictly rigid polymers, R_G is easily calculated by:

$$R_G = aN/\sqrt{12} \tag{2.5}$$

For semi-rigid polymers such as xanthans and scleroglucans, the number of statistical segments N is always small enough to assume the effects of excluded volume to be negligible. Under these conditions, the worm-like chain model can be used to calculate chain dimensions:

$$R_0^2 = 2qL_c[1 - (1/x)(1 - e^{-x})] \tag{2.6}$$

and

$$R_G^2 = (qL_c/3)\{1 - 3/x + (6/x^2)(1 - (1/x)(1 - e^{-x})\} \tag{2.7}$$

where L_c is the contour length, q the persistence length and $x = L_c/q = 2N$.

For polyelectrolytes, a reasonable approximation consists of considering persistence length q (or the length of statistical segment a, since $a = 2q$) as the sum of an intrinsic part q_i and of an electrostatic part q_e. Over a wide range of salinity C_s, the effects of the screening of electrostatic repulsions can be accounted for by assuming $q_e \propto C_s^{-1/2}$. Typical values for q, R_G and L_c are respectively 140, 500 and 2000 nm for xanthan and scleroglucan and 1250 and 25 000 nm for HPAM in 20 g/liter NaCl.

2.3.2 *Dynamic properties*

The direct derivation of solution viscosity from macromolecule conformation may be carried out theoretically by integrating the elementary viscous friction due to each segment over the entire macromolecule. This elementary friction f on a monomer of radius r is proportional to the solvent viscosity η_s and to the difference in velocity V between this monomer and the solvent according to Stokes law:

$$f = 6\pi\eta_s rV \tag{2.8}$$

However, such a calculation of viscosity is beyond the capacity of the most powerful computers when $N > 20$. In practice, viscosity measurements are used to estimate macromolecule dimensions using the Fox–Flory equation:[11]

$$[\eta]_0 = \Phi' R_G^3/M \tag{2.9}$$

where Φ' is an universal constant, M is the molecular weight and $[\eta]_0$ is zero shear rate intrinsic viscosity. M must be determined by light scattering and $[\eta]_0$ is deduced from viscosity measurements at different polymer concentrations. Indeed, if we except the case of polyelectrolytes at very low salinity, the viscosity of polymer solutions at low concentrations increases with the polymer concentration C according to the Huggins equation:

$$\eta_{r0} = 1 + C[\eta]_0 + k'C^2[\eta]_0^2 \tag{2.10}$$

where η_{r0} is relative viscosity, i.e. the polymer-to-solvent viscosity ratio, and k' is the Huggins constant, which characterizes intermolecular interactions. Typical values for $[\eta]_0$ are in the range of 6000–9000 cm^3/g for xanthan and scleroglucan and for HPAM increases from 4000 cm^3/g to 12 000 cm^3/g when salinity decreases from 20 to 2 g/liter NaCl. In the low shear rate limit, for macromolecules that are highly permeable to solvent with short range repulsive interactions, k' is equal to 0·4 whatever their conformation may be.

However, for highly flexible molecules in good solvents, polymer size decreases significantly as concentration increases, leading to k' values lower than 0·4. In poor solvents, macromolecule permeability decreases and k' increases beyond 0·4. Values of k' higher than 0·4 for rigid polymers and higher than 1 for flexible ones are indicative of attractive interactions.[7,8]

In practice, the direct measurement of M by light scattering is not easy and the empirical Mark–Houwink equation is often used to estimate M from viscosity:

$$[\eta]_0 = KM^{\alpha} \tag{2.11}$$

where K and α are parameters depending on the polymer–solvent system. For very large N, $\alpha = 0{\cdot}5$ and $0{\cdot}8$, for θ and good solvents, respectively. For small to medium values of N, such as for polyelectrolytes in low salinity waters, $0{\cdot}8 < \alpha < 1$, and for strictly rigid polymers $\alpha = 1{\cdot}8$.

Some other macromolecular characteristics, such as relaxation times τ and diffusion coefficients D, are also needed to predict rheological and transport properties. Most equations that may be used to estimate these characteristics are found in Refs 7, 10 and 11. Among these equations, the most useful is probably that of Rouse, which gives the longest relaxation time τ of isolated flexible macromolecules:

$$\tau = (6/\pi^2)\eta_s[\eta]_0 M/\mathrm{RT} \tag{2.12}$$

where R is the gas constant and T the absolute temperature. For HPAM in 20 g/liter NaCl, $\tau \approx 10$ ms and $D \approx 2 \times 10^{-8}\ \mathrm{cm^3/s}$.

2.3.3 *Concentration regimes*

Both static conformation and dynamic properties reported above are only valid for very dilute solutions in which macromolecules are isolated. However, as the concentration increases, macromolecules begin to interact and their size decreases progressively. Above a critical concentration C^* usually estimated by $C[\eta]_0 = 1$, they begin to overlap, and then form an entangled network in which polymer segment density becomes nearly uniform. However, the transition between the dilute and semi-dilute regime is smooth for high $\bar{M}_w$ macromolecules dissolved in good solvents. In this later case, the semi-dilute regime is reached only for $C[\eta]_0 > 3$–5. Since the concentration of solutions used in polymer flooding is frequently such that $C[\eta]_0 < 5$, the properties of semi-dilute solutions are not discussed here but can be found in the literature.[10,11]

2.4 Polymer solution rheology

The rheological behavior of polymer solutions, and particularly the variations of their viscosity with deformation rate, depend strongly on the type of flow. In this section, shear, elongational and complex or mixed flows will be examined successively. The deformation rate range of interest is very wide, typically from 1 to $10^4\ s^{-1}$, corresponding to the values prevailing in the bulk of the reservoir and near the well bores, respectively.

2.4.1 *Shear flows*

In pure shear flows where the velocity gradient $\dot{\gamma}$ (or shear rate) is perpendicular to the flow direction, macromolecules rotate and are thus subjected to periodic extensional and compressional stresses. As $\dot{\gamma}$ increases, hydrodynamic stresses increase and their effects are non-negligible compared with those of Brownian motion when rotation time becomes smaller than the longest rotational relaxation time τ_r, thus explaining that:

(1) At low shear rates ($\dot{\gamma}\tau_r < 1$), macromolecules rotates at a constant angular velocity ($\omega = \dot{\gamma}/2$) without significant conformational change. Hence, the viscosity remains constant and the regime is Newtonian.
(2) At higher shear rates ($\dot{\gamma}\tau_r > 1$), the macromolecules are either deformed for flexible polymers or oriented for rigid rods into the flow direction. Their space and time-averaged dimensions in the direction perpendicular to flow decreases, and thus shear viscosity decreases. The regime is shear-thinning.

Typical plots of viscosity versus shear rate are shown in Fig. 2.4 for xanthan solutions. Usually, such curves are effectively fitted by using the empirical Carreau model over a wide range of shear rates:

$$\eta_r = \eta_{r_0}/[1 + (\dot{\gamma}\tau_r)^2]^m \qquad (2.13)$$

where η_{r_0} is the Newtonian relative viscosity, and $m = (1 - n)/2$ where n is the standard power law exponent.

The analysis of viscosity–shear rate curves gives access to most of the macromolecular characteristics (polymer size, Huggins constant, relaxation time, rigidity) that are necessary to predict the behavior in porous media flows.[12]

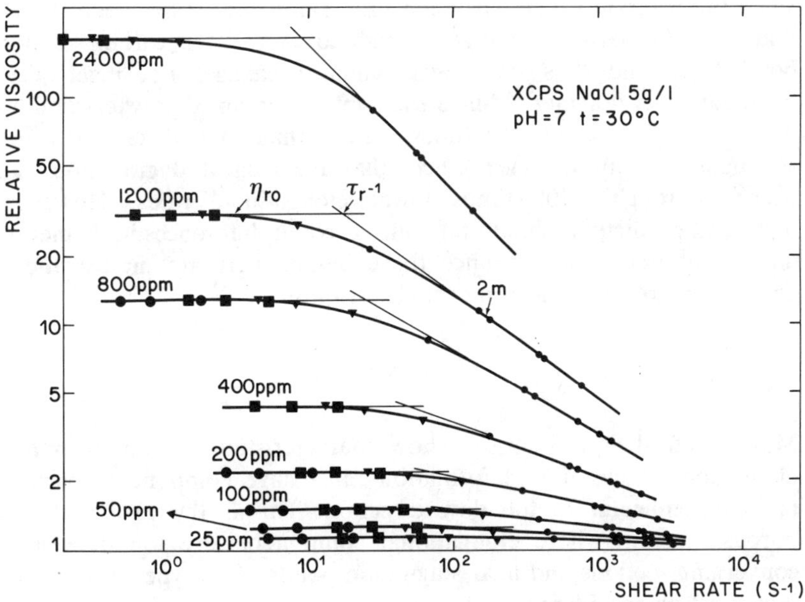

Fig. 2.4 Viscosity of xanthan solutions in pure shear flows. (From Ref. 12, reproduced by permission of The Society of Rheology, Inc.)

2.4.2 *Elongational flows*

In pure elongational flows where the velocity gradient $\dot{\varepsilon}$ (or elongation rate) is parallel to flow direction, the condition to be respected for the deformation or orientation of macromolecules is also $\dot{\varepsilon}\tau > 1$.[11,13] However, the effective deformation that occurs in the flow direction may be very large for flexible polymers and a full stretching is achieved only if the macromolecule remains in the elongational flow over a sufficiently long time.

The first consequence of macromolecule stretching in elongational flows is a strong increase in viscosity. For high molecular weight fiexible polymers such as HPAM used in EOR, the elongational viscosity may be as high as 10^4 times the zero shear rate viscosity.[13] The second consequence is that the macromolecule is subjected to an internal tensile force F which is given by:

$$F \propto \eta_s \dot{\varepsilon} L_c^2 \tag{2.14}$$

when the macromolecule has been fully elongated. It has been shown that F could possibly be high enough to exceed the carbon–carbon bond force and thus may cause chain breakage, i.e. mechanical degradation of polymer.[14] Since the contour length of polyacrylamides used in EOR is about ten times greater than that of xanthans and scleroglucan, eqn (2.14) predicts that mechanical degradation can occur at roughly 100 times lower elongational rates. However, improved predictions should take into account differences in diameters and in internal cohesion since these biopolymers are multistranded and thus more resistant to mechanical rupture.

2.4.3 *Complex flows*

Most practical flow situations show characteristics of both shear and elongational flows, i.e. deformation rates have components parallel and perpendicular to the flow direction. This is the case within a porous medium where elongational components are found both in converging sections and near stagnation points. Two types of complex flow must be considered:

(a) *Flows where the only elongational zones are converging/diverging sections.* The rheology of both polyacrylamide and xanthan solutions in such flows has been studied by using capillary systems of varying cross-section. A typical example of the variation of the effective viscosity of a HPAM solution with the maximum wall shear rate $\dot{\gamma}$ in constricted sections is shown in Fig. 2.5 (curve with solid squares). From the results of experiments carried out with such capillary systems,[7] the following conclusions can be drawn:

(i) The effective viscosity remains nearly equal to shear viscosity up to a critical shear rate $\dot{\gamma}^*$. Beyond $\dot{\gamma}^*$, the elongational rate in converging sections is high enough to induce macromolecule elongation. As a consequence, viscous friction in the converging zones increases drastically and produces significant thickening behavior, although viscosity in the shearing zones is still decreasing (curve with open squares in Fig. 2.5).

(ii) The criterion for the onset of thickening behavior ($\dot{\gamma}^*\tau = 10$, i.e. $\dot{\varepsilon}\tau \approx 1$ in converging sections) is valid for polyacrylamides in good solvents, and independent of solvent nature and viscosity.

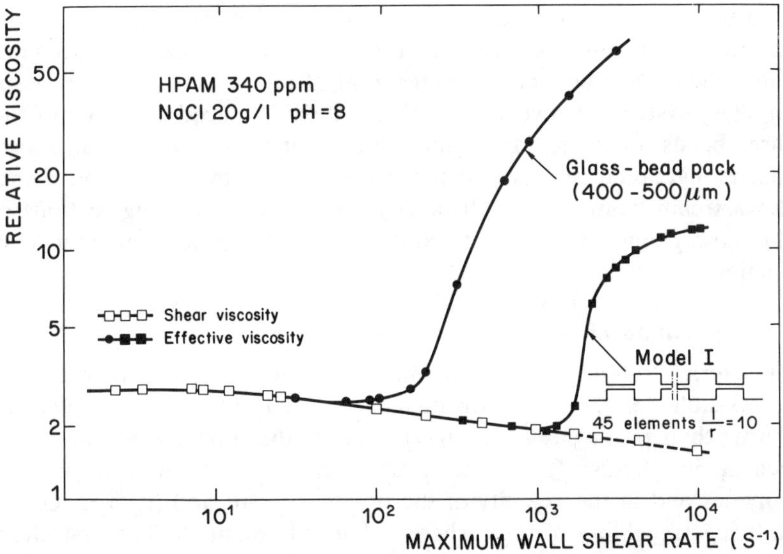

Fig. 2.5 Influence of pore geometry on the thickening behavior of polyacrylamide solutions in complex flows. (From Ref. 7, reproduced by permission of The Society of Petroleum Engineers.)

(iii) The value of $\dot{\gamma}^*$ is independent of the entrance shape, the contraction ratio, the distance between two successive contractions and of the number of contractions.

(iv) The maximum effective viscosity at the highest $\dot{\gamma}$ increases with contraction ratio and molecular weight, is nearly insensitive to salinity for HPAM and is very small for xanthan, as expected from its rigidity and small contour length.

(v) The increase in viscosity beyond $\dot{\gamma}^*$ is sharper when the persistence length decreases, for instance when salinity increases with HPAM.

(vi) The effective viscosity just beyond $\dot{\gamma}^*$, increases more sharply when there are successive contractions separated by distances too short to achieve complete relaxation of macromolecules. Under these conditions, elongation and thus viscosity increase with the number of contractions.

(b) *Flows which are elongational both in converging sections and stagnation points.* The behavior of HPAM solutions in such flows was studied in small glass bead arrangements.[7] For the converging flow towards pore throats between three beads, the regime becomes shear-thickening beyond the same critical shear rate ($\dot{\gamma}\tau = 10$) as in capillary systems. However, if a fourth bead is placed upstream of the three beads, thickening behavior is found for $\dot{\gamma}\tau \approx 1$. This result shows that elongational rate are 10 times higher near the stagnation points downstream from the fourth bead than in the converging sections in this flow geometry, which is close to that of a porous medium (see Section 2.7.1).

2.5 Retention of polymers

The term 'retention' is used here to cover all the mechanisms responsible for the reduction of mean velocity of polymer molecules during their propagation through porous media. Polymer retention has two main effects: (i) a decrease in the concentration of flowing polymer and in the velocity of the polymer front, and (ii) a reduction in the permeability of rock where polymer is retained. The first effect is always detrimental to polymer flooding efficiency whereas the second is expected to be mainly beneficial since it contributes to the reduction of the mobility of both the polymer slug and chase brine.

Three distinct mechanisms can lead to the retention of macromolecules in flows through porous media: (i) adsorption on the surface of pores, (ii) mechanical entrapment in pores surrounded by constrictions smaller than macromolecules, and (iii) hydrodynamic retention, i.e. the local accumulation of macromolecules by a diffusion–convection controlled mechanism. According to the retention mechanism, macromolecules are retained in different locations within the pore space, as shown in Fig. 2.6.

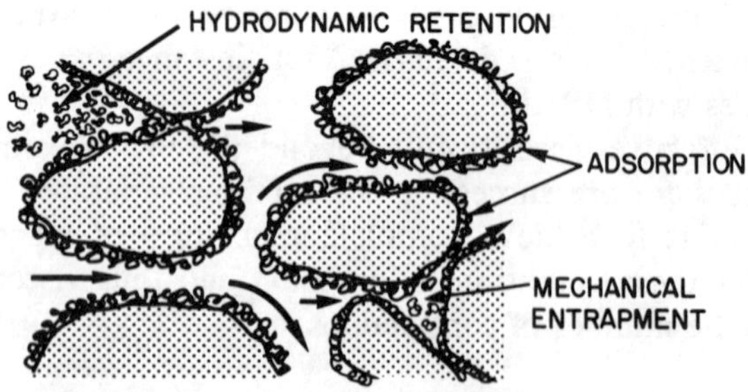

Fig. 2.6 Location of the various types of polymer retention in porous media flows.

2.5.1 *Specific features of polymer adsorption*

Most specific features of polymer adsorption such as the high affinity character of isotherms, its pseudo-irreversibility, the thickness of adsorbed layer, etc., depend on the conformation of the adsorbed molecules.

(a) *Conformation of adsorbed macromolecules.* When macromolecules are adsorbed, some portions of the polymer chains are in direct contact with the surface as trains, and the remaining portions extend into bulk solution as loops and tails. The extension perpendicular to the surface of loops and tails, which determines the thickness of the adsorbed layer, increases with molecular weight for flexible polymers. Semi-rigid polymers, i.e. low molecular weight flexible chains or high molecular-weight rigid chains, such as xanthan and scleroglucan, adopt a flat conformation when adsorbed on a plane surface.

(b) *Thermodynamics of polymer adsorption.* The driving factor for polymer adsorption is the change in free enthalpy ΔG of the system which includes all the species that are competing for adsorption, i.e. water, small ions and polymer. The condition for polymer adsorption is that all enthalpic and entropic terms including contributions from various origins such as van der Waals and electrostatic forces, ion–dipole interactions, solvation, dilution, etc., must lead to a negative value of ΔG:

$$\Delta G = \Delta H - T\Delta S < 0 \tag{2.15}$$

where H and S are enthalpy and entropy per mole, respectively.

The sign and magnitude of the enthalpic term depend on the interactions between all the species. This term can be strongly negative if polymer and surface carry charges of opposite signs, particularly at low salinities. Such situations, which may lead to very high levels of adsorption, are unusual in polymer flooding. Relatively high negative values of enthalpic term are also encountered if direct hydrogen bonds between the polymer and the surface occurs, while they are much smaller when polymer/surface interactions are due only to van der Waals forces or ion-dipole interactions.

The adsorption of polymer causes a slight loss of chain conformational entropy while the removal of water molecules or small ions from the vicinity of the surface or from the hydration layer of the polymer significantly increases the entropy of the system. On the whole, therefore, the contribution of entropy is often favorable to adsorption, thus explaining why polymer adsorption is sometimes an endothermic process. This means that polymer can be adsorbed even if its binding energy with the surface is less than that of the solvent.

(c) *Adsorption isotherms.* When high molecular weight polymers are adsorbed on a surface, there is always a large number of segments in contact with the surface and the overall adsorption energy per macromolecule is often high even if the adsorption energy per segment is small. As a consequence, isotherms have a high-affinity nature, i.e. at low concentrations, the adsorption density, Γ, rises sharply, while at higher concentrations it reaches a pseudoplateau. The adsorption density at the plateau Γ_m depends strongly on polymer flexibility. For a rigid polymer such as xanthan, which is flattened on the surface, Γ_m does not depend on molecular weight and increases when polymer–polymer electrostatic repulsions decrease, from low values (100 $\mu g/m^2$) up to reach a maximum (600–800 $\mu g/m^2$) corresponding to a monolayer at low pH or high salinities.[15] For flexible polymers such as polyacrylamides, the adsorption density increases with molecular weight mainly at high adsorption energy and may vary strongly from low values (80 $\mu g/m^2$) corresponding to 0·3 monolayer when the adsorption energy is nearly zero, as is the case for the anionic HPAM on a negative surface like sand just above critical salinity,[16] up to very high values (8000 $\mu g/m^2$ or 30 monolayers) when adsorption occurs on a positively-charged surface.[17]

(d) *Kinetics of polymer adsorption.* In most situations, adsorption is delayed by the time of access to the surface, which is often controlled by diffusion, thus favoring faster access of low molecular weight fractions. However, even after the macromolecules reach the surface, the following further changes are expected to occur: (i) a change in conformation from bulk to adsorbed state; (ii) a reconformation due to the competition with newly arriving macromolecules; (iii) exchanges due to preferential adsorption of high molecular weight fractions; (iv) a retrogradation of polymer and surface ionization due to a too high charge density at the vicinity of the surface. All of these phenomena usually contribute to the delay in the attainment of equilibrium, which may vary between some minutes to some weeks.

2.5.2 *Adsorption on mineral surfaces*

A wide variety of minerals is found in oil reservoirs, e.g. quartz, feldspars, carbonates, clays (kaolinite, montmorillonite, illite, etc.).

The adsorption density Γ of polyacrylamides on quartz surfaces is always low ($\Gamma < 500\ \mu g/m^2$), because of the absence of strong interactions with surface silanols. Γ is independent of pH and salinity for PAM, whereas it increases with salinity and as pH decreases for HPAM.[16] The variations of xanthan adsorption with salinity and pH are similar to that of HPAM.[15]

The adsorption on kaolinite has been studied in detail for both polyacrylamides[18] and xanthans.[19] For PAM, Γ is high on lateral faces ($3500\ \mu g/m^2$) due to hydrogen bonds between carboxyl groups and surface aluminols, but low on the hydroxyde aluminium basal face ($\sim 500\ \mu g/m^2$) and zero on siloxan basal face (Fig. 2.7). In contrast, xanthan adsorbs only on basal faces. For both polymers, adsorption decreases very sharply for pH > 10. For ionic HPAM and xanthans, Γ

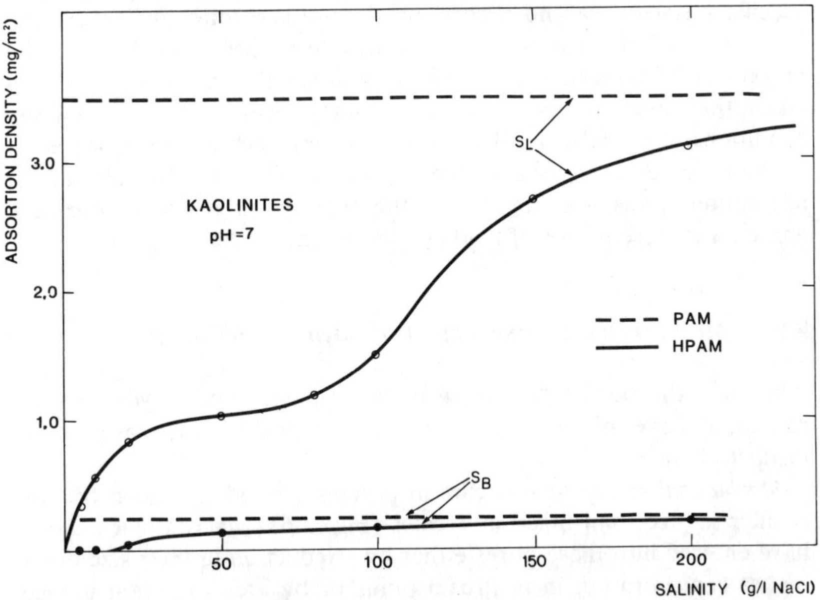

Fig. 2.7 Influence of salinity on the adsorption of polyacrylamides respectively on lateral (S_L) and basal (S_B) surfaces of kaolinites. (From Ref. 18, reproduced by permission of The Academic Press, Inc.)

increases with salinity as expected from the effects of both charge screening and ion condensation and reaches that of PAM at high salinities.

The potential adsorption on montmorillonite is considerable, as for instance, for PAM on Na-montmorillonite at low salinity and high polymer concentration.[19] However, many studies[19] have shown that the effective adsorption of anionic polyelectrolytes on clays is much smaller than that of neutral polymers and decreases when salt concentration is increased due to particle coagulation, which reduces accessibility to clay surface.

The adsorption density of PAM on calcium carbonate is low ($\Gamma \approx 400\ \mu g/m^2$) and close to that found on TiO_2 and siliceous minerals, while HPAM adsorption is only weakly dependent on pH and salinity.[21]

In practice, the determination of adsorption in natural cores always requires direct measurement by material balance in a core flow experiment. Indeed, the assumption that overall adsorption can be estimated from the weight content of each mineral multiplied by its specific surface area and adsorption density is erroneous because the major part of clay surface is not accessible to polymers. Indeed, the proportion of accessible surface may be as small as a few per cent and adsorption levels rarely exceed 20–30 $\mu g/g$, even under high salinity conditions for well-consolidated sandstones. As a consequence, if reservoir cores are disaggregated when received in the laboratory, adsorption levels will depend on the subsequent packing procedure and data representative of field conditions cannot be obtained.

2.5.3 *Mechanical entrapment and hydrodynamic retention*

Although the terms 'mechanical entrapment' and 'hydrodynamic retention' have often been used interchangeably, they are in fact distinct phenomena.

Mechanical entrapment occurs in pores surrounded by constrictions smaller in size than macromolecules (Fig. 2.6). Macromolecules may have entered into these pores either by a reduction in their size due to higher concentration in upstream pores or by a deformation induced by a high elongational rate at the pore entrance. The trapping of polymer occurs either when polymer concentration decreases at the rear of the slug or when the flow rate is reduced. Retention by mechanical entrapment is thus partly irreversible. The amount of polymer retained by this mechanism is usually low.

Hydrodynamic retention results from a flow-induced increase in polymer concentration in the converging zones upstream from pore constrictions having a size smaller than that of macromolecules (Fig. 2.6). In these zones, polymer concentration, and thus retention, increases as long as the convection flux remains higher than the sum of the fluxes of diffusion and of that resulting from the passage through pore constriction. For flexible polymers, the flux through pore constriction increases continuously as polymer concentration increases, due to the correlative reduction in polymer size and thus polymer concentration and retention reach a maximum.[22] Hydrodynamic retention is expected to increase with flow rate at low flow rates and then to decrease when the size of macromolecules is so reduced by concentration effects that the volume of pore space where the retention occurs decreases significantly. In addition, at the highest flow rates, macromolecules can be elongated and oriented (see Section 2.4.2) and in this way can pass easily through even the smallest constrictions.

The existence of hydrodynamic retention was shown for xanthan[23] and polyacrylamides[24] flowing through natural porous media. For polyacrylamides, full reversibility could be observed, as shown in Fig. 2.8. Although the pore fraction in which polymer can be retained by hydrodynamic retention is usually small, the concentration in the retention zones can be very high so that the amount of polymer retained by this mechanism may be higher than that due to adsorption.

When microgel-free polymers well-suited to polymer flooding are injected in porous media of medium to high permeabilities ($k \geqslant 0{\cdot}1\ \mu m^2$ [100 mD]), a maximum level of hydrodynamic retention is reached for a given flow rate leading to a moderate and thus nondetrimental increase in pressure drop. However, if k is too small or if the polymer sample contains microgels, a gel-like cake can be formed on the core inlet and even inside pore structure, possibly leading to severe core plugging.[7]

2.5.4 *Retention kinetics and polymer propagation*

Several phenomena related to the restructuring of adsorbed polymer layers as discussed in Section 2.5.1(d) may explain why the completion of adsorption requires a relatively long time even in batch

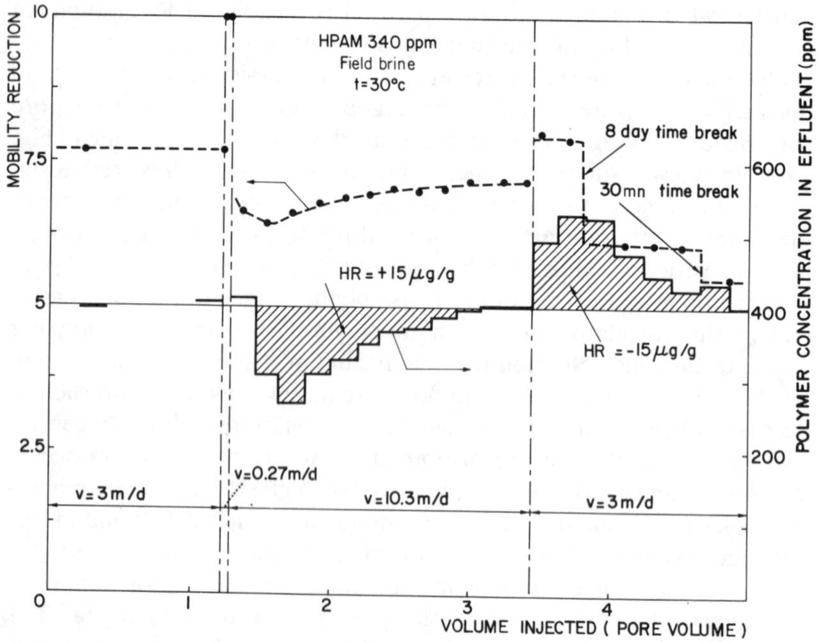

Fig. 2.8 Reversibility of hydrodynamic retention (HR) of polyacrylamides with flow rate. (From Ref. 24, reproduced by permission of The Society of Petroleum Engineers.)

experiments where agitation ensures rapid access to solid surfaces. However, in porous media flows, only a fraction of the pore surface is accessible by hydrodynamic convection. On this part of the surface, the time of access to the surface is short and independent of molecular weight. However, other parts of the surface are accessible only by diffusion and thus the low molecular weight fractions are first adsorbed. This explains why, in laboratory experiments, the polymer samples that arrive first at the core outlet are depleted in low molecular weight fractions and thus have higher molecular weights even in adsorbent porous media, i.e. in the absence of surface exclusion chromatography.[25] On the contrary, in field tests, the exchanges between flowing polymer and the adsorbed layer which result in preferential adsorption of the highest molecular weight molecules have sufficient time to be completed, and low molecular weight components arrive first in producing wells.[26]

2.6 Polymer stability

When polymers are used in oil recovery, it is clearly important that their physical properties are not rapidly degraded. The main property of interest in this respect is generally polymer solution viscosity. Polymer degradation refers to any process that will change the molecular structure of the macromolecule and the main degradation pathways concern chemical, mechanical and biological degradations.

2.6.1 *Chemical stability of polymers*

A large number of laboratory studies have been carried out to investigate the chemical stability of polymers in the context of EOR. A typical example of such applied studies is that of Davison and Mentzer[27] who evaluated over 140 polymers of several different types for viscosity retention and porous medium flow performance under high temperature (90°C) and sea water conditions. In their study, scleroglucan and polyvinylpyrrolidone had the best thermal stabilities, xanthan had 'intermediate' stability. Polyacrylamides were found to be particularly bad under these conditions and precipitated at 90°C within 60 days. A detailed study of HPAM behavior[28] showed that this polymer does precipitate from solution in hard brines, but was much more stable, under conditions of lower salinity/hardness and lower temperature. More basic research works contributed to the understanding of the main degradation mechanisms, which may be considered under two headings: short-term free-radical attack, and long-term hydrolysis of the sidegroups and backbone of the macromolecules.

(a) *Short-term free-radical attack.* Both xanthan[29,30] and HPAM[31,32] are susceptible to free radical attack, which leads to degradation of the macromolecule. The initial free radicals may be formed by reaction between reducing species in solution and oxygen. Activators that may assist this process are chemicals such as peroxides or catalysts such as transition metals (e.g. iron) present in trace quantities in solution. Radical species will then react very rapidly with oxygen, producing hydroperoxide radicals. These then cause propagation reactions which may proceed until oxygen has been fully consumed. The hydroperoxide produced may decompose rapidly or over a longer term producing more free radical species. Some of the subsequent reactions with these radicals may result in backbone scission within the polymer, which reduces the polymer viscosity. In order to remove the oxygen from solution, reducing agents such as sulfite and bisulfite have been proposed but their reaction with oxygen also creates free radicals which degrade the polymer.[24] Thus oxygen must be removed before the polymer is added to the injection water. However, the removal of oxygen does not always prevent polymer degradation. Addition of acrylamide monomers has been found to improve the stability of HPAM[32] but the stability is often determined by the nature of impurities in the samples, thus requiring specific packages for the efficient protection of the macromolecules.

(b) *Long-term hydrolysis of polymers.* The longer term hydrolysis of the molecule is largely unavoidable. The hydrolysis rate is a function of several parameters; the most important of which being pH and temperature for both polyacrylamides[31] and xanthans.[30]

The effect of pH on the stability of oxygen-free xanthan solutions was studied in some detail by Seright and Henrici.[30] They found that the pH of unbuffered xanthan solutions dropped within the first day when they were aged at high temperatures, due to the hydrolysis of acetyl groups and that this drop in pH was responsible for an increase in the degradation rate. Indeed, when the solutions are buffered, the total viscosity decay constant t_d, shows the behavior illustrated in Fig. 2.9. Figure 2.9 also indicates that xanthan is more stable at a pH between about 7 and 8 and the findings also indicate that the stability of all the commercial xanthans was quite comparable when the solutions were buffered. Thus, as reservoir systems are often strongly buffered, xanthan samples from different commercial sources could give similar degradation time constants.

Hydrolysis also plays a key role in the longer term degradation of HPAM.[31] The rate of hydrolysis, which involves the acrylamide side groups and not the main chain, is also at a minimum at neutral pH and increases with temperature. The higher the equilibrium hydrolysis level, the lower the salinity beyond which HPAM precipitates, which explains why the temperature stability of polyacrylamides is severely reduced in hard brines.

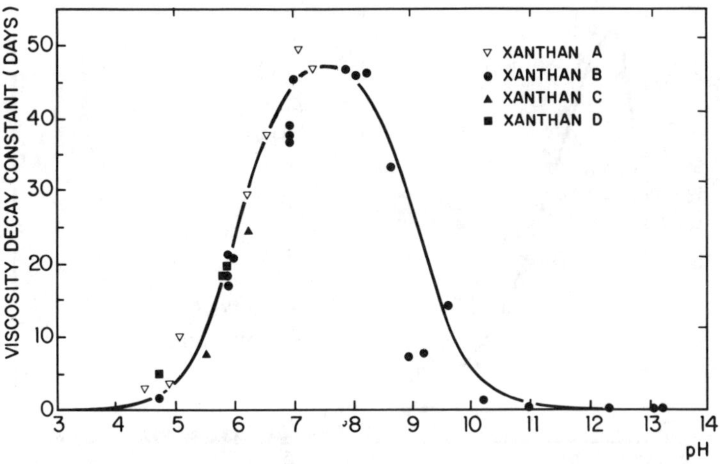

Fig. 2.9 Effect of pH on xanthan stability at 120°C. (From Ref. 30, reproduced by permission of The Society of Petroleum Engineers.)

2.6.2 *Mechanical degradation*

Mechanical degradation occurs almost exclusively in elongational flows for dilute solutions. This process, which occurs fairly rapidly in the high flow rate region close to the wellbore, essentially breaks the large molecules into fragments and thus changes the molecular weight distribution of the polymer. This shear-damaged polymer, which has a lower average molecular weight than the original polymer, may, however, still have satisfactory properties for a polymer flood.

One of the first experimental approaches to polyacrylamide degradation was to measure the viscosity of a polymer solution as a function of the number of passages N_p through capillaries of different lengths.[14] These simple experiments showed that mechanical degradation: (i) does not occur in shear flows up to $5 \times 10^4\,s^{-1}$; (ii) can be observed only when macromolecules have been stretched in the entrance elongational flow; (iii) increases with strain rates and reaches a maximum only after a large number of passages ($N_p > 500$); (iv) decreases the elongational viscosity more strongly than shear viscosity by preferentially affecting high molecular weight fractions.

A striking feature concerning the mechanical stability of the polymers commonly used in EOR is that xanthan appears to be

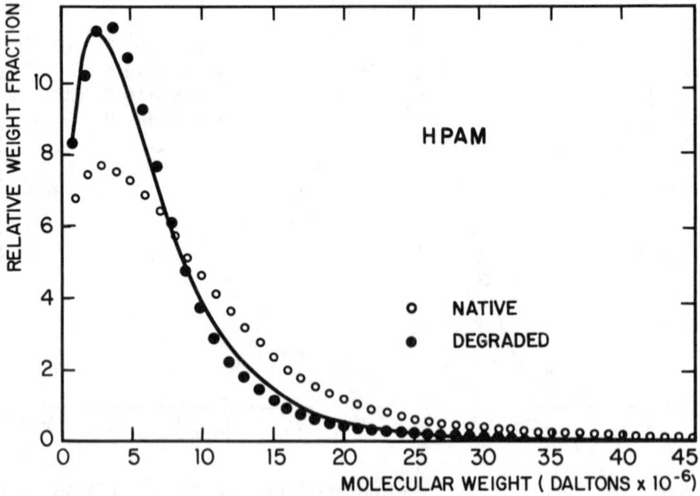

Fig. 2.10 Effects of the mechanical degradation on the molecular weight distribution of a polyacrylamide. (From Ref. 33, reproduced by permission of The American Chemical Society.)

extremely stable while polyacrylamides appear to be very degradable. Such observations are consistent with the predictions that can be derived from eqn (2.14) as discussed in Section 2.4.2.

Similar conclusions have been drawn from experiments in natural rocks and correlations were developed for the extent of mechanical degradation of HPAM as a function of a mean elongational rate and of the length of porous media. Seright *et al.*[33] directly measured the change in molecular weight distribution induced by flow through porous media and showed clearly (Fig. 2.10) the preferential degradation of the highest molecular weight fractions. In addition, mechanical degradation was shown to increase as permeability decreases and as salinity increases.

2.6.3 *Biological degradation*

Biological degradation due to the presence of bacteria may occur either during storage of the polymer before injection, or in the cooler regions of the reservoir. The almost universal solution to this problem is to use a biocide. Early tests concerning the use of biocides[24] concluded that most of these products, currently used in small quantities could be adsorbed on minerals or dissolved in oil and consequently lose their efficiency in EOR conditions while formaldehyde used at higher concentrations (500–5000 ppm) remained active. This conclusion was confirmed by a more extensive study including very convincing field tests.[34]

2.7 Rheology in porous media

The rheological behavior of polymer solutions in porous media is expected to differ from that observed in model systems, such as those considered in Section 2.4, for two main reasons: (i) the pore geometry is more complex and (ii) the dimensions of macromolecules are usually non-negligible compared with pore dimensions.

2.7.1 *Effects of pore geometry*

When pore dimensions are very large compared with the size of macromolecules, the rheology is determined only by the complex geometry of the pore channels. This complexity makes it impossible to calculate the mean value of the maximum wall shear rate in pore throats $\dot{\gamma}_{PT}$ which must be determined experimentally. The value of $\dot{\gamma}_{PT}$ has been estimated by fitting the onset of shear rate dependence in both viscometers and porous media flows assuming that:

$$\dot{\gamma}_{PT} = 4\alpha V_f / r \tag{2.16}$$

with

$$r = (8\,k/\Phi)^{1/2} \tag{2.17}$$

where V_f is frontal velocity, r is the pore throat radius, k is the permeability, Φ is the porosity and α is a geometrical factor characterizing the pore structure. The value of α, which is minimum in calibrated sphere packs ($\alpha = 1{\cdot}7$), increases up to 2·5 for sandpacks and can reach much higher values in natural heterogeneous media.[12]

By using eqn (2.16) with $\alpha = 1{\cdot}7$, a comparison was made between the rheological behavior of a flexible polymer solution in a glass bead pack and in a capillary system with successive constrictions (Fig. 2.5). The onset of thickening behavior is observed in the porous medium when the wall shear rate in pore throats $\dot{\gamma}_{pt}$ is such as $\dot{\gamma}_{pt}\tau = 1$, i.e. when the elongational rate in converging sections is too small ($\dot{\varepsilon}\tau \approx 0{\cdot}1$) to induce a coil–stretch transition. Thus, the elongation of macromolecules responsible for the thickening behavior occurs near the stagnation points downstream from the grains constituting the porous medium and not in converging sections towards pore throats. Thus, the maximum elongational rate must be estimated by an equation like $\dot{\varepsilon}_{max} \approx V/D_g$ where D_g is the grain diameter.

Semi-rigid macromolecules are also oriented first downstream from the particles, but the increase in viscous dissipation due to this orientation is quite small and does not induce a shear thickening behavior but only a reduced shear thinning.[12] The rheological behavior shown in Fig. 2.5 is typical of a highly flexible polymer. If polymer flexibility is decreased, for instance by reducing salinity, the shear thinning is more pronounced and shear thickening is less sharp.

2.7.2 *Influence of pore size*

In the vicinity of the interface between a polymer solution and another phase (gas, liquid or solid), the viscosity properties are altered either by an excess of polymer in the case of adsorption or by the depletion in polymer if the interface is repulsive. In both cases, these interfacial properties are expected to affect the rheological properties in porous media, more particularly in the presence of oil and in porous media where the size of macromolecules is not negligible compared with that of pores.

(a) *Non-adsorbing walls.* Near a non-adsorbing wall, the probability for the macromolecule center of mass to be at a distance x from the interface decreases strongly when x becomes smaller than half of the end-to-end distance R_0 (eqns (2.1) and (2.3)). This wall exclusion has two important effects. (i) The macromolecules are excluded from the slowest stream lines in the porous medium and thus move faster than the solvent. In addition, since wall exclusion distance scales with R_0, macromolecule velocity increases with molecular weight. For polydispersed polymers this leads to a fractionation commonly called hydrodynamic chromatography, which could be used to determine molecular weight distribution of EOR grade xanthans.[35] (ii) The polymer concentration near the pore wall is smaller than in bulk for

$x < R_0$. As a consequence, in this depletion layer, the viscosity is reduced progressively from bulk viscosity for $x = R_0$ down to that of water for $x = 0$.

The effects of this depletion layer on the relative effective viscosity in pores η_{rp}, which is commonly called the mobility reduction or resistance factor were investigated in various types of porous media by using a microgel-free xanthan sample.[12] A depletion model has been proposed, giving the viscosity reduction factor η_{rb}/η_{rp} (where η_{rb} is the relative viscosity in the bulk solution) as a function of macromolecule-to-pore size ratio. For $C < C^*$ and $\delta/r \ll 1$, the model can be simplified and becomes:[36]

$$\eta_{rb}/\eta_{rp} = 1 + 4(1-\beta)(\delta/r)C[\eta]_0 \tag{2.18}$$

where β and δ are respectively equal to 0·64 and 0·7 L_c for rigid rods and to 0·3 and 1·24 R_G for flexible coils.

In the dilute regime, the linearity in δ/r and $C[\eta]$ predicted by eqn (2.18) has been experimentally verified for both xanthan and HPAM flowing through various types of porous media. Figure 2.11 shows that, for xanthan solutions, agreement between depletion model predictions

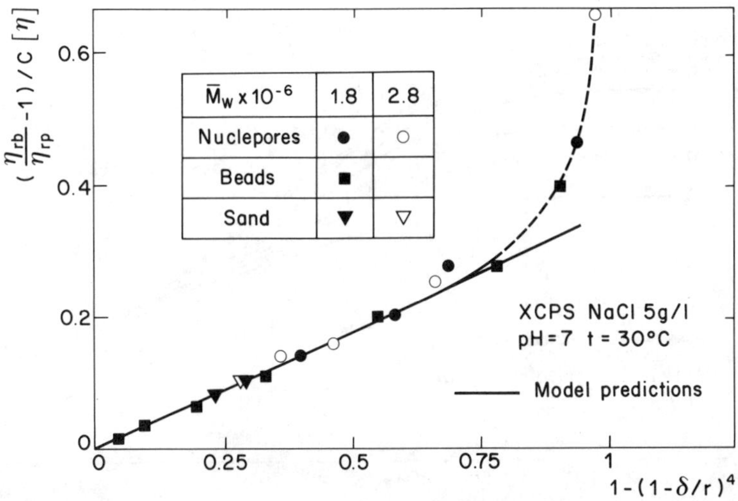

Fig. 2.11 Decrease in Newtonian effective viscosity with pore size: comparison between depletion model predictions and experimental data for two microgel-free xanthan samples. (From Ref. 36, reproduced by permission of The Academic Press, Inc.)

and experiments is good as long as δ/r remains smaller than 0·75. In the semi-dilute regime, up to $(C[\eta]_0 \approx 10)$, eqn (2.18) may still be used with $\delta = \text{constant}$ for rigid rods[7] and $\delta \propto R_G(C/C^*)^{3/4}$ for flexible polymers in good solvents.[7,36]

The validity of eqn (2.18) is limited to the Newtonian regime. Indeed, as soon as hydrodynamic forces are large enough to change macromolecule conformation, various phenomena not included in the model are expected to affect effective viscosity: (i) the decrease in depletion layer thickness due to an increased orientation in flow direction as flow rate increases; (ii) an increase in viscous dissipation in elongational zones, and (iii) a change in flow distribution through pores of different size.[37] These three effects contribute to a decrease in shear thinning at intermediate shear rates as shown in Fig. 2.12.

(b) *Adsorbing walls.* The effect of adsorption on flow properties is strongly dependent on the conformation of adsorbed macromolecules.

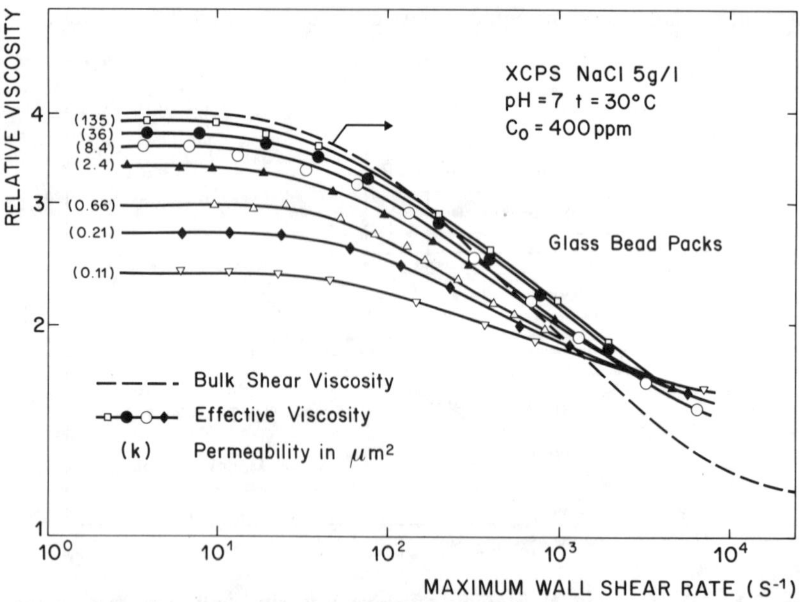

Fig. 2.12 Shear thinning behavior of a xanthan solution in the flow through porous media of various permeabilities. (From Ref. 12, reproduced by permission of The Society of Rheology, Inc.)

For semi-rigid polymers, if we except the case of low permeability media ($k < 0{\cdot}1\ \mu m^2$ [100 mD]), where macromolecules are too long to accommodate entirely flat on the pore surface, the hydrodynamic thickness of the adsorbed layer is negligible and does not affect flow properties. By contrast, for flexible polymers, loops and tails can extend far from the wall and the pore radius is reduced by a value ε_h—the hydrodynamic thickness. Thus the permeability of the porous medium is reduced by a factor R_k—the permeability reduction. The use of capillary model for porous media leads to:

$$R_k = (1 - \varepsilon_h/r)^{-4} \tag{2.19}$$

Typical values of ε_h for microgel-free polyacrylamides vary between 0·2 and 0·4 μm. Under good solvent conditions, which usually prevail in polymer flooding, interactions between the flowing macromolecules and the adsorbed layer are repulsive, and a depleted layer, similar to the one existing near a non-adsorbing wall, is expected on top of the adsorbed layer. In this case, the depletion model can be extended to take into account the effects of the adsorbed layer, and eqn (2.18) for δ and $\varepsilon_h \ll r$ may be modified to the following form:[36]

$$\eta_{rb}/\eta_{rp} = \frac{1}{R_k}\left[1 + 4(1-\beta)\frac{\delta}{r - \varepsilon_h} C[\eta]_0\right] \tag{2.20}$$

Experimental data concerning the flow of hydrolyzed polyacrylamides at two different concentrations in the dilute regime through various types of porous media are plotted in Fig. 2.13 according to the coordinates suggested by eqn (2.20). The data concerning adsorbent and non-adsorbent media lie exactly on the same lines showing that the thickness of the depleted layer on top of the adsorbed layer is the same as near a non-adsorbing wall.

Of course, the validity of eqn (2.20) is also restricted to the Newtonian regime. Indeed, at higher deformation rates, the prediction of effective viscosity would need to include in the model not only the complex effects on flowing macromolecules as discussed in Section 2.7.2(a), but also the effects of flow-induced changes in the adsorbed macromolecule conformation, i.e. the rheology of adsorbed polymer layers.

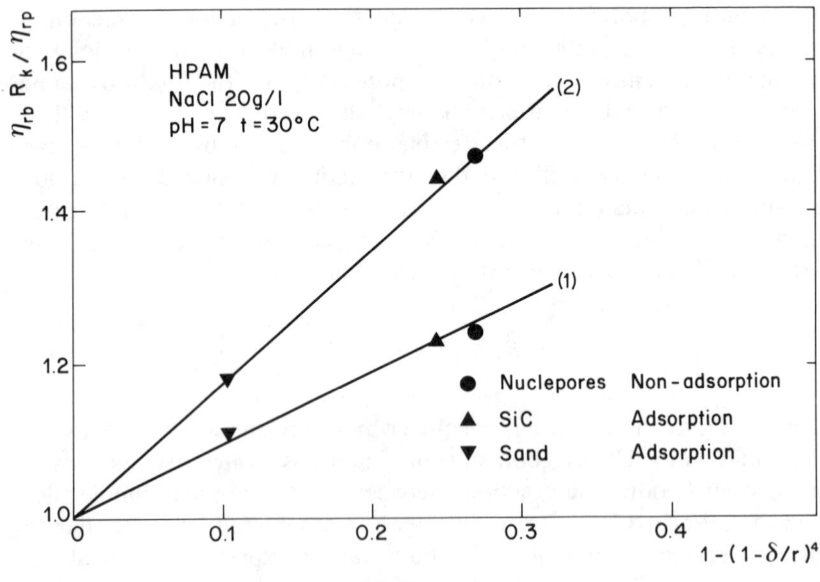

Fig. 2.13 Newtonian effective viscosity of two dilute polyacrylamide solutions (curve 1: $c = 170$ ppm; curve 2: $c = 340$ ppm) in adsorbing and non-adsorbing pores showing the existence of a depletion layer on top of adsorbed layer. (From Ref. 36, reproduced by permission of The Academic Press, Inc.)

2.7.3 *Rheology of adsorbed polymer layers*

The flow-induced changes in the conformation of adsorbed macromolecules under good solvent conditions result essentially from hydrodynamic forces exerted by the solvent flowing through and above adsorbed layer.

In pure shear flows of solvent above a polymer adsorbed on a flat surface, ε_h is expected to be constant at low shear rate and thus to decrease continuously. However, in flow through porous media, the mean hydrodynamic thickness of the adsorbed layer ε_h as derived from permeability reduction through eqn (2.19) varies in a more complex way as shown in Fig. 2.14:[38]

(i) At low shear rates, ε_h is constant and thus the regime is Newtonian.
(ii) Beyond a critical shear rate, $\dot{\gamma}_c$, practically equal to that characterizing the onset of shear rate dependence for the

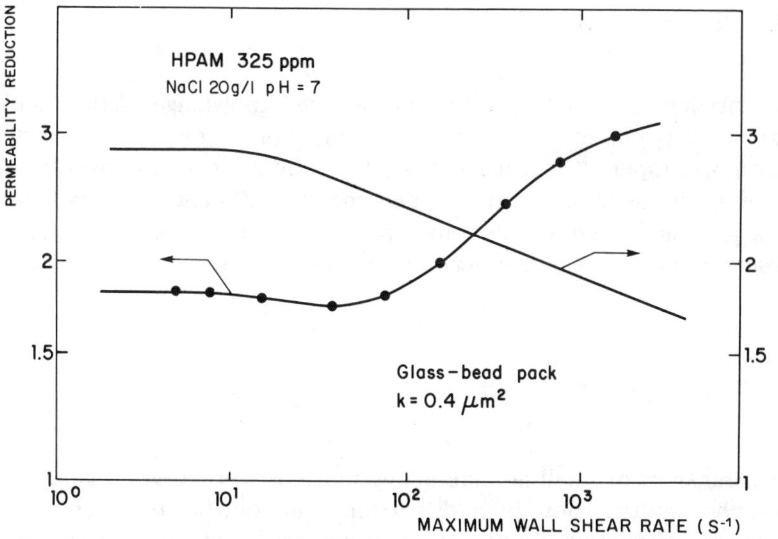

Fig. 2.14 Rheological behavior of a layer of a flexible polymer adsorbed in a porous medium subjected to solvent flow. (From Ref. 38, reproduced by permission of The American Chemical Society.)

viscosity of a dilute solution of the adsorbed polymer ($\dot{\gamma}_c \approx 10\,s^{-1}$, see Figs 2.5 and 2.14), ε_h is found to decrease as expected in a pure shear flow.

(iii) At higher shear rates, for $\dot{\gamma} > \gamma^*$, ε_h increases, showing that its decrease in shearing zones is overcome by its significant increase in elongational zones. Interestingly, the thickening behaviors of adsorbed (Fig. 2.14) and flowing (Fig. 2.5) macromolecules are observed beyond the same critical shear rate. This observation is a confirmation that the highest elongational rates in porous media flows are located downstream from the particles and not in converging zones.

(iv) At the highest shear rates, the mean thickness of the adsorbed layer tends towards stabilization, which probably corresponds to a maximum extension of adsorbed layer without desorption since full reversibility in $R_k = f(\dot{\gamma})$ curve was observed.

From the above interpretation, with more rigid polymers or with HPAM dissolved in lower salinity water, the shear thinning behavior is expected to be more pronounced and the shear thickening to be less sharp.

2.8 Transport in porous media

The prediction of polymer flooding requires knowledge of the rheology at all points in the reservoir as a function of time, and also the transport properties for each molecular weight fraction. In this section, we first consider the various phenomena that may affect polymer transport and then we describe the effects that have been clearly observed and the way they have been taken into account.

2.8.1 *Transport parameters*

The transport of small noninteracting molecules or tracers is generally described by the convection–dispersion equation, but to describe the transport of macromolecules, several extensions to this equation are required to take into account some phenomena which are either specific or particularly important in the case of polymers. These phenomena are as follows:

(i) Small molecular diffusion: diffusion coefficients are very low, molecular weight-dependent and are further reduced when the radius of gyration becomes non-negligible compared with pore size.
(ii) Dominant hydrodynamic dispersion which is, even at low flow rates, is controlled by the effects of effective viscosity ratio, non-Newtonian behavior and permeability reduction.
(iii) Non-equilibrium behavior[25,39,40] coming both from low diffusion-controlled access to some parts of pore surface and from slow adsorption kinetics, as discussed in Section 2.5.4.
(iv) Inaccessible pore volume, first described by Dawson and Lantz[41] and analyzed later as being mainly due to a surface exclusion mechanism.[12,35,42]
(v) Adsorption–retention described in Section 2.5 and possible desorption in the case of salinity or pH variations.
(vi) Permeability heterogeneities (see Section 2.9.1).
(vii) Viscous instabilities at the rear end of the slug which can be very severe in two- and three-dimensional floods.

2.8.2 *Viscous slug propagation*

A typical effluent profile in a polymer/tracer coreflood using a chlorine-36 labeled viscous polymer in brine is shown for a non-adsorbing sandstone core in Fig. 2.15. The main points to note from this figure are that although a 'square wave' slug of tracer was injected:

(i) the slug is spread at the frontal edge by hydrodynamic dispersion and surface exclusion chromatography, and it turns out that the dispersion coefficient for the polymer is larger than that of the tracer by a factor of between 2 and 3;[39,40]

(ii) the polymer front reaches the outlet of the core before the tracer front, as expected from the surface exclusion mechanism;

(iii) there is a tailing of the polymer (and tracer) effluents in the rear edge of the breakthrough curve due to viscous fingering since, in this displacement, a low viscosity brine is displacing a higher viscosity polymer solution.

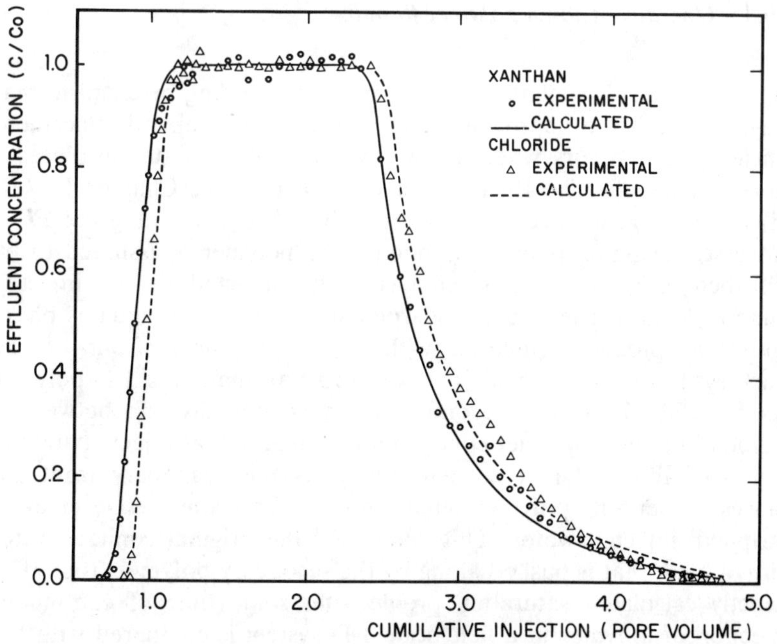

Fig. 2.15 Effluent concentration for polymer and tracer (two-pore volume polymer slug injected in a sandstone core). (From Ref. 39, reproduced by permission of The Society of Petroleum Engineers.)

The inclusion in the convection–dispersion equation of terms simulating these effects leads to a satisfactory description of polymer transport[39,40] in that they reproduce the main features observed in Fig. 2.15.

2.9 Oil displacement mechanisms

The polymer may operate by increasing the microscopic displacement efficiency or by improving the areal or vertical sweep efficiency. That is, the application of polymers may be considered for both unfavorable mobility ratio waterfloods or in reservoirs where there is excessive heterogeneity.

2.9.1 *One-dimensional polymer flooding*

For 'linear' displacements, the only polymer flooding mechanism that may operate is the improvement in local or microscopic displacement efficiency that occurs when a lower viscosity drive water displaces a more viscous oil. The Buckley–Leverett theory (see Chapter 1, Ref. 25) has been generalized[43] to describe this effect for a range of EOR processes including polymer flooding. The polymer is assumed to be adsorbed onto the rock which has been contacted by the injected solution, resulting in two self-sharpening fronts in the aqueous phase saturation profile, compared with the single shock front of the Buckley–Leverett theory as is shown for an example of a 1-D polymer flood in Fig. 2.16. The second front marks the division between a region of uniform polymer concentration equal to that injected and the region of zero polymer concentration. As this secondary front advances, it deposits polymer on the rock at its leading edge creating 'stripped' injected water. This water and the original connate water form a bank that is pushed along by the secondary polymer front. The directly calculated saturation profile after some time of continuous injection of polymer in a field-scale 1-D system is compared with the finite difference numerical solutions for the saturation profile in Fig. 2.16 and good agreement is observed.

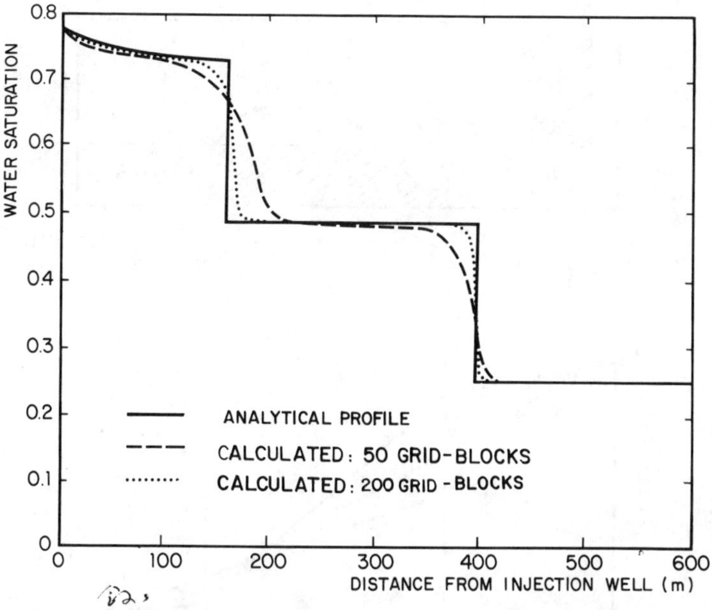

Fig. 2.16 Aqueous phase saturation profiles predicted by generalized Buckley–Leverett theory for 1-D polymer flooding.

2.9.2 *Improvement in areal sweep efficiency*

The schematic ideas about improved areal sweep are illustrated by the study of Patton *et al.*[44] who presented simplified methods for calculating polymer flood recoveries in both linear and five-spot systems. They compared the oil recovery for different sized polymer slugs with the waterflood performance. Particular points relating to areal polymer flooding processes emerged from this work: (i) a slug-type polymer flood involving a relatively modest increase in drive water viscosity may lead to appreciable improvements in the areal sweep efficiency; (ii) there appears to be a diminishing return in terms of incremental oil as the polymer slug is increased in size. This implies that there is an optimum slug size in terms of volume of incremental oil per unit weight of polymer injected.

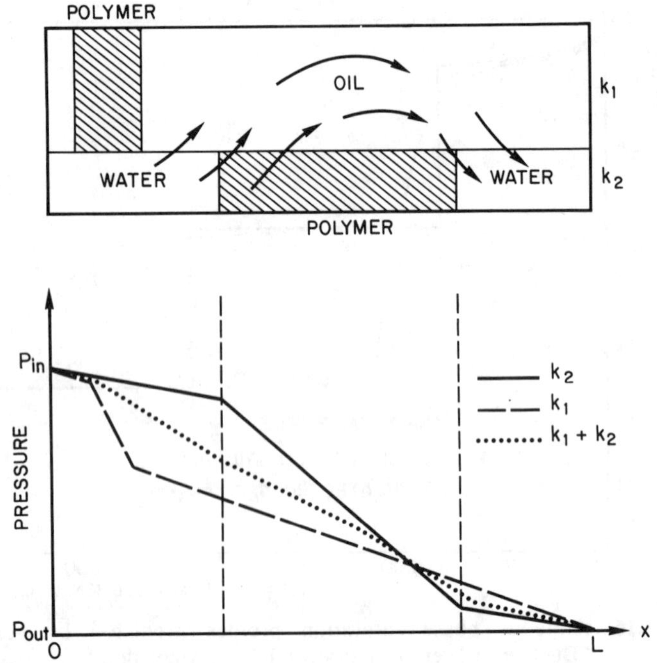

Fig. 2.17 Crossflow effects in polymer flood of layered systems ($k_1 < k_2$). (From Ref. 45, reproduced by permission of The Society of Petroleum Engineers.)

2.9.3 *Flow mechanisms in flooding layered systems*

An important application of polymers in several areas of the world is in the improvement of waterflooding performance in highly heterogeneous reservoirs. In particular, polymers can reduce the harmful effects of high permeability layers in a vertically stratified system and so reduce watercut and improve vertical sweep efficiency. Polymers can act through a combination of two mechanisms:

(i) mobility control in which the polymer slug viscosity changes fluid flow patterns within the reservoir;
(ii) adsorption, leading to local reductions in permeability, which again alter fluid flow.

For applications of this type, it has been shown that, even for relatively low vertical permeability, the polymer flood will be dominated by the crossflow of fluids (oil, water and polymer solution) between layers of different permeability.[45,46] Crossflow is a principal mechanism both in oil recovery enhancement and in its effects on the integrity and location of the polymer slug.

Polymer behavior in a layered system is often described as follows: an injected polymer slug predominantly enters a high permeability layer, which it 'blocks' due to its high viscosity, thus diverting chase water into layers of lower permeability. Reality is even more complicated, particularly in reservoirs where the layers are in good vertical communication, and a schematic view of the situation is pictured in Fig. 2.17. The polymer slug acts as a 'plug' in the high permeability layer, but its effect on flow is localized rather than global in the sense that the fluid flow ratio in high-to-low permeability layers is altered only in the neighborhood of the slug. Chase water is diverted out of the high into the low permeability layer at the rear of the slug, carrying some polymer with it. In this case, the slug will be lost from the high permeability layer by progressive crossflow at its rear end. Flow in the low permeability layer adjacent to the slug is enhanced, and this additional flow crosses back into the high permeability layer head of the slug. Therefore, in the time before the polymer has approached the production well, the oil that it sweeps from the low permeability layer actually emerges from the high permeability layer. It should also be noted that there will be a delay between the start of polymer injection and the first appearance of incremental oil. The delay period is the time taken for crossflowed oil to flow from injector to producer, ahead of the slug, along the high permeability layer.

The outcome of polymer flooding in a reservoir with noncommunicating layers is quite different. In this latter case, the polymer causes a gross change in the flow ratio between the layers. Fluid diversion in the absence of crossflow, and hence incremental oil production, begins as soon as polymer is injected and the incremental oil is produced directly from the low permeability layers.

In conclusion, the polymer slug breakdown due to reservoir heterogeneity is one of the most important aspects of polymer flooding.

2.10 Numerical simulation of polymer flooding

2.10.1 *Polymer properties to be included in simulators*

The many factors described above, including the transport properties of the polymer, the interactions with the rock matrix, possible chemical reactions and the non-Newtonian rheological behavior, combine to make polymer flow through porous media very complex. The main features that should be included in a polymer simulator for both field and laboratory uses are as follows:

(i) Polymer apparent viscosity in porous media: the viscosity of a polymer solution is strongly dependent on its concentration and on its flow rate and thus a model for in-situ rheological behavior must be included in the simulator.

(ii) Adsorption–retention and permeability reduction: virtually all types of polymer are adsorbed/retained in the porous medium, which can have a very important effect on the outcome of a polymer flood. The resulting permeability reduction of rock to the aqueous phase must also be included. Adsorption may be described by an equilibrium isotherm or via a non-equilibrium process.

(iii) Transport properties: in porous media, polymer molecules show dispersion behavior, wall exclusion/inaccessible pore volume and viscous fingering effects which are quite different to those of tracers as discussed above.

(iv) Chemical reactions: polymers will gradually degrade chemically in reservoir environments, causing a corresponding loss of solution properties. Polymers may also react chemically with other species such as trivalent metal ions to crosslink and form gel-like materials in the reservoir.

(v) Thermal effects: polymer properties such as viscosity, degradation rate, etc., are dependent on temperature. If large amounts of cool water are injected into a hot reservoir there may be significant temperature changes, and calculations should take account of this.

(vi) Salinity and pH effects: polymers are often injected in a water having lower salinity or higher pH than that of water initially in place and these two factors strongly affect polymer properties.

(vii) Polydispersity effects: all the phenomena cited above are strongly dependent on molecular weight and there may be some improvement in the predictive capability of simulators by considering the polymer as *n* discrete components transported in the aqueous phase to represent the molecular weight distribution.

The full polymer transport equation in 3-D is usually taken to be a generalized convection–dispersion equation. In addition to the mass conservation and pressure equations, a heat balance equation may also be included in the simulator in order to calculate temperature distributions within the reservoir as a function of time.[45–47]

2.10.2 *Simulations including temperature-dependent degradation*

In systems where there is local cooling, for example in high permeability streaks, several features of the polymer flood may be affected, e.g. chemical degradation time scales. As an example of the application of a simulator to model such effects in a polymer flood, we consider chemical degradation of polymer and its dependence on temperature. Although, as discussed in Section 2.6, the process of chemical degradation may be fairly complex, the ultimate result is that a desirable property (e.g. viscosity) is eroded with time. If we assume a simple first-order reaction for the polymer degradation term, then we may define a polymer half-life, $\tau_{1/2}$, at a given temperature. Thus, we may calculate the effects of chemical degradation on a polymer flood. Figure 2.18 shows the cumulative oil recovery for a waterflood, chemically stable polymer flood and a polymer flood with $\tau_{1/2} = 6$ months in a simple isothermal two-layer reservoir system with the properties specified elsewhere.[45] Practically, local cooling will occur and its consequence will be a much lower chemical degradation rate for the polymer. If the degradation rate is characterized by an activation energy E_a, then its rate depends on the absolute temperature T, according to the Arrhenius equation. With this assumption, the calculation is repeated for the stable and degrading polymer floods in the two-layer system. The half-life is again taken to be 6 months at 93°C but now increases (i.e. degradation rate decreases) at lower temperatures. Figure 2.18 also shows the calculated cumulative oil recovery for this case along with some other cases for comparison. It is clearly seen that the recovery efficiency for this case where the degradation rate depends on temperature is very close to that for the stable polymer.[45]

2.10.3 *Simulation of adsorption effects*

In most cases, adsorption reduces the efficiency of polymer flooding but some compensation can be gained by the accompanying per-

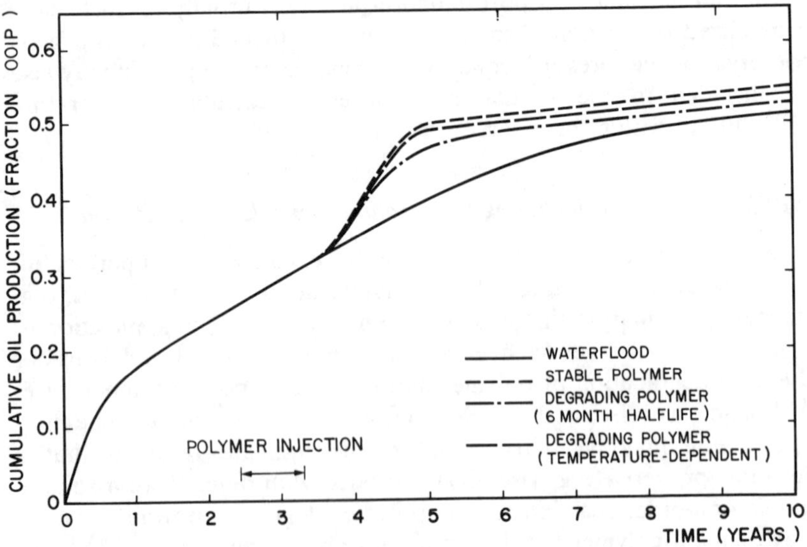

Fig. 2.18 Influence of polymer degradation on oil recovery by polymer flooding. (From Ref. 45, reproduced by permission of The Society of Petroleum Engineers.)

meability reduction and the quantitative effects of adsorption on oil recovery can only be predicted using numerical simulations. This quantitative prediction is of importance to decide if a method for reducing adsorption is profitable. Another reason for simulating adsorption effects concerns the selection of cores for laboratory studies. Indeed, adsorption may vary greatly from one part of a reservoir to another and it is not practical to determine adsorption–retention values from many core floods. Simulation studies have shown that oil production is more sensitive to the adsorption levels in the higher permeability regions for both stratified and areally heterogeneous reservoirs, and thus cores from these regions should be selected preferentially for laboratory adsorption measurements.

2.11 Status and perspectives of polymer flooding

Several hundred polymer flooding tests, including industrial applications, have been implemented to date (see also Ref. 25 in Chapter 12).

Many of these field tests are considered as technical and economic successes. Among the field tests that have been described in some detail in the oil literature are: the polyacrylamide injection in the low-salinity, low-temperature Chateaurenard field,[26] the xanthan flood in the high-salinity, low-temperature Loudon field,[34] and the polyacrylamide flood in the medium-temperature, high-salinity Hankensbuettel field, where a nonionic, low molecular weight, polyacrylamide dissolved in fresh water was used successfully for preconditioning the reservoir.[48] At present, for many reservoirs, polymer flooding can be used without excessive risk. A thorough discussion of the applicability of polymer flooding may be found in Ref. 49.

The understanding of phenomena related to polymer behavior in the field of EOR has been greatly improved during the last two decades, and the process may now be economically attractive. However, progress still remains to be achieved in order to decrease the technical risks, to improve profitability, and to extend the range of applicability of the process.

This progress concerns fundamental aspects of polymer behavior in porous media. Improved physical laws describing hydrodynamic retention and viscous fingering are required for ensuring better reliability of numerical simulators used to optimize field operations. These phenomena, not sufficiently controlled to date, are of primary importance for the cost effectiveness of the process. Some aspects of in-situ rheology will also have to be clarified.

But the main progress to be made certainly concerns industrial products. This will be a challenge for applied chemistry. The polymer flooding process needs high-quality polymers, free of microgels, to provide good injectivity and to reduce the risks of plugging. It also needs polymers having both improved long-term bacterial and thermal stability. The alternative could be either to improve the methods of synthesis and the manufacturing process itself for presently-used polymers, or to develop new polymers, such as modified polyacrylamides, through sulfonation for instance, or new multistranded polysaccharides, by making use of recent advances in biotechnologies. In other respects, new additives could replace conventional stabilizers to ensure better polymer stability against bacteria and oxidation.

Whatever the polymers may be, their conditioning has to be improved. For instance, the transportation of polysaccharide broths should be abandoned to the benefit of more concentrated solutions. The use of polymers in emulsion should be developed rather than those in powders, even if emulsion stability, especially thermal stability, is sometimes questionable and needs to be fully mastered.

Experience gained due to past Research and Development activity leads us to anticipate that polymer flooding has a high potential for oil production. This will be especially true if the process is implemented together with other chemicals, such as surfactants and alkaline agents, and/or in combination with new production techniques, such as horizontal drilling.

2.12 References

1. Willhite, G. P., *Waterflooding,* SPE Textbook Series No. 3, SPE, Richardson, Texas, 1986.
2. Pye, D. J., Improved secondary recovery by control of water mobility. *J. Petroleum Technol.,* **16**(8) (August 1964) 911–16.
3. Sandiford, B. B., Laboratory and field studies of waterfloods using polymer solutions to increase oil recovery. *Petroleum Trans. AIME,* **231** (1964) 917–29.
4. Gogarty, W. B., Mobility control with polymer solutions. *Soc. Petroleum Engineers J.,* **7**(2) (1967) 161–70.
5. Szabo, M. T., Laboratory investigations of factors influencing polymer flood performance. *Soc. Petroleum Engineers J.,* **15**(4) (1975) 338–46.
6. Kolodziej, E. J., Mechanisms of microgel formation in xanthan biopolymer solutions. SPE Paper No. 16730, 62nd Annual Techn. Conf. and Exhibition of Soc. Petroleum Engineers, Dallas, 27–30 September 1987.
7. Chauveteau, G., Fundamental criteria in polymer flow through porous media and their relative importance in the performance differences of mobility control buffers. In *Water Soluble Polymers—Beauty with Performance,* ed. J. E. Glass. Adv. Chem. Series, American Chem. Soc., Washington, **213** (1986) 227–68.
8. Rivenq, R., Donche, A. & Noïk, C., Improved scleroglucan for polymer flooding under harsh reservoir conditions. SPE Paper No. 19635, 64th Annual Techn. Conf. and Exhibition of Soc. Petroleum Engineers, San Antonio, 8–11 October 1989.
9. von Halasz, S. P., Vinyl sulfonate/vinyl amide copolymers and different surfactants as suitable systems in EOR at higher temperature and salinities. *Revue de l'Institut Français du Pétrole,* **43**(4) (1988) 545–53.
10. Flory, P., *Principles of Polymer Chemistry,* Cornell University Press, Ithaca, 1971.
11. de Gennes, P. G., *Scaling Concepts in Polymer Physics,* Cornell University Press, Ithaca and London, 1979.
12. Chauveteau, G., Rodlike polymer solution flow through fine pores: influence of pore size on rheological behavior. *J. Rheology,* **26**(2) (1982) 111–42.

13. Acierno, D., Titomanlio, G. & Nicodemo, L., Elongational flow of dilute polymer solutions. *Rheologica Acta,* **13** (1974) 532–7.
14. Ghoniem, S., Chauveteau, G., Moan, M. & Wolff, C., Mechanical degradation of semi-dilute polymer solutions in laminar flows. *Canadian J. Chem. Engineering,* **59** (1981) 450–4.
15. Chauveteau, G., Lecourtier, J. & Lee, L. T., Reduction of polymer adsorption on reservoir rocks. *Revue de l'Institut Français du Petrole,* **43**(4) (1988) 533–43.
16. Lecourtier, J., Lee, L. T. & Chauveteau, G., Adsorption of polyacrylamides on siliceous minerals. *Colloids and Surfaces,* **47** (1990) 219–31.
17. Pefferkorn, E., Jean-Chronberg, A. C., Chauveteau, G. & Varoqui, R., Adsorption of hydrolyzed polyacrylamides onto amphoteric surfaces. *J. Colloid Interface Science,* **137**(1) (1990) 66–74.
18. Lee, L. T., Rahbari, R., Lecourtier, J. & Chauveteau, G., Adsorption of polyacrylamides on the different faces of kaolinites. *J. Colloid Interface Science,* in press.
19. Rahbari, R., Lee, L. T., Lecourtier, J. & Chauveteau, G., Adsorption of hydrosoluble polymers on the different faces of kaolinites. Paper presented at the 200th ACS National Meeting, Washington, 26–31 August 1990.
20. Bottero, J. Y., Bruant, M., Cases, J. Canet, D. & Fiessinger, F., Adsorption of nonionic polyacrylamide on sodium montmorillonite. *J. Colloid Interface Science,* **124**(2) (August 1988) 515–27.
21. El Attar Sofi, Y., Carquille, C., Lemarche, J. M. & Foissy, A., Influence of calcium and phosphate ions on the adsorption of partially hydrolyzed polyacrylamides on TiO_2 and $CaCO_3$. *Progress in Colloid and Surface Science,* **82** (1990) 43–51.
22. Argillier, J. F., Filtration de macromolécules flexibles à travers des micropores modèles. Thèse de doctorat, University of Paris VI, France, 1989.
23. Maerker, J. M., Dependence of polymer retention on flow rate. *J. Petroleum Technol.,* **15** (1973) 1307.
24. Chauveteau, G. & Kohler, N., Polymer flooding: the essential elements for laboratory evaluation. SPE Paper No. 4745, 2nd Improved Oil Recovery Symp. of Soc. Petroleum Engineers, Tulsa, 22–24 April 1974.
25. Chauveteau, G. & Lecourtier, J., Propagation of high molecular weight polymer solutions slugs through adsorbent porous media. In *Water Soluble Polymers for Petroleum Recovery,* ed. C. G. Stahl & D. N. Schulz, Plenum Press, New York, 1988, pp. 53–8.
26. Putz, A., Lecourtier, J. & Bruckert, L., Interpretation of high recovery obtained in a new polymer flood in the Chateaurenard field. SPE Paper No. 18093, 63rd Annual Techn. Conf. and Exhibition of Soc. Petroleum Engineers, Houston, 2–5 October 1988.
27. Davison, P. & Mentzer, E., Polymer flooding in North Sea reservoirs. *Soc. Petroleum Engineers J.,* **22**(3) (1982) 353–62.
28. Moradi-Araghi, A. & Doe, P. H., Stability of polyacrylamides in hard brines at elevated temperatures. *SPE Reservoir Engineering,* **2**(2) (1987) 189–98.

29. Wellington, S. L., Biopolymer solution viscosity stabilization: polymer degradation and antioxydant use. *Soc. Petroleum Engineers J.,* **23**(6) (1983) 901–12.
30. Seright, R. S. & Henrici, B. J., Xanthan stability at elevated temperatures. *SPE Reservoir Engineering,* **5**(1) (1990) 52–60.
31. Muller, G., Thermal stability of polyacrylamide solutions: effects of residual impurities in the molecular-weight-degradation process upon heating. *Polymer Bulletin,* **5** (1981) 39–45.
32. Kheradmand, H., François, J. & Plazanet, V., Effects of additives on the degradation of acrylamid–acrylic acid copolymer in aqueous solutions. *J. Applied Polymer Science,* **39** (1990) 1847–57.
33. Seright, R. S., Maerker, J. M. & Holzwarth, G., Mechanical degradation of polyacrylamide induced by flow through porous media. *Polymer Preprint,* **22** (1988) 30–3.
34. Bragg, J. R., Maruca, S. D., Gale, W. W., Gall, L. S., Wernau, W. C., Beck, D., Goldman, I. M., Laskin, A. I. & Naslund, L. A., Control of xanthan degrading organisms in the Loudon pilot: approach, methodology, results. SPE Paper No. 11989, 58th Annual Techn. Conf. and Exhibition of Soc. Petroleum Engineers, San Francisco, 5–8 October 1983.
35. Lecourtier, J. & Chauveteau, G., Xanthan fractionation by surface exclusion chromatography. *Macromolecules,* **17** (1984) 1340–3.
36. Chauveteau, G., Tirrel, M. & Omari, A., Concentration dependence of the effective viscosity of polymer solutions in small pores with repulsive or attractive walls. *J. Colloid Interface Sci.,* **100**(1) (1984) 41–54.
37. Sorbie, K. S., Clifford, P. & Jones, E. R. W., The rheology of pseudo-plastic fluids in porous media using network modeling. *J. Colloid Interface Science,* **130** (1989) 508–34.
38. Bagassi, M., Chauveteau, G., Lecourtier, J., Englert, J. & Tirrell, M., Behavior of adsorbed layers in shear and elongational flows. *Macromolecules,* **22** (1989) 262–6.
39. Sorbie, K. S., Parker, A. & Clifford, P. J., Experimental and theoretical study of polymer flows in porous media. *SPE Reservoir Engineering,* **2**(3) (1987) 281–304.
40. Kolodziej, E. J., Transport mechanisms of xanthan biopolymer solutions in porous media. SPE Paper No. 18090, 63rd Annual Techn. Conf. and Exhibition of Soc. Petroleum Engineers, Houston, 2–5 October 1988.
41. Dawson, R. & Lantz, R. B., Inaccessible pore volume in polymer flooding. *Soc. Petroleum Engineers J.,* **12**(5) (1972) 448–52.
42. Brown, W. B. & Sorbie, K. S., Dispersion and polydispersity effects in the transport of xanthan in porous media. *Macromolecules,* **22** (1989) 2835–45.
43. Pope, G. A., Application of fractional flow theory to enhanced oil recovery. *Soc. Petroleum Engineering J.,* **20**(3) (1980) 191–205.
44. Patton, J. T., Coats, K. H. & Colegrove, G. T., Prediction of polymer flood performance. *Soc. Petroleum Engineers J.,* **11**(1) (1971) 72–84.
45. Clifford, P. J. & Sorbie, K. S., The effects of chemical degradation on polymer flooding. SPE Paper No. 13586, International Symposium on Oil

Field and Geothermal Chemistry of Soc. Petroleum Engineers, Phoenix, 9–11 April 1985.

46. Sorbie, K. S. & Clifford, P. J., The simulation of polymer flow in heterogeneous porous media. In *Water Soluble Polymers for Petroleum Recovery,* ed. G. A. Stahl & D. N. Schulz, Plenum Press, New York, 1988, pp. 69–99.
47. Scott, T., Sharpe, S. R., Sorbie, K. S., Clifford, P. J., Roberts, L. J., Foulser, R. W. S. & Oakes, J., A general purpose chemical flood simulator. SPE Paper No. 16029, Ninth Symp. on Reservoir Simulation of Soc. Petroleum Engineers, San Antonio, 1–4 February 1987.
48. Maitin, B., Application of chemical flood processes in the oil field Hankensbuettel. *Proc. Third European Meeting on Improved Oil Recovery,* Rome, 16–18 April 1985. Agip Spa, pp. 283–93.
49. Sorbie, K. S., *Polymers in Improved Oil Recovery*. Blackie and Sons, Glasgow, 1991.

3 Reduction of capillary forces by surfactants

Gary A. Pope
University of Texas at Austin, Texas, USA

and

Marc Bavière
Institut Français du Pétrole, Rueil-Malmaison, France

3.1	**Introduction: the need for low interfacial tensions**	89
3.2	**Physicochemistry of surfactant solutions**	90
3.2.1	The surfactants	91
3.2.2	Properties of surfactant solutions	93
3.2.3	Phase behavior and phase properties in EOR systems	96
3.2.4	Thermodynamics: phase behavior modeling	102
3.3	**Mechanisms involved in oil displacement by surfactant solutions**	104
3.3.1	Immiscible displacement under low interfacial tension conditions	104
3.3.2	Mass transfer mechanisms	106
3.3.3	Fluid/rock interactions	107
3.3.4	Modeling of surfactant flooding	112
3.4	**Process design**	114
3.4.1	Fundamentals	114
3.4.2	Salinity gradient concept	115
3.4.3	Field operations and perspectives	117
3.5	**References**	119

3.1 Introduction: the need for low interfacial tensions

Capillary forces limit the microscopic displacement efficiency of water (see Chapter 1). This phenomenon was known early and gave rise to the idea of adding chemicals to reduce these forces by altering interfacial properties of the oil/brine/rock systems: alkaline agents (see Chapter 4), surfactants and solvents (see Chapters 6 and 7). As early as the 1920s, the main parameters that influence oil recovery were identified, especially interfacial parameters, i.e. wettability and oil/brine interfacial tension.

Originally, surface active agents were primarily thought of as detergents or wetting agents.[1] Wettability was early recognized as being an important factor regarding oil entrapment and distribution in the pore space, but investigations in this field appear to be very delicate. At the same time, experimenters were aware of the need to lower interfacial tension. However, for a long time the results were not consistent, probably because of adsorption preventing the progression of surfactants through the formation, and also because of the poor adaptation of the surfactant to the oil/brine system.

The displacement of residual phases is related to the competition between viscous and capillary forces, usually expressed by the capillary number N_c (see Section 1.2.1).

Residual oil cannot be displaced until a critical value of N_c has been reached, which depends on the rock properties.[2] For typical waterflood conditions, N_c crit. $=10^{-6}$. It has to be increased by three or four orders of magnitude (even more for the displacement of the wetting phase) in order to have a near-zero oil saturation. So, considering that injection rates cannot be increased to a great extent, interfacial tension between oil and brine, which is usually around 30 mN/m, needs to be reduced by a factor of 10^3 or 10^4. This was shown with gas/liquid mixtures without any additives[3] and confirmed with surfactant-containing systems in the early 1970s.[4]

There is now agreement that lowering interfacial tension will have a predominant effect, even if other mechanisms may also occur, depending on reservoir conditions. The lowering to values around 10^{-3} mN/m can be obtained with surfactants. This is the principle of surfactant flooding, also known under the names of micellar flooding, low-tension water flooding and microemulsion flooding. The degree of lowering is closely related to the physicochemical behavior of the oil/brine/surfactant system. This behavior is examined below.

3.2 Physicochemistry of surfactant solutions

This section presents the main concepts related to the phase behavior of surfactants and its interpretation. It aims at defining the rules of formulation leading to ultra-low interfacial tensions for given fluids, temperatures and pressures.

3.2.1 *The surfactants*

(a) *Definition and types.* Surfactants are amphiphilic molecules, i.e. they consist of two segregated portions called (when water is the solvent) hydrophilic and hydrophobic groups. The latter is most often a long-chain hydrocarbon radical, from 8 to 18 carbon atoms, straight or branched, saturated or not, possibly associated with naphthenic or aromatic structures. Depending on the hydrophilic group, surfactants are classified in four categories.

—*Anionics.* The surface-active part of the molecules in aqueous solutions is negatively charged. They are, for example, sodium alkylsulfates or sodium alkylarylsulfonates. They are the most commonly used surfactants in EOR, especially sulfonates, because of their lower adsorption on reservoir rock compared to other types and also because of their availability and relatively low cost.
—*Cationics,* such as quaternary ammonium salts. These molecules are not used in EOR because they extensively adsorb onto anionic mineral surfaces (silica, clays).
—*Nonionics.* These molecules are not ionized in aqueous solutions. Their water solubility is limited by temperature (cloud point). They are more salt-tolerant than most anionics.
—*Zwitterionics.* They result from a combination of anionic and cationic moieties. They have not been used in EOR because of their high cost and adsorption level.

(b) *Surfactants used in EOR.* Petroleum sulfonates have been the most widely used in EOR applications. They are generally manufactured by sulfonating aromatic refinery products, such as lube oils. The resulting alkylaryl sulfonic acid is then neutralized by adding alkali, usually sodium hydroxide. The final surfactant is a complex mixture of components having different structures, molecular weights, and degrees of sulfonation (number of sulfonate functions per molecule). Average molecular weights typically range from 350 to more than 500. Since refinery products contain polyaromatic components, petroleum sulfonates consist of very water-soluble polysulfonates (green acids) in addition to rather oil-soluble molecules (mahogany sulfonates). Hence petroleum sulfonates are better characterized by an average equivalent weight, which is the average molecular weight divided by the sulfonation degree.

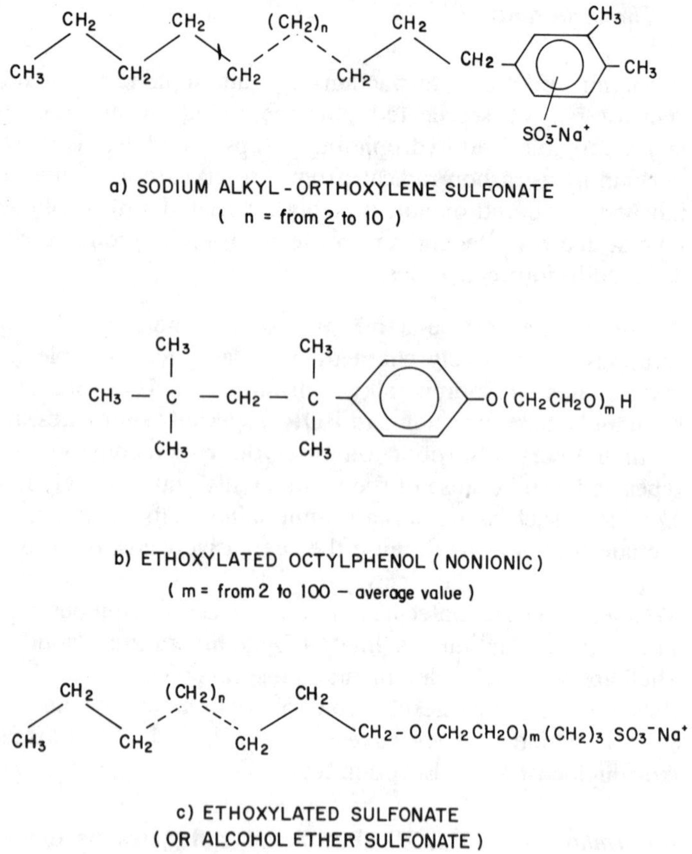

Fig. 3.1 Representative structure of surfactant molecules.

Alkylbenzene or alkylxylene sulfonates (Fig. 3.1(a)), referred to as synthetic sulfonates, have also been used. They are reported to be more easily tailored for a given application and more efficient.

Other surfactant families have been used or proposed either as cosurfactants or as primary surfactants, especially for medium-to-high salinity (i.e. for sea-water salt content and above). These particularly include nonionics (such as ethoxylated alcohols or alkylphenols (Fig. 3.1(b)), alkanolamides), ethoxylated sulfates (their use is limited to low temperatures because of their sensitivity to hydrolysis), ethoxylated sulfonates (Fig. 3.1(c)) and propoxylated–ethoxylated sulfonates, alpha–olefin sulfonates (a mixture consisting mainly of alkene sulfonates, hydroxyalkane sulfonates and disulfonates), and sulfonated esters.

Anionic surfactants employed in EOR are usually sodium salts, and contain impurities such as water, inorganic salts, and unreacted feedstock oil.

Prior to interfacial efficiency examination, the long-term chemical stability of the surfactant molecule has to be checked, particularly the effects of pH, temperature, and the presence of oxygen.[5]

3.2.2 *Properties of surfactant solutions*

(a) *Surface activity, micellization, and solubilization.* When dissolved in water, surfactant molecules accumulate at the air/water interface as an oriented monolayer, where only the hydrophilic part remains in contact with the solvent. This adsorption reduces the free energy increase due to surfactant/water interactions and prevents the surfactant from becoming insoluble. The monolayer formed lowers surface tension as a function of the excess surfactant concentration, defined as the difference between interface and bulk concentrations.

At saturation adsorption, and if the temperature is high enough (Krafft point), surfactant molecules aggregate into micelles (a few nanometers in diameter), with hydrocarbon parts turned inward, out of contact with water, and in dynamic equilibrium with nonassociated molecules (monomers). Micelles form at a concentration called critical micelle concentration (CMC) and are identified by a change in physical properties of the solution, such as surface tension (Fig. 3.2). The CMC value decreases with the surfactant affinity for water. The micellization process leads to a great increase in the surfactant solubility.

Micellar solutions are able to solubilize oil in the interior of micelles. These are the solutions named S_1 by Winsor (Fig. 3.3).[6,7] If the oil content exceeds the solubilization capacity of micelles, an excess oil phase forms which contains only traces of water and surfactant. The system is then called a type I system. Similarly, surfactants may form inverted micelles in oil, with their hydrophilic part directed toward the interior of the aggregate. Solubilization of water gives rise to S_2 solutions, and type II systems if an excess water phase forms.

By changing one of the variables of the system (see below), there is

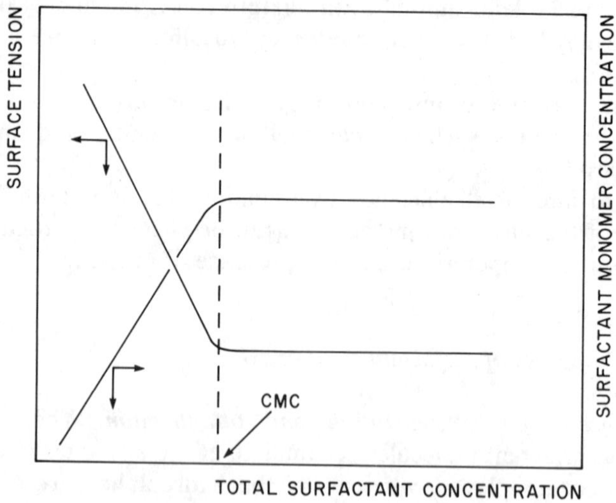

Fig. 3.2 Schematic of surface tension versus surfactant concentration.

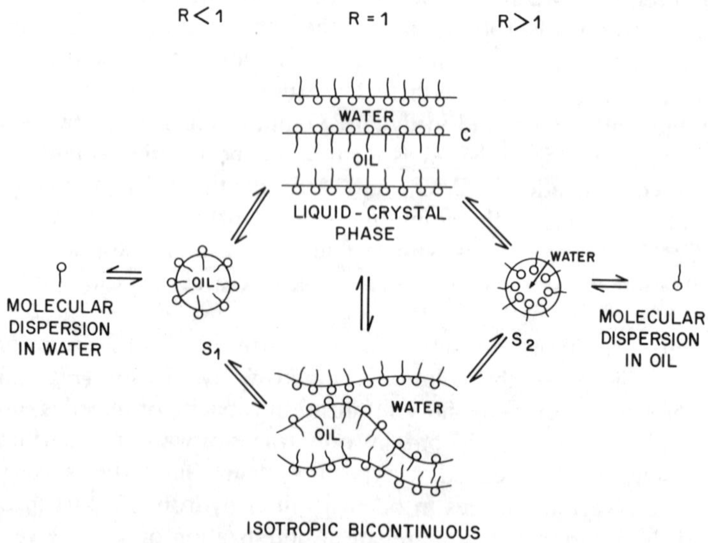

Fig. 3.3 Intermicellar equilibrium. (Adapted from Ref. 7, reproduced by permission of Butterworth-Heinemann.)

a tendency to pass gradually from S_1 to S_2 (or vice versa) through an intermediate state in which both oil and water are continuous phases. This is the bicontinuous structure suggested by Scriven.[8] According to the rigidity of the amphiphilic membrane separating the two bulk phases, this structure can be fluid and isotropic, or rigid and anisotropic (lamellar liquid–crystal mesophase), as shown in Fig. 3.3. In the presence of an excess of oil and water, a three-phase system forms, called type III by Winsor, with a micellar phase (middle-phase) coexisting with both oil and water excess phases. A gradual anisotropic–isotropic transition can be obtained by creating some disorder in the amphiphilic membrane, either by increasing temperature or by adding other species of amphiphilic molecules, such as cosurfactants or cosolvents (alcohols).

Isotropic micellar phases of all three types are called swollen or solubilized micellar solutions, or microemulsions.[9]* Single-phase oil/water/surfactant mixtures are called type IV, the structure being S_1, S_2 or bicontinuous.

For our purpose, solubilization is an essential phenomenon because interfacial tensions are closely related to it.

(b) *Winsor's R-theory of solubilization.*[6,7] In water/oil/surfactant mixtures, the surfactant forms an interfacial (amphiphilic) region C between bulk water and oil phases (Fig. 3.3). On both sides of C, cohesive intermolecular forces (van der Waals, hydrogen bonding, electrostatic forces) promote associations, and therefore solubilization. These interactions are mainly between oil and the hydrophobic part of the surfactant (A_{co}) and between water and the hydrophilic part of the surfactant (A_{cw}). Interactions between members of the same species promote separation, e.g. interactions A_{oo} between oil molecules and between the lipophilic parts of the surfactant, or interactions A_{ww} between water molecules and between the hydrophilic parts of the surfactant.

Winsor defined R as the ratio of solvent attraction between surfactant and oil to solvent attraction between surfactant and water. So,

$$R = (A_{co} - A_{oo})/(A_{cw} - A_{ww}) \tag{3.1}$$

* The term is misleading; contrary to ordinary emulsions, microemulsions are thermodynamically stable (see Section 3.2.4).

The higher A_{co} and A_{cw}, the stronger the solvent effect of C, therefore the higher the solubilization per unit of interfacial area. If $R < 1$, i.e. the surfactant is preferentially water soluble, C tends to be convex toward water by preferential solvation (Bancroft's rule), giving rise to a S_1 solution, or a type I system. If $R > 1$, C tends to be convex toward oil, giving rise to a S_2 solution, or a type II system. When $R = 1$, the radius of the natural curvature of C is infinite. This leads to intermediate structures, or a type III system.

Any change of a variable which increases R values favors surfactant partitioning in oil, thus tending to promote the I–III–II transition. For instance, increasing water salinity induces both a decrease in A_{cw} and an increase in A_{ww} (interactions between surfactant polar heads); then R increases and a I–III–II transition is expected. The salinity effect is particularly strong with an ionic surfactant.

As shown in Table 3.1, there are many variables to consider because EOR systems consist of many components. Numerous studies have investigated their effects on phase transition. A comprehensive review by Bourrel and Schechter details the various active mechanisms.[10] Table 3.2 summarizes the main features. Two variations of opposite trend may compensate one another so that the R value remains unchanged.

3.2.3 *Phase behavior and phase properties in EOR systems*

(a) *Pseudoternary representation of phase behavior.* The phase behavior of EOR systems is often represented in a pseudoternary diagram, where apexes represent respectively the amphiphilic com-

Table 3.1 Usual components of micellar solutions for EOR

Solution	*Components*
Oil	• Crude oil (usually stock-tank oil)
Brine	• Formation water (or water from other sources, such as sea, lakes and rivers) containing inorganic ions: Na^+, Ca^{++}, Mg^{++}, Cl^-, SO_4^{--}, etc. • Hydrosoluble polymers, bactericides, sacrificial agents, etc.
Amphiphiles	• Surfactant (primary) • Cosurfactant • Cosolvent (alcohols from C_3 to C_5)

Table 3.2 Variations promoting the I–III–II transition

Oil	—Decrease in molecular weight —Decrease in branching (or aromaticity)
Brine	—Increase in salinity —Increase in divalent/monovalent cation ratio —Decrease in pH (carboxylated surfactants)
Amphiphiles	—Increase in molecular weight (or branching) of surfactant-hydrophobic part —Decrease in polarity of surfactant-hydrophilic part —Increase in long-chain alcohol concentration —Decrease in short-chain alcohol concentration
External variables	—Decrease in temperature (anionic surfactants) —Increase in temperature (nonionic surfactants) —Decrease in pressure

ponents, the brine, and the oil. Within the diagram, a demixtion curve separates regions of one- and multi-phase behavior. Strictly, this representation does not permit us to draw tie lines because pseudo-components do not partition as a single component. For instance, electrolytes may partition independently of water. However, in a first approximation, this will be neglected to keep visualization easy.

Figure 3.4 illustrates the main configurations. For example, increasing brine salinity gives the sequences a–b–c–d–e, where the overall composition OC separates into a type I, II or III system according to the R value. Each two-phase region exhibits a plait point, unless it terminates on a diagram boundary as the two-phase region in c. The three-phase region originates from the split of a critical tie line, as shown in b and d.

Inside the type I and II lobes (c in Fig. 3.4), the structure of the micellar phase is similar to that which exists in the three-phase region (tie triangle), so it is considered as being in a type III environment.[11] Any overall composition located in a tie triangle separates into invariant compositions B, M and O. As salinity increases, the invariant point M migrates from B to O.

The amount of surfactant required to form a single phase from equal amounts of oil and water is minimal when the amounts of oil and brine solubilized in the middle phase M are equal ($R = 1$). This configuration is defined as an optimum, and salinity, or some other equivalent variable, is called optimum. Oil and brine solubilization can be expressed in terms of solubilization parameters SP_o and SP_w, which

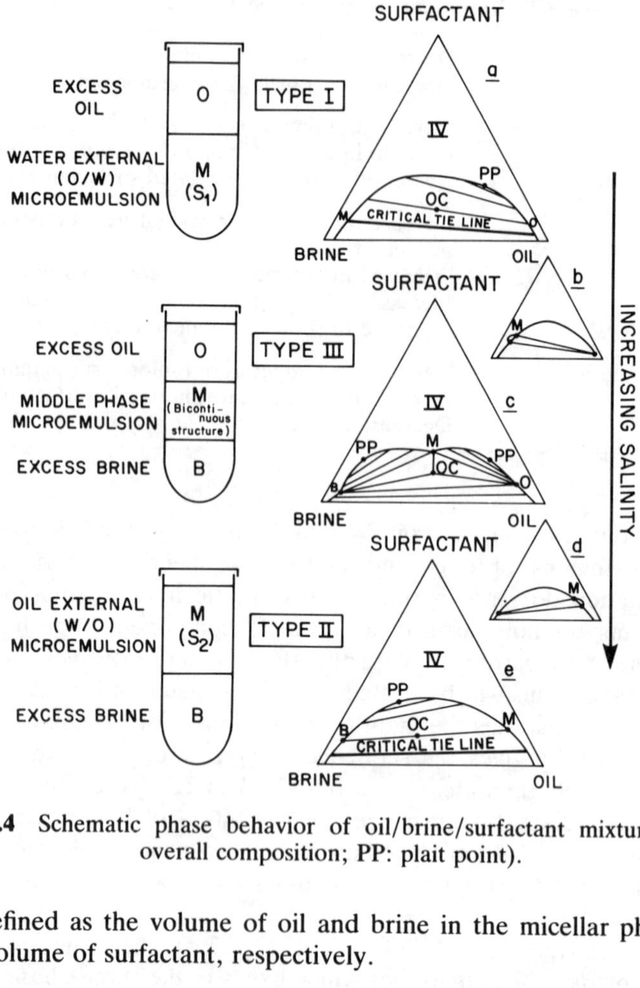

Fig. 3.4 Schematic phase behavior of oil/brine/surfactant mixtures (OC: overall composition; PP: plait point).

are defined as the volume of oil and brine in the micellar phase per unit volume of surfactant, respectively.

(b) *Solubilization and interfacial tensions.* The correlation between solubilization and interfacial tensions was clearly stated by Reed and Healy:[12] interfacial tension between the oil (or brine) excess phase and the micellar phase decreases as oil (or brine) solubilization increases (Fig. 3.5). This gives a fundamental basis for designing the process.

In the type III environment, interfacial tensions are very low—about 10^{-3} mN/m. As shown in Fig. 3.6, interfacial tensions between the middle phase and excess phases are lower than in type I and II

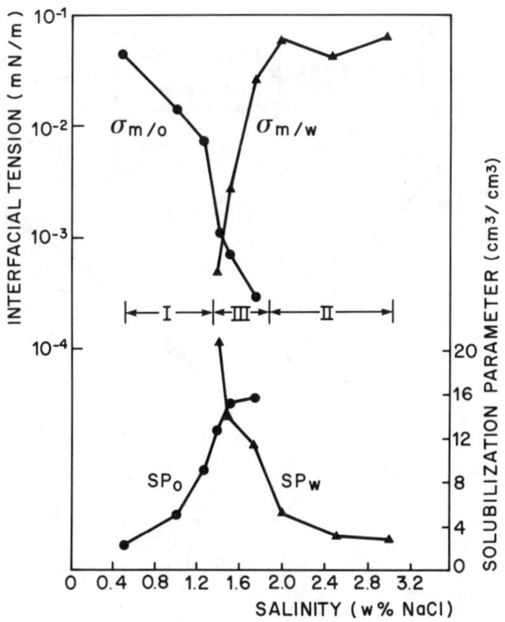

Fig. 3.5 Interfacial tensions and solubilization parameters versus salinity. (From Ref. 12, reproduced by permission of The Academic Press, Inc.)

systems. The higher value is associated with a high surface pressure of the surfactant film. The lower value decreases toward zero at the boundary between two-phase and three-phase systems and is associated with the nearness of critical points.[13]

Usually, interfacial tension curves intersect approximately at the same abscissa as solubilization curves. A relationship between solubilization parameter *SP* and interfacial tension σ has been empirically established.[12] Huh developed a model layer for describing the middle phase that leads, at the optimum, to the following equation:

$$\sigma = \mathrm{k}/(SP)^2 \tag{3.2}$$

where k is a constant depending on the surfactant.[14] This relation has been reasonably verified with pure hydrocarbons. Even though the measurement of σ has been facilitated by the use of the spinning drop method,[15] the correlation between *SP* and σ is very useful from a practical standpoint, because solubilization is more easily measured than interfacial tension, especially when crude oils are used.

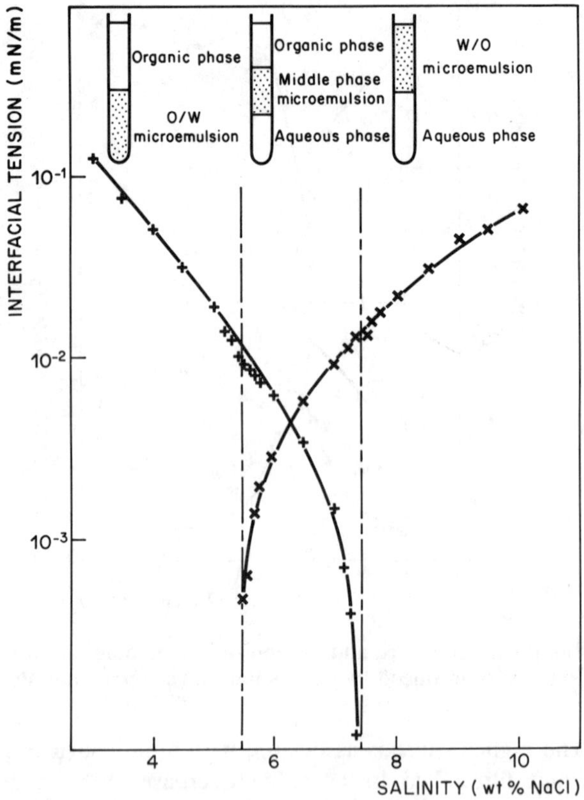

Fig. 3.6 Interfacial tension variation with salinity. +, Tensions between microemulsion and top phases; ×, tensions between microemulsion and bottom phases; —-—-, limits of the three-phase domain. (From Ref. 13.)

(c) *Other phase properties.* The micellar phase viscosity and the coalescence rate of the emulsion formed by mixing the micellar phase with excess phases may have important implications in the process, and this can be correlated with the phase behavior. Changes in microemulsion viscosity during a I–III–II transition are very sensitive to the composition. Many authors have observed a maximum slightly before the I–III transition and again after the III–II transition, indicating a certain ordering (sometimes the micellar phase is anisotropic and non-Newtonian), and a minimum close to the optimal configuration. This minimum is more or less pronounced according to the fluidity of the interfacial region, for instance, depending on alcohol concentration or temperature. It is not always present and may disappear for mesophase formation.

The coalescence rate behaves in a similar way,[10,16] probably for the same reasons. A minimum of interfacial tension has also been correlated with other properties, such as electrophoretic mobility[17] and a minimum of interfacial viscosity.[18]

(d) *Partitioning of pseudocomponents*. Alcohol (or other component) partitioning can be responsible for changes in the optimal conditions when the overall composition varies. This is especially noticeable with the amphiphilic pseudocomponents (surfactant plus alcohol). For instance, if the concentration of amphiphiles decreases by dilution, the optimum salinity shifts because of the change in alcohol concentration. If the water/oil ratio (WOR) is changed, the behavior depends on whether alcohol partitioning favors oil or brine.[19] These and similar changes can be correlated with the R value. An alcohol with approximately equal partitioning between the brine and oil under reservoir conditions of interest will cause the least variation in optimal conditions and will usually be the preferred choice for this reason. Current practice favors not using any alcohol, but this can be difficult at low temperatures and under a variety of other conditions, so this decision must be made carefully to avoid an unfavorable tradeoff.

A change in phase behavior may result from salt partitioning. With NaCl brine, the electrolyte is partially excluded from the micellar phase.[20] However, the opposite trend is observed with $CaCl_2$ brine because of the strong association of anionic surfactant with divalent cations (see Section 3.3.3(c)). Therefore, decreasing surfactant concentration reduces interactions between the interfacial region and brine, then optimal salinity decreases.[21]

The partitioning of hydrocarbon mixtures between micellar and excess oil phases tends to enrich the oil in the micellar phase in polar species because of their preferential association with the amphiphilic membrane.[10] This can be detrimental for interfacial properties when these species have a high molar volume, with a consequent decrease in the interfacial density of the surfactant.

An appropriate choice of pseudocomponents reduces the dependence of the phase behavior on the overall composition and can be used for modeling purposes. The salinity gradient design concept aims at minimizing the impact of the nonideal phase behavior or to take advantage of it (see Section 3.4.2).

(e) *Optimization.* The three-phase environment with $R = 1$ is assumed to be the best. However, not all optimized systems are equivalent. According to Winsor's theory, variables have to be adjusted so that the R-ratio numerator and denominator are as high as possible to maximize solubilization.[10] This also corresponds to the minimum in the size of the multiphase region.[22]

Pressure, temperature and oil composition are not typically used as design variables, but salinity can usually be changed to improve the process efficiency (see Section 3.4.2).

Oil composition is of course an essential parameter. To make laboratory research easier, especially when using gas-containing crude oils, it has been proposed that live oil be replaced by a model oil having the same phase behavior.[23] However, this concept also requires that the crude oil and the model oil have equal molar volumes and solubilization parameters and, if possible, the same viscosity and middle-phase environment.[24] This emphasizes that the concept still has to be used carefully. In particular, it is difficult to model the effect of heavy components such as paraffins, resins and asphaltenes.

The optimization of the surfactant involves increasing the interactions between surfactant and both oil and brine as much as possible. This is achieved by increasing both the molecular weight of the hydrophobic part and the strength of polar heads, and also by minimizing the area per surfactant molecule in the interfacial region, for example by using a straight rather than branched alkyl chain. Liquid–crystal phases resulting from compact ordering are usually prevented by using alcohols. However, since alcohols diminish surfactant interfacial density and decrease solubilization because of much smaller interactions between alcohol molecules and both oil and brine, it seems to be more advisable to use a mixture of surfactants to reach both an optimum configuration and an acceptable phase condensation. In other respects, alcohols improve the compatibility between the surfactant and the polymer used for controlling mobility[25] (see Section 3.3.1). They may also reduce the loss of surfactant by precipitation or adsorption on the rock. So the selection of a micellar composition requires a compromise when taking into account all the phenomena.

3.2.4 *Thermodynamics: phase behavior modeling*

The interpretation of surfactant flooding tests by simulators requires accurate knowledge of phase behavior based upon considerable laboratory work. Another way is to create models able to predict the stability of oil/brine/amphiphile mixtures, i.e. the location of demixtion curves in the phase diagram, and the nature and composition of phases that may form from the overall mixture.

Since Schulman's work on the stability of microemulsions,[9] it has been shown that the calculation of the free enthalpy of mixing should take into account the increase of entropy resulting from the high degree of division of the system, and that the stability of the mixture can be reached if interfacial tension has been lowered enough by the surfactant.[26]

Several models of microemulsion have been proposed, together with corresponding equations giving the free enthalpy of mixing. The mathematical treatment of these equations becomes more complex as the number of components increases. This number can be reduced by considering pseudocomponents. One way of doing this is to associate the pseudophase assumption with a thermodynamic model leading to free-enthalpy determination.[27,28]

In the pseudophase model, microemulsion phases formed from oil/brine/surfactant/alcohol mixtures consist of three microdomains, called pseudophases: an aqueous pseudophase W' (brine + alcohol), an oleic pseudophase O' (oil + alcohol) and an interface pseudophase M' (surfactant + alcohol). Brine and oil excess phases, if any, have the

→	POLYMER DRIVE 50 to 100% PV	SURFACTANT SLUG 5 to 60% PV	OIL BANK 35 to 70% OIL CUT	INITIAL CONDITIONS	→

Typical range of surfactant slug composition:

- Aqueous slug

0.1	to	10%	surfactant
0.05	to	0.2%	polymer
0	to	10%	cosurfactant

Balance is brine

- Microemulsion

1	to	20%	surfactant
10	to	90%	oil
1	to	20%	cosurfactant

Balance is brine

Typical range of polymer drive composition:

0.05 to 0.2% polymer in brine

Fig. 3.7 Schematic cross-section of a typical micellar-polymer flood.

same compositions as W' and O', respectively. The pseudophase compositions are calculated from the overall composition and from the partition coefficient of alcohol between pseudophases. Thus, the composition of microdomains is taken into consideration, especially the interfacial region, and, by considering the plane defined by the pseudophase compositions, the system is treated as a ternary in which pseudophases are true pseudocomponents. The possible existence of excess phases is determined by a thermodynamic model using physicochemical parameters characterizing the system, namely (1) the entropy of dispersion of oil and brine, (2) interactions promoting the solubilization and those opposed to it, and (3) the curvature energy of the interface. The model still needs to be improved, in particular for simulating the region of the phase diagram near critical points.

3.3 Mechanisms involved in oil displacement by surfactant solutions

3.3.1 *Immiscible displacement under low interfacial tension conditions*

There are at least four distinct regions of immiscible flow during a typical displacement of oil by surfactant solutions in a porous medium (Fig. 3.7). The first of these is the two-phase flow of oil and water under the initial conditions of the reservoir. This region is often, but not always, close to residual oil saturation to water and in laboratory corefloods may be exactly at this endpoint. The second region, going upstream toward the injector, is the oil bank. Assuming a stable displacement of the oil by the surfactant solution, then a region of oil flowing at higher oil saturation than the initial condition must form because of overall material balance considerations, and this two-phase flow region is called the oil bank. The third region is characterized by the two- or three-phase flow of oil, brine and microemulsion, depending on phase behavior and fractional flow considerations. This is the low interfacial tension region. The fourth region is the polymer solution. This is a high interfacial tension region again and will usually be characterized by the single-phase flow of the aqueous phase.

(a) *Role of relative permeabilities.* Each of these immiscible flow regions will be governed in part by the relative permeabilities of the phases which form in them. In addition to the usual petrophysical factors that affect the oil and water relative permeabilities in porous media, the effect of changing interfacial tension must be considered. This can be done using the capillary number given by eqn (1.14) in Chapter 1. Although such a dependence of the two- and three-phase flow relative permeabilities on the capillary number is complex and few measurements have been reported, several observations can be made.[29] (1) The endpoint relative permeabilities of each of the three phases increase as the capillary number increases, and the residual saturations decrease until each endpoint reaches ~1 at very high capillary numbers. (2) The curvatures of the relative permeability versus saturation curves change very slowly with the capillary number and are still significantly different from straight lines even at the highest capillary numbers measured to date (about 0·1). Also, the curvature for the microemulsion phase was found to be the opposite of the usual positive curvature for oil and water. (3) There is little difference between the relative permeability of each phase under two-phase flow conditions and three-phase flow conditions at the same saturation and capillary number, i.e. the relative permeability of each phase depends only on its own saturation rather than on two saturations.

The relative permeabilities together with the viscosities of each fluid can be used to calculate the fractional flow of each phase as a function of the saturations and capillary number. Fractional flow theory[30,31] has been found to be very useful to gain insight into flow in permeable media, neglecting dissipative effects such as dispersion and capillary pressure. This theory helps us understand which banks form and how they propagate. This in turn helps us understand certain design requirements, such as for mobility control.

(b) *Mobility control.* The fundamental criterion of a stable displacement is that each bank is less mobile than the one it is displacing. The total relative mobility of a given bank is defined as the sum of the mobilities (relative permeability divided by viscosity) of each fluid in the bank. For an oil bank assumed free of chemicals, this is simply the sum of the mobilities of the oil and water phases. The total relative mobility curves for four rock samples are shown in Fig. 3.8. The most conservative design approach, as suggested by Gogarty *et al.*,[32] is to take the minimum of the lowest curve as the basis for the design of the surfactant and polymer solutions displacing the oil bank. For this example, that means the total relative mobility of these solutions needs to be less than 0·03 $(mPa/s)^{-1}$ or the equivalent viscosities need

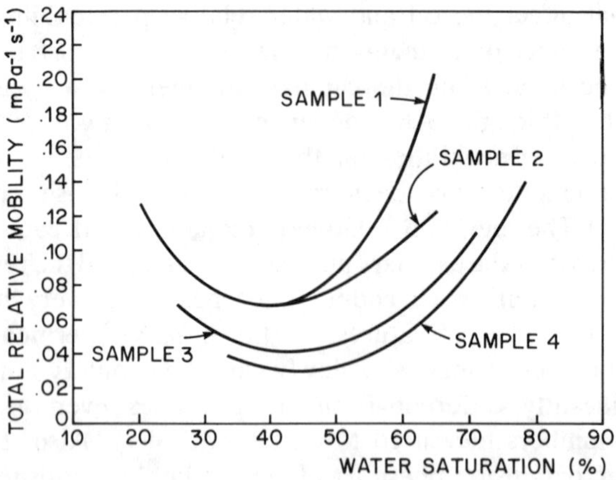

Fig. 3.8 Total relative mobilities for samples from the same reservoir. (From Ref. 32, reproduced by permission of The Society of Petroleum Engineers.)

to be at least 33 mPa/s. Most surfactant solutions in brine do not have such a high viscosity, so either oil or polymer must be added to the surfactant slug to make it more viscous (but interactions with surfactant and non-Newtonian behavior have to be taken into account (see Chapter 2)).

Decreasing the mobility ratio increases areal, vertical and displacement sweep efficiency in surfactant floods as it does in other floods. In addition, the mobility control in a surfactant flood affects the distribution of the slug areally and vertically into the reservoir in a way that improves as the mobility ratio decreases. Good mobility control is essential to protect the integrity of a small slug against dissipation because of fingering, channeling, mixing, crossflow and other reservoir mechanisms. Fingering is greater in a low-tension process than in most others because at low tension the capillary forces are low and thus there is little or no stabilization of the fingers by capillary-driven crossflow across the fingers. The microscopic or displacement sweep efficiency is improved as the mobility ratio decreases because the fractional flow curve becomes steeper, i.e. more favorable for a piston-like displacement.

3.3.2 *Mass transfer mechanisms*

Surfactant flooding is also affected by a variety of displacement mechanisms beside those governing immiscible flow. The most important of these is the mass transfer resulting from the change in phase behavior as the fluids propagate and mix in the reservoir. Emulsification and wettability changes can also affect surfactant displacements to various degrees, although less is known quantitatively about these effects. They are generally thought to be of less importance than the equilibrium phase behavior.

The mass transfer that actually occurs in a particular displacement will depend on the phase behavior environment and the composition of the injected solutions. For example, if an aqueous surfactant solution is injected and mixes with oil and brine in the composition region corresponding to the tie triangle of a type III phase diagram, then it will solubilize large amounts of oil and form a middle-phase microemulsion in equilibrium with both excess brine and excess oil. If the composition is such that the phase behavior is type II, then an upper-phase microemulsion will form, and this microemulsion will be swollen compared to the oil phase free of surfactant. Thus, some oil recovery can occur even if there is no decrease in interfacial tension, since this swollen phase will have a greater saturation than the pure oil.

The properties of the in-situ microemulsion, such as viscosity, relative permeability and specific gravity, rather than the properties of the injected solution, will then determine its behavior and displacement of oil and brine. Since the temperature and crude oil composition are constant or nearly so in most displacements, one of the key reservoir variables that determines this outcome is the electrolyte composition of the brine. The electrolyte composition tends to change because of variations in both the salinity (usually expressed as the total dissolved anions in the brine) and the amounts and distributions of the monovalent and divalent cations (hardness) in the brine. These variations result from nonuniform cation salinity in the reservoir at the start of the flood and also from the fact that the injected preflood (if any), surfactant slug, polymer drive and chase brine are all in general of different salinity. The differences can be used advantageously to create a favorable salinity gradient as discussed in Section 3.4.

3.3.3 *Fluid/rock interactions*

Fluid/rock interactions that affect surfactant flooding are adsorption, cation exchange, precipitation–dissolution phenomena, capillary phenomena, and dispersion. All of these directly or indirectly affect the retention of surfactant. The tendency for surfactants to be adsorbed on mineral surfaces has been extensively studied and widely recognized to be important for the economics of the process, since only small losses of the expensive surfactant can be tolerated. There has also been much concern about the precipitation of surfactants if brines of high hardness are encountered, but this is generally not a problem in practice because most if not all of the surfactants used in surfactant flooding will partition in the oil or in the upper-phase microemulsion formed under these conditions, rather than precipitate. Phase trapping caused by microscopic capillary forces in the pores of the rock has received much less attention but is of potentially more importance.

(a) *Capillary phenomena.* Residual microemulsion can form in much the same way as residual oil forms and can be correlated by the same capillary number approach as discussed earlier with respect to oil trapping. Actually, the trapping of microemulsion can be greater than that of oil.[33,34] This is because the oil is nonwetting and the microemulsion is the wetting phase, and wetting phases are always held more tightly than nonwetting phases. Figure 3.9 shows that the residual microemulsion saturation can be high even at very high capillary numbers (field average capillary numbers, even assuming ultra-low interfacial tensions of about 0·001 mN/m, rarely exceed 0·01

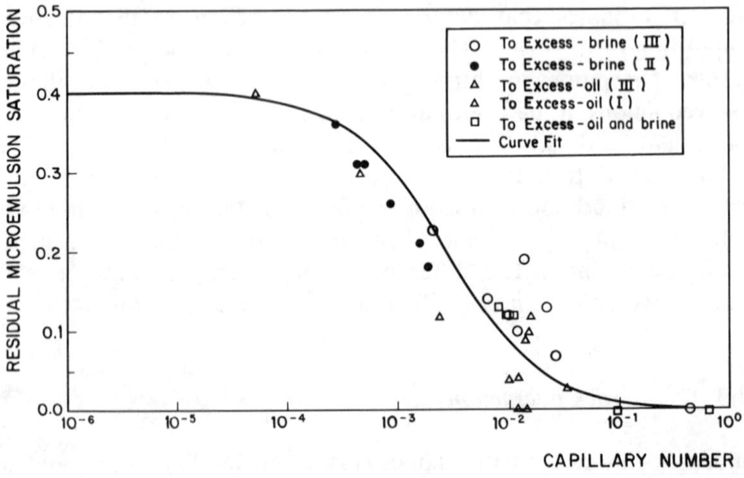

Fig. 3.9 Microemulsion capillary desaturation curve. (From Ref. 34.)

and are often about 0·001). A residual microemulsion saturation of even 10% will result in a greater loss than that due to typical adsorption in clean sandstones.

Fortunately, the trapping of microemulsion can often be eliminated entirely by designing the process in such a way that the phase behavior environment is type I upstream from the surfactant slug. The simplest and least expensive way to do this is to reduce the salinity of the polymer drive. As the surfactant slug mixes with the polymer drive, it will change from type II or type III to type I. No trapping will occur with type I behavior because the lower phase microemulsion will be miscibly displaced by the brine in the polymer drive.

(b) *Adsorption.* Surfactant adsorption is caused by the attraction (van der Waals and electrostatic forces, hydrogen bonding) between the surfactant molecules and the mineral surfaces. The forces differ greatly between nonionic and anionic surfactants and with the nature of the mineral surface. Adsorption also depends on the temperature, electrolyte composition and pH of the brine, the surfactant concentration, the cosurfactant or cosolvent concentration if any, the polymer concentration and even the oil concentration, since the oil affects both the phase behavior and the micellar structure of the surfactant solution. Thus, it is difficult to describe the adsorption behavior of an equilibrium fluid/rock system. The dynamic adsorption of a surfactant flowing in a porous medium would depend upon not just the level of adsorption but also the selective adsorption from mixtures of surfactants, since this may lead to chromatographic separation of the surfactants.[35] For this reason, here the discussion is restricted to consideration of what controls the adsorption of a particular surfactant under reservoir conditions, and it is assumed that a surfactant with favorable adsorption characteristics has already been selected from screening experiments. It should be noted that the same conditions that give high solubilization of oil and brine in the microemulsion phase also tend to result in lower adsorption.[36]

For a particular surfactant and a given set of reservoir conditions (temperature, crude oil, clay type and content), the primary remaining variables are the concentrations of the surfactant, electrolytes, cosolvent, and alkaline agents if any. Surfactants increase in adsorption as the surfactant concentration increases until a plateau value is reached at the CMC, as illustrated in Fig. 3.10.[37] Since injected surfactant concentrations are usually at least two orders of magnitude greater

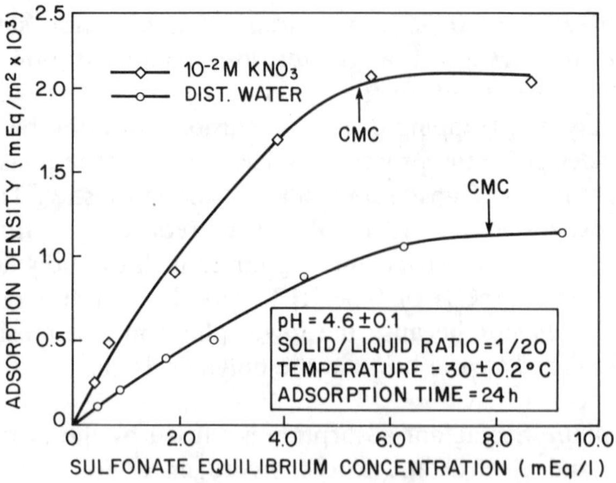

Fig. 3.10 Adsorption isotherms of sodium dodecylsulfonate on kaolinite. (From Ref. 37, reproduced by permission of The Academic Press, Inc.)

than the CMC of effective surfactants, this plateau is reached very quickly and can be approximated as a very steep Langmuir-type isotherm. With mixtures the increase occurs more gradually, but the plateau value is still reached at very low concentrations of surfactant. Thus as a first approximation the plateau adsorption value can be taken as the adsorption that will occur during the surfactant flood. This leaves the electrolyte composition of the brine as the key variable that controls the adsorption for a particular reservoir.

Polymer adsorption has already been considered in Chapter 2, but new aspects come into play when a polymer is used with a surfactant. For example, surfactant–polymer incompatibility (SPI) can increase the retention of both surfactant and polymer.[25] SPI is a more important factor than polymer adsorption when large polymer slugs are injected.

(c) *Cation exchange.* Almost all sandstone oil reservoirs contain clay with significant cation exchange capacity. Cation exchange will occur when one or more injected fluids differ in electrolyte composition from the initial fluid saturating the permeable medium. This is always the case when using an anionic surfactant. So cation exchange will typically occur, and the only question is how important it is in a given situation. Actually, there will be two cation exchange reactions, since exchange will occur both with the clay and with the double layer surrounding the micelles.[38] The net effect of these reactions has thus to be considered.

To simplify the discussion, only sodium and calcium ions will be mentioned even though many other cations will be present. The ion exchange reaction of these cations with a negative charge site M on either clay or micelle can be represented as follows:

$$2Na^{+} + CaM_2 \rightleftarrows Ca^{++} + 2NaM \tag{3.3}$$

The same concentration of calcium in mEq/liter has a greater effect on anionic surfactants such as petroleum sulfonates than sodium does. Thus, if the optimum salinity is low and a large amount of exchange occurs, great changes in the interfacial properties of the surfactant will result and must be taken into account in the surfactant flood design. However, if the optimum salinity is very high and/or the exchange relative to this optimum salinity is low, cation exchange may not be very important in the design.

Simple analytical calculations can be performed to estimate the exchange with clay during a coreflood. An example is shown in Fig. 3.11.[39] Calculations taking into account exchange with the micelles, dispersion, adsorption and other fluid/rock interactions have been

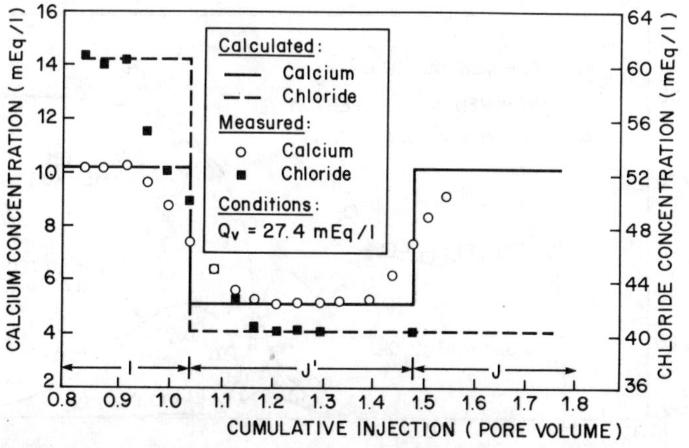

Fig. 3.11 Comparison between theory and experiment for displacement with cation exchange. *I* and *J* are initial and injected brine composition, respectively; *J'* is the plateau region. (From Ref. 39, reproduced by permission of The Society of Petroleum Engineers.)

made using numerical simulators.[40] Under some conditions, even a very small amount of dissolution of a slightly soluble mineral such as calcite is enough to keep the clays partially loaded with calcium. For example, the solubility of calcite at 25°C is only 20 ppm as calcium, but this is enough to affect the cation exchange reaction with clay in low salinity brines, and therefore to increase the adsorption of the surfactant. More soluble minerals such as anhydrite occur less often in sandstone reservoirs but have a larger and more direct effect on anionic surfactants when present.

3.3.4 *Modeling of surfactant flooding*

The most comprehensive and versatile approach to modeling the macroscopic displacement of oil by surfactants is the compositional reservoir simulator. It will be briefly reviewed here.

A compositional simulator is one that is based upon the mass conservation of each component. This leads to a set of partial differential equations often called material balance equations. These equations are nonlinear for the multiphase flow process of interest, so in general they must be solved numerically, usually by a finite

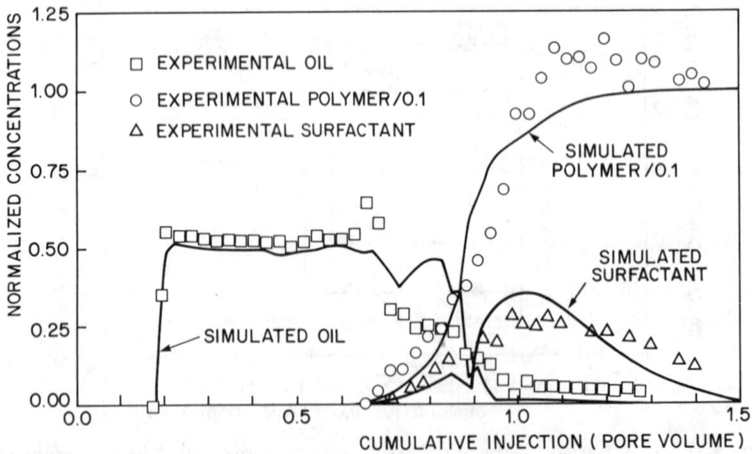

Fig. 3.12 Comparison between simulated and experimental total oil, surfactant and polymer histories. (From Ref. 43, reproduced by permission of The Society of Petroleum Engineers.)

difference method. Simulators of this type have been reported by Datta-Gupta *et al.*,[40] Van Quy and Labrid[41] and Scott *et al.*,[42] among others.

In addition to the material balance equations, Darcy's law and a large number of equations describing the physical properties of interest such as phase behavior, interfacial tension, relative permeability, capillary desaturation, adsorption, dispersion and others are needed. One of the key assumptions usually made is that the fluids are at local thermodynamic equilibrium. Most of the other assumptions and approximations deal with the detailed behavior of each property, such as the character of the adsorption isotherms, and can be improved or refined if justified.

Despite the numerous simplifications and approximations in such a simulator, reasonably good agreement for engineering purposes has been achieved both with laboratory corefloods and with field pilot data. Partial results from a comparison between simulation and typical Berea coreflood are shown in Fig. 3.12,[43] and between simulation and the Big Muddy surfactant pilot data in Fig. 3.13.[44]

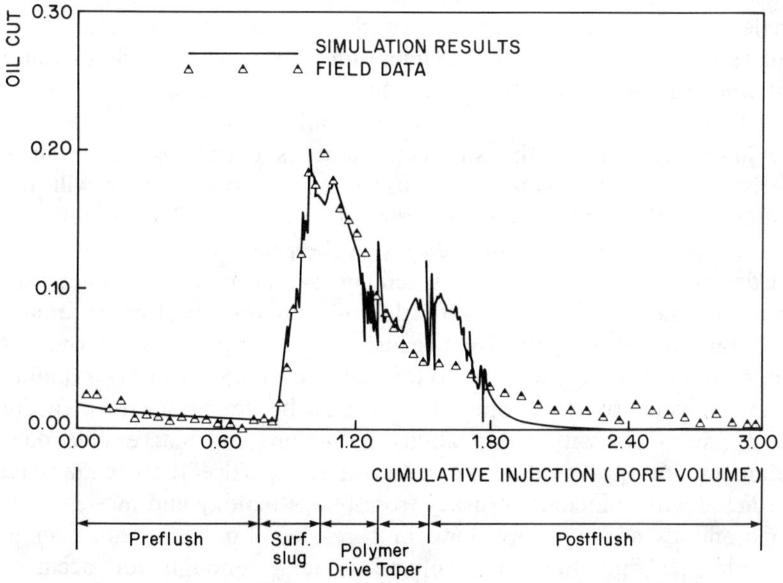

Fig. 3.13 Comparison between simulated and field oil cuts. (From Ref. 44, reproduced by permission of The Society of Petroleum Engineers.)

3.4 Process design

3.4.1 *Fundamentals*

The four fundamental requirements for a successful surfactant flood are: (1) the surfactant must mobilize the residual oil and reduce its saturation essentially to zero microscopically by lowering the interfacial tension by several orders of magnitude; (2) the retention of the surfactant must be sufficiently low that it can be propagated across the entire oil reservoir using only a small amount of injected surfactant; (3) the mobility control must be sufficiently good that high sweep efficiency occurs in the reservoir so that the mobilized oil will be banked up and efficiently displaced and captured by the producing wells; and (4) the reservoir characterization and reservoir engineering must take into account the reservoir characteristics by appropriate well spacing, well control, profile control and so forth.

The first step in the process design is to screen and select a surfactant taking these fundamental requirements into account. The physicochemistry of surfactant solutions described earlier in this chapter should be used as the basis for this screening process. The single most important screening criterion is high solubilization parameters (in the order of 10 or higher) under reservoir conditions and optimum salinity, since this invariably means the interfacial tension will then be very low. Molecular considerations, such as the length of the lipophilic part of the surfactant and its branching, were summarized above, so all subsequent discussion of process design will just assume that this step has been successfully completed. Then corefloods must be done to evaluate the efficacy of the process.

Corefloods should be conducted in reservoir cores. They are important not only to evaluate the oil recovery performance and surfactant retention, but also because of potential problems such as poor transport of polymer due to low core permeability or poor quality of the polymer solution. Special tests must be performed to measure the apparent viscosity, permeability reduction, and inaccessible pore volume to the polymer in the reservoir rock, since these cannot be predicted with sufficient accuracy from bulk rheology and models.

Corefloods are typically done in cores 0·3–1 m long and 5 cm in diameter so that the pore volume is large enough for accurate measurements of produced fluids. Tests in shorter cores can be very misleading because of mixing phenomena (for example, emulsion formation depends on the residence time of the surfactant solution) and capillary end effects. Such short corefloods are so poorly scaled that the results are very difficult to interpret. An example of a successful coreflood is shown in Fig. 3.12. Such well designed corefloods produce 80–90% of the oil as a clean oil bank (before surfactant breakthrough) and 90–100% of the oil by 1·2 to 1·5 pore volumes. In addition to oil recovery, the complete effluent composition and the pressure drop across two or more sections of the core should be measured. Additional analysis of the core, which is also desirable, includes the final oil saturation and surfactant retention along the length of the core. These measurements are valuable both for a general assessment of the experimental results and for comparisons of compositional simulations needed for scale-up to reservoir conditions during the field design process.

Such high oil recovery in corefloods when done with small slugs (e.g. 10% of the pore volume at 6% active chemical) and realistic operating conditions (e.g. 0·02 MPa/m pressure gradient) generally requires not only a good surfactant, which produces ultra-low interfacial tension and good mobility control,[36] but also several other equally important factors that apply to both corefloods and reservoir floods. These factors are most easily discussed in terms of the salinity gradient concept.

3.4.2 *Salinity gradient concept*

The fundamental premise of the salinity gradient concept is that the most efficient design of a surfactant flood is not one that operates at a constant salinity corresponding to the optimum salinity, but rather a design that imposes a salinity gradient to take advantage of the favorable properties of the surfactant first at high salinity and then at low salinity.[11,45,46] This may at first seem like a contradiction, since the optimum salinity of a surfactant would appear to be the best salinity for the surfactant flood. But the optimum salinity, as conventionally defined in Section 3.2.3, is not the point of lowest interfacial tension between the microemulsion and the oil to be mobilized and displaced. The minimum in the latter's σ value occurs at the upper critical salinity, at which the phase behavior changes from type III to type II (Fig. 3.6). This minimum σ value produces the maximum capillary number for a microemulsion displacing oil, hence the lowest residual oil saturation. Moreover, the optimum salinity is not the point at which the lowest retention occurs, nor the point of optimal mobility control with polymer, both of which occur at lower salinities. Clearly, a design that takes advantage of the most favorable behavior first at a high salinity and then a low salinity is superior to one that considers only one salinity. High and low salinity in this context are strictly relative terms; if the optimum salinity of the surfactant is, say, 100 000 ppm, then low may mean 80 000 ppm.

It turns out that the salinity gradient design not only takes advantage of these changes in salinity, but it is also a more stable and robust design from a reservoir engineering point of view. Nor is a preflood required to impose this salinity gradient; in fact it may even be detrimental. A certain formation salinity is desirable, so that a lower salinity can be applied in the surfactant slug, then a still lower salinity applied in the polymer drive. Mixing in the reservoir will nicely take care of modifying these steep changes into a gradually decreasing salinity gradient. This gradient will also make the surfactant less sensitive to cation exchange.

At the rear of the microemulsion, the most important interfacial tension is the one between the polymer drive brine and the microemulsion, which will be a middle-phase microemulsion if the salinity environment is type III. Interfacial tension will likely not be low enough to prevent trapping of the microemulsion by the brine and thus very high retention of surfactant.[25] Figure 3.9 shows that the capillary numbers required to reduce the microemulsion saturation to zero are very high. Thus, the solution is to reduce the polymer drive salinity below the lower critical salinity at which the type III behavior changes to type I. Since the polymer slug is much larger than the surfactant slug, this lower salinity will then eventually allow the polymer drive to miscibly displace the lower-phase microemulsion, so no trapping of surfactant will occur. The adsorption of the surfactant will also decrease at this point, assuming it is reversible with salinity. With some polymers, there will also be a decrease in adsorption, an increase in viscosity, and a decrease in SPI. All of these factors lead to a more focused surfactant slug that is less affected by dispersion.

The salinity gradient also makes use of the variation and poor predictability of reservoir and fluid properties across the reservoir. For example, not only do many waterflooded reservoirs have spatial variations in salinity at the start of the surfactant flood, but they may have variations, known and unknown, in a variety of other conditions such as pressure, dissolved gas in crude, temperature, clay composition (and thus calcium), and soluble minerals. Designing a slug that is optimum for all of these conditions is very difficult, but imposing a salinity gradient that passes through the various optimum surfactant conditions for these variable properties will often be relatively easy. Furthermore, the surfactant may change properties because of a change in plant specifications or changes induced by the reservoir after it is injected. For example, the molecular weight might shift slightly or the surfactant-to-cosolvent (e.g. alcohol) ratio may change. A salinity gradient produces a window of opportunity for the surfactant to pass through its optimum salinity as these phenomena occur.

For all these reasons, the probability of an efficient displacement is much greater with a gradient than without. Under some conditions, however, a salinity gradient will not be feasible because of the location of the project. Other alternative gradients should then be studied. For example, a gradient in the surfactant or cosolvent properties is also possible. Cation exchange can be used to create a favorable calcium concentration gradient under relatively fresh water conditions. Finally, the process can be made to work with no imposed gradient (there will always be uncontrolled gradients of some kind in the reservoir), but the demands on the surfactant are greater, e.g. it needs a wider type III region of behavior.

3.4.3 *Field operations and perspectives*

Many surfactant floods have failed or performed far below expectations simply because of inadequate attention to balancing well patterns, lack of areal or vertical confinement, poor or unknown reservoir characteristics including fractures, channels, and bottom water, unsuitable well completions, or injectivity profiles. Good reservoir engineering practice should either correct or, if not feasible to correct, then rule out the reservoir as a surfactant flood candidate. All of these and other similar factors are much more important to the economics of a surfactant flood than they are to a waterflood.

As far as polymer is concerned, the most important consideration is the total amount of polymer injected. For economic reasons, this amount is minimized by using concentrations in the mobility buffer, which are reduced gradually in such a way that the concentration reaches zero in the chase water. Figure 3.14 shows the recovery

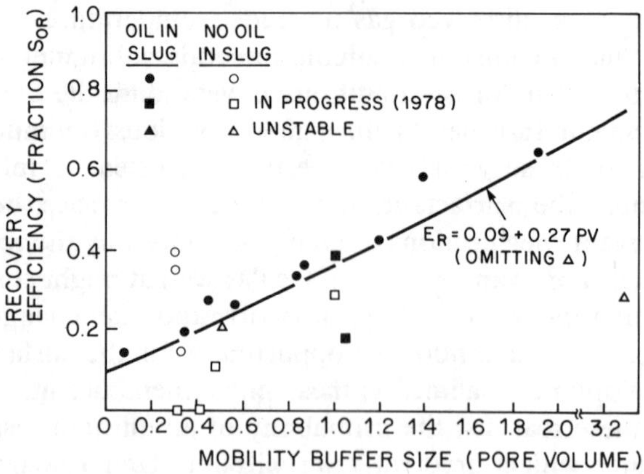

Fig. 3.14 Recovery efficiencies from micellar–polymer field tests. (From Ref. 47, reproduced by permission of Edgell Communications, Inc.)

efficiency (ultimate oil produced divided by oil in place at the start of the surfactant flood) from a survey of 20 field tests correlated as a function of mobility buffer size in pore volumes.[47] This was the most important factor studied—even more important than field capillary number or slug size.

Some research and pilots have focused on combining the alkaline and surfactant processes (see Chapter 4, Refs 9 and 10).

Only two particular field tests will be cited here based upon their success, but also because they used a recent technology with considerable laboratory research reported and because they represent very different reservoir and design conditions: the Loudon Field test reported by Exxon[48] and the Chateaurenard Field test reported by Elf Aquitaine.[49] Both of these surfactant floods recovered about 68% of the oil in place at the start of the flood. The Loudon test was in a low-permeability, high-salinity sandstone reservoir containing a 5 mPa/s crude. A dilute aqueous surfactant slug containing xanthan gum polymer followed by a polymer drive was used. The Chateaurenard test was in a high-permeability fresh-water sand containing a 40 mPa/s crude. A high concentration microemulsion slug with oil used to generate viscosity followed by a hydrolyzed polyacrylamide polymer drive was used. Clearly these tests show that surfactant flooding can be technically successful when implemented by using appropriate chemicals, reservoir characterization, and numerical simulation models.

Attempts to reduce the cost of the process, either by coinjecting surfactant and polymer at low concentration followed by a mobility control buffer[50] or by using the previously mentioned combinations with alkaline chemicals, look promising. Both of these processes reduce the surfactant retention to very low levels and thus reduce the chemical requirements. With new surfactants that can be used at almost any reservoir salinity (see Section 3.2.1) and polymers stable at high temperatures (see Section 2.2 in Chapter 2), technical barriers to the use of the process have been removed.

As with polymer flooding, many physicochemical aspects still remain to be investigated, such as surfactant phase behavior, surfactant/polymer interactions and rock/surfactant solution interactions. But there is also a highly-important chemical aspect. Considering that the share of chemicals represents about two-thirds of the technical cost of oil produced by surfactant flooding, the process would certainly benefit from the conjugate efforts of petroleum and chemical industries to promote appropriate structures and manufacturing processes. More efficient chemicals combined with better reservoir characterization and process implementation are the keys to technical and economic success in the future.

3.5 References

1. Nutting, P. G., Chemical problems in the water driving of petroleum from oil sands. *Industrial and Engineering Chemistry,* **17** (1925) 1035–6.
2. Taber, J. J., Dynamic and static forces required to remove a discontinuous oil phase from porous media containing both oil and water. *Soc. Petroleum Engineers J.,* **9**(1) (1969) 3–12.
3. Wagner, O. R. & Leach, R. O., Effect of interfacial tension on displacement efficiency. *Soc. Petroleum Engineers J.,* **6**(4) (1966) 335–47.
4. Foster, W. R., A low-tension waterflooding process. *J. Petroleum Technol.,* **25** (1973) 205–10. *Trans. Soc. Petroleum Engineers AIME,* **255**.
5. Bavière, M., Bazin, B. & Labrid, J., Chemical stability of surfactants for micellar flooding under North Sea conditions. *In Situ,* **13**(3) (1989) 101–20.
6. Winsor, P. A., Hydrotropy, solubilisation and related emulsification process. *Trans. Faraday Soc.,* **44** (1948) 376–98, 451–71; **46** (1950) 762–72.
7. Winsor, P. A., *Solvent Properties of Amphiphilic Compounds,* Butterworth, London, 1954.
8. Scriven, L. E., Equilibrium bicontinuous structures. *Nature,* **263** (1976) 123–5.

9. Schulman, J. H. & Montagne, J. B., Formation of microemulsions by amino alkyl alcohols. *Annals of the New York Academy of Science,* **92**(2) (1961) 366–71.
10. Bourrel, M. & Schechter, R. S., *Microemulsions and Related Systems,* Marcel Dekker, New York, 1988.
11. Nelson, R. C. & Pope, G. A., Phase relationships in chemical flooding. *Soc. Petroleum Engineers J.,* **18**(5) (1979) 325–38.
12. Reed, R. L. & Healy, R. N., Some physicochemical aspects of microemulsion flooding: a review. In *Improved Oil Recovery by Surfactant and Polymer Flooding,* ed. D. O. Shah & R. S. Schechter, Academic Press, New York, 1977, pp. 383–437.
13. Langevin, D., Meunier, J. & Cazabat, A.-M., Correlations between interfacial tensions and microemulsion structure in Winsor equilibria. Role of surfactant film curvature properties. *Colloids and Surfaces,* **19** (1986) 159–70.
14. Huh, C., Interfacial tensions and solubilizing ability of a microemulsion phase that coexists with oil and brine. *J. Colloid Interface Sci.,* **71**(2) (1979) 408–25.
15. Cayias, J. L., Schechter, R. S. & Wade, W. H., The measurement of low interfacial tension via the spinning drop technique. In *Adsorption at Interfaces,* ACS Symposium Series, Vol. 8, ed. J. R. Gould, American Chemical Society, Washington, 1975, p. 234.
16. Vinatieri, J. E. Correlation of emulsion stability with phase behavior in surfactant systems for tertiary oil recovery. *Soc. Petroleum Engineers J.,* **20**(5) (1980) 402–06.
17. Sharma, M. K. & Shah, D. O., Use of surfactants in oil recovery. In *Enhanced Oil Recovery, II Processes and Operations,* ed. E. C. Donaldson, G. V. Chilingarian & T. F. Yen, Elsevier, Amsterdam, 1989, pp. 255–315.
18. Wasan, D. T., Shah, S. M., Aderangi, N., Chase, M. S. & McNamara, J. J., Observations on the coalescence behavior of oil droplets and emulsion stability in enhanced oil recovery. *Soc. Petroleum Engineers J.,* **18**(6) (1978) 407–17.
19. Bavière, M., Schechter, R. S. & Wade, W. H., The influence of alcohols on microemulsion compositions. *J. Colloid Interface Sci.,* **81**(1) (1981) 266–79.
20. Tosch, W. C., Jones, S. C. & Adamson, A. W., Distribution equilibria in a micellar solution system. *J. Colloid. Interface Sci.,* **31**(3) (1969) 297–306.
21. Glover, C. J., Puerto, M. C., Maerker, J. M. & Sandvik, E. L., Surfactant phase behavior and retention in porous media. *Soc. Petroleum Engineers J.,* **19**(3) (1979) 183–93.
22. Holm, L. W., Use of soluble oils for oil recovery. *J. Petroleum Technol.,* **23** (1971) 1475–83.
23. Cayias, J. L., Schechter, R. S. & Wade, W. H., Modeling crude oils for low interfacial tension. *Soc. Petroleum Engineers J.,* **16**(6) (1976) 351–7.
24. Puerto, M. C. & Reed, R. L., A three-phase parameter representation of surfactant/oil/brine interactions. *Soc. Petroleum Engineers J.,* **23**(4) (1983) 669–82.

25. Trushenski, S. P., Micellar flooding: sulfonate–polymer interaction. In *Improved Oil Recovery by Surfactant and Polymer Flooding*, ed. D. O. Shah & R. S. Schechter, Academic Press, New York, 1977, pp. 555–75.
26. Ruckenstein, E. & Chi, J. C., Stability of microemulsions. *J. Chem. Soc.*, **71** (1975) 1690–707.
27. Biais, J., Clin, B. & Lalanne, P., Phase diagrams and pseudophase assumption. In *Microemulsions: Structure and Dynamics*, ed. S. E. Friberg & P. Bothorel, CRC Press, Boca Raton, Florida, 1987, pp. 1–31.
28. Trouilly, J. L., Contribution à l'étude de la stabilité thermodynamique des microémulsions dans le cadre du modèle des pseudophases. PhD Dissertation, University of Bordeaux, France, 1986.
29. Delshad, M. Delshad, M., Pope, G. A. & Lake, L. W., Two- and three-phase permeabilities of micellar fluids. *SPE Formation Evaluation*, **2**(3) (1987) 327–37.
30. Helfferich, F. G. Theory of multicomponent, multiphase displacement in porous media. *Soc. Petroleum Engineers J.*, **21**(1) (1981) 51–62.
31. Hirasaki, G. J., Application of the theory of multicomponent, multiphase displacement to three-component, two-phase surfactant flooding. *Soc. Petroleum Engineers J.*, **21**(2) (1981) 91–204.
32. Gogarty, W. B., Meabon, H. P. & Milton, Jr, H. W., Mobility control design for miscible-type waterfloods using micellar solutions. *J. Petroleum Technol.*, **22** (1970) 141–7.
33. Delshad, M., Delshad, M., Bhuyan, D., Pope, G. A. & Lake, L. W., Effect of capillary number on the residual saturation of a three-phase micellar solution. SPE/DOE Paper No. 14911, Fifth Symposium on Enhanced Oil Recovery of Soc. Petroleum Engineers and US Department of Energy, Tulsa, 20–23, April 1986.
34. Delshad, M., Trapping of micellar fluids in Berea sandstone. PhD dissertation, University of Texas, Austin, USA, 1990.
35. Harwell, J. H., Surfactant adsorption and chromatographic movement with application in enhanced oil recovery. PhD dissertation, University of Texas, Austin, USA, 1983.
36. Vernet, M. R. M., Truchetet, R. & Schechter, R. S., Compositional effect on the adsorption of surfactant from aqueous phases in equilibrium with a middle phase microemulsion. SPE Paper No. 11212, 57th Annual Fall Techn. Conf. and Exhibition of Soc. Petroleum Engineers, New Orleans, 26–29 September 1982.
37. Somasundaran, P. & Hanna, H. S., Physico-chemical aspects of adsorption at solid/liquid interfaces. In *Improved Oil Recovery by Surfactant and Polymer Flooding*, ed. D. O. Shah & R. S. Schechter, Academic Press, New York, 1977, pp. 205–74.
38. Hirasaki, G. J., Ion exchange with clays in the presence of surfactant. *Soc. Petroleum Engineers J.*, **22**(2) (1982) 337–46.
39. Pope, G. A., Lake, L. W. & Helfferich, F. G., Cation exchange in chemical flooding, Part 1: Basic theory without dispersion. *Soc. Petroleum Engineers J.*, **18**(6) (1978) 418–34.
40. Datta-Gupta, A., Sepehrnoori, K., Thrasher, R. L. & Pope, G. A., A symmetric, positive definite formulation of a three-dimensional

micellar/polymer simulator. *SPE Reservoir Engineering,* **1**(6) (1986) 622–32.
41. Van Quy N. & Labrid, J., A numerical study of chemical flooding—comparison with experiments. *Soc. Petroleum Engineers J.,* **23**(3) (1983) 461–74.
42. Scott, T., Sharpe, S. R., Sorbie, K. S., Clifford, P. J., Roberts, L. J., Foulser, R. W. S. & Oakes, J. A., A general purpose chemical flood simulator. SPE Paper No. 16029, Ninth SPE Symposium on Reservoir Simulation of Soc. Petroleum Engineers, San Antonio, 1–4 February 1987.
43. Camilleri, D., Fil, A., Rouse, B., Sepehrnoori, K. & Pope, G. A., Comparison of an improved compositional micellar/polymer simulator with laboratory corefloods. *SPE Reservoir Engineering,* **2**(4) (1987) 441–51.
44. Saad, N., Sepehrnoori, K. & Pope, G. A., Simulation of Big Muddy surfactant pilot. *SPE Reservoir Engineering,* **4**(1) (1989) 24–34.
45. Nelson, R. C., Chemically enhanced oil recovery: the state-of-the-art. *Chem. Eng. Proc.,* (1989) 50–7.
46. Hirasaki, G. J., van Domeslaar, H. R. & Nelson, R. C., Evaluation of the salinity gradient concept in surfactant flooding. *Soc. Petroleum Engineers J.,* **23**(3), (1983) 486–500.
47. Lake, L. W. & Pope, G. A., Status of micellar-polymer field tests. *Petroleum Engineer Int.,* **51** (1978) 38–60.
48. Reppert, T. R., Bragg, J. R., Wilkinson, J. R., Snow, T. M., Maer, Jr, N. K. & Gale, W. W., Second Ripley surfactant flood pilot test. SPE/DOE Paper No. 20219, SPE/DOE Seventh Symposium on Enhanced Oil Recovery of Soc. Petroleum Engineers and US Department of Energy, Tulsa, 22–25 April 1990.
49. Chapotin, D., Lomer, J.-F. & Putz, A., The Chateaurenard (France) industrial microemulsion pilot design and performance. SPE/DOE Paper No. 14955, Fifth Symposium on Enhanced Oil Recovery of Soc. Petroleum Engineers and US Department of Energy, Tulsa, 20–23 April 1986.
50. Kalpakci, B., Arf, T. G., Barker, J. W., Krupa, A. S., Morgan, J. C. & Neira, R. D., The low-tension polymer flood approach to cost-effective chemical EOR. SPE/DOE Paper No. 20220, SPE/DOE Seventh Symposium on Enhanced Oil Recovery of Soc. Petroleum Engineers and US Department of Energy, Tulsa, 22–25 April 1990.

4 The use of alkaline agents in enhanced oil recovery processes

Jean Labrid
Institut Français du Pétrole, Rueil-Malmaison, France

4.1	**Historical**	123
4.2	**Chemicals**	125
4.3	**Fluid–fluid interactions**	127
4.3.1	Lowering of the water–crude oil interfacial tension	127
4.3.2	Emulsification	131
4.3.3	Precipitation of salts and hydroxides	134
4.3.4	Effect of alkalis on the chemical stability of surfactants and polymers	136
4.4	**Rock–fluid interactions**	138
4.4.1	Ion exchange	138
4.4.2	Wettability alteration	139
4.4.3	Dissolution of minerals by alkalis	141
4.4.4	Effect of alkalis on the retention of surfactants and polymers	147
4.5	**Process design**	149
4.6	**References**	152

4.1 Historical

The use of alkaline chemicals, also called alkalis, for improving oil recovery dates back to the early 1920s with the issuing of a patent to Squires who claimed that oil displacement may be more complete by introducing an alkali into the water.[1] Nutting[2] prescribed using alkaline salts, such as sodium carbonate, instead of sodium hydroxide, to diminish the adverse effects of rock dissolution. The acting mechanisms were not clearly understood at that time, but the hypothesis of wettability change combined with oil–water interfacial tension (IFT) lowering was emitted. As a matter of fact, Reisberg and

Doscher measured interfacial tensions between a California crude and sodium hydroxide solutions, and they observed very significant IFT lowerings.[3] In other respects, it was shown that an improvement in oil recovery could result from changes in relative permeability curves due to a reversing of rock wettability from oil-wet to water-wet.[4]

Laboratory flooding tests demonstrated that emulsification is probably a major mechanism by which caustic waterflooding may improve oil recovery.[5] If interfacial tension is not quite low enough, entrapment occurs with a reduced water mobility which improves both vertical and areal sweep efficiency.

Many field tests have been performed during the last thirty years, which have revealed that alkaline flooding is not a simplistic method but requires careful project design and monitoring techniques.[6] The profitability of these field tests has not been clearly established. One reason why conventional alkaline floods have been so discouraging is the lack of mobility control during alkali injection. Another reason is that the low caustic soda concentrations required for obtaining low interfacial tensions were not capable of propagating alkali because of the consumption by ion exchange and dissolution–precipitation processes.

To circumvent such technical problems and to take into account deep changes in the economic situation, the trend in recent years has been to replace caustic soda by some less reactive chemicals, such as sodium carbonate or sodium bicarbonate, and to benefit from some synergistic effects stemming from combining surfactants, polymers and alkalis in a chemical flood.

In this way, different approaches have been under investigation. Alkaline chemicals can be injected as a sacrificial agent ahead of a micellar-polymer slug to reduce the ion hardness level and the retention of chemicals.[7] With the same technical objectives, alkalis have been added to polymer solutions or to low surfactant concentration systems to improve oil recovery, while using smaller amounts of chemicals.[8] In such applications, no particular properties of the crude oil are required concerning its chemical composition, i.e. its acid number.

In other respects, the great number of studies carried out in chemical flooding have led to procedures for selecting chemicals which have been applied to alkaline flooding.[9] Just as in chemical flooding, a cosurfactant can be used to raise the salinity requirement of an alkaline flood. When the system is properly designed, adding

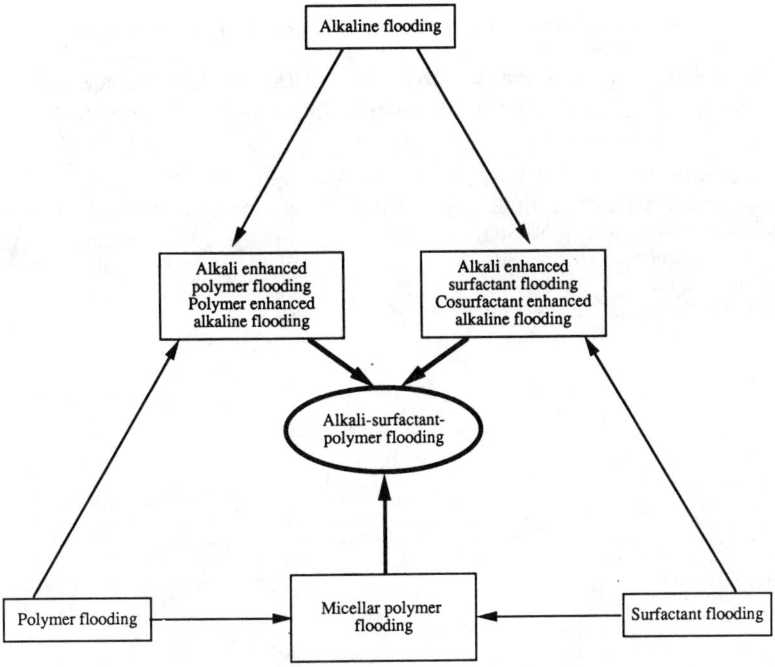

Fig. 4.1 Evolution of the chemical EOR process.

surfactant to the alkaline solution can increase oil recovery significantly. On the other hand, adding polymers to an alkaline slug is essential to speeding up oil production.

Finally, all strategies converged when investigators studied the feasibility of replacing part of the high cost chemicals, i.e. surfactants and polymers, with lower cost alkali, while maintaining high oil recoveries.[10] In a sense, all these processes can be viewed as variants of a single process, which takes advantage of combining in a synergistic way the specific properties of the different chemicals (Fig. 4.1).

4.2 Chemicals

The selection of alkaline chemical systems is quite complex due to the many interactions that can occur with chemicals injected simultaneously and with the reservoir itself. For these reasons, an absolute

Table 4.1 Alkaline agents commonly used in EOR processes

Alkali	*Chemical formula*	*Molecular weight*	*Alkalinity $Na_2O(K_2O)$ (% wt)*	*Relative cost (base = 100 for NaOH)*
Sodium hydroxide	NaOH	40	77·5	100
Potassium hydroxide	KOH	56	83·9	150
Sodium orthosilicate	Na_4SiO_4	184	67·4	135
Sodium metasilicate	Na_2SiO_3	122	50·8	105
Sodium carbonate	Na_2CO_3	106	58·5	65
Potassium carbonate	K_2CO_3	138	68·1	130

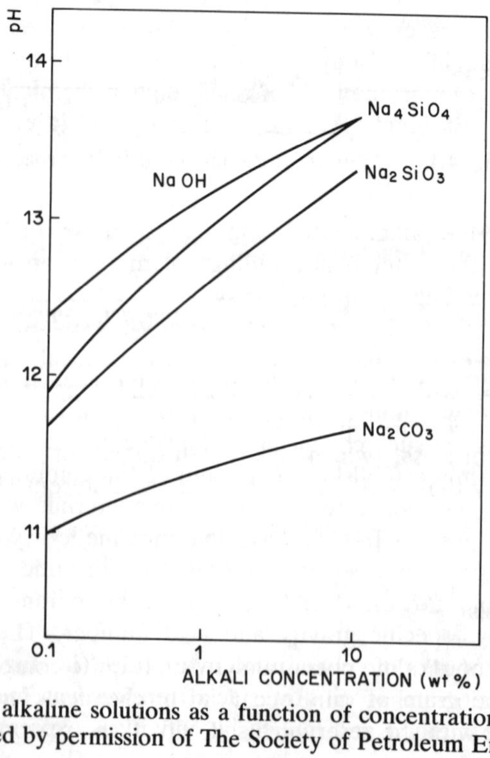

Fig. 4.2 pH of alkaline solutions as a function of concentration. (From Ref. 6, reproduced by permission of The Society of Petroleum Engineers.)

classification is obviously meaningless. Each alkali has its own properties and cost, and the real objective is to obtain the greatest efficiency/cost ratio. As a matter of fact, the current trend of using alkalis combined with surfactants and polymers adds greater complexity to the design of the different processes.

The alkalis that are found to be useful in such processes are carbonates, silicates and sodium hydroxide. Table 4.1 provides some data concerning alkalinity and relative costs of the different chemicals. Alkalinity is defined as the amount of acid required to neutralize the alkali. It is expressed as the percentage of Na_2O (K_2O) contained in each chemical. In addition to ortho and metasilicate solutions, solutions with SiO_2/Na_2O ratios varying from 1·6 to 3·3 are also used.

A graph of pH versus alkali concentration is given in Fig. 4.2. Of course, the pH of the solutions varies with the salt content. For instance, the pH of caustic solutions decreases from 13·2 to 12·5 when the salinity increases from 0 to 1%wt NaCl. By comparison, the pH of sodium carbonate solutions is less dependent on salinity.

Price ranges are given in Table 4.1. They refer to chemicals of a technical grade in an anhydrous form. Potassium-based alkalis are also included, though they are of much higher chemical cost, because they must be considered when clay swelling and injectivity problems are expected.

4.3 Fluid–fluid interactions

4.3.1 *Lowering of the water–crude oil interfacial tension*

The lowering of the crude oil–water interfacial tension by using alkaline agents was noted a long time ago, and this observation is the basis of the use of such chemicals to improve oil recovery.

It is now firmly established that some of the components contained in crude oils may generate surface active materials when they react with alkaline agents. In this way, Jennings made systematic investigations on a great number of crude oils to determine the factors that are responsible for interfacial activity.[11] The parameters examined were viscosity, specific gravity and acid number. The acid number gives the amount of potassium hydroxide (in mg) required to neutralize one gram of oil. Interfacial tension was measured by the pendant drop method and interfacial activity is expressed in terms of an empirical parameter called the caustic coefficient and calculated from the interfacial tension–caustic concentration curve. The caustic coefficient is the number of square cycles of log–log scales bounded by the interfacial curve between interfacial tension values of 0·01 and 1·0 mN/m and caustic concentrations between 0·001 and 1·0% wt. Results showed that nearly 80% of the 160 oils examined possessed a marked surface activity against caustic solutions with acid numbers in the range 0–2·5 mg/g (Fig. 4.3)[11,12] and that some correlation can be found between acid number and caustic coefficient (Fig. 4.4). But while the presence of some acid species appears to be a requirement, no direct correlation has been observed between the acid number of the oil and the oil recovery factor.

Attempts made to identify surface active components present in the crude oil–water interfacial region revealed the complexity of the molecular structures involved in the process. However, it has been demonstrated that interfacial activity is related to the content of heavy constituents, i.e. asphaltenes and resins, which associate to form a micellar type structure.[13,14] These high molecular weight constituents are obtained by various separation procedures, resulting in slight differences in the composition of the fractions.

Asphaltenes consist of condensed polynuclear aromatic rings bearing short alkyl side chains and including heteroelements such as

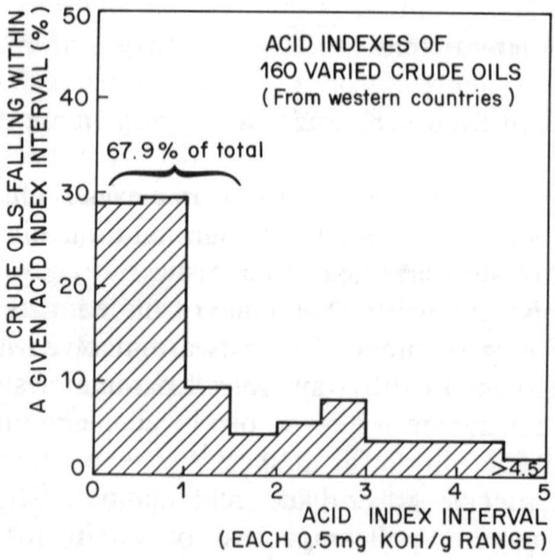

Fig. 4.3 Histogram of acid numbers. (From Ref. 12, reproduced by permission of Editions Technip.)

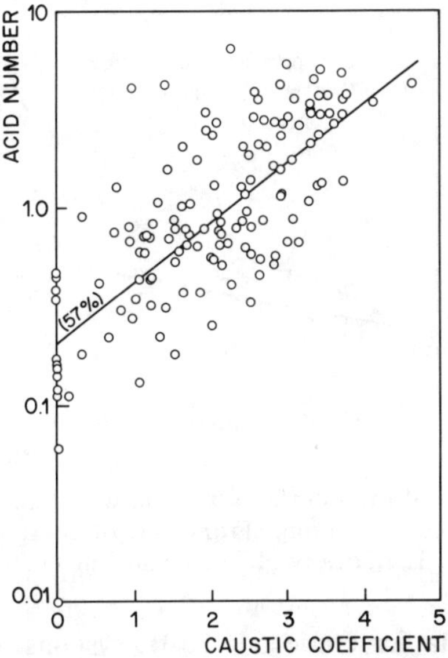

Fig. 4.4 Relationship between acid number and caustic coefficient. (From Ref. 11, reproduced by permission of The Society of Petroleum Engineers.)

nitrogen, oxygen and sulfur.[15] Resins are made up of molecules, the structures of which are comparable to asphaltenes, but with a reduced number of aromatic nuclei. They contain high molecular weight carboxylic acids from naphthenic and naphtheno aromatic type structures with carbon numbers from C_{10} to C_{30}. Carboxylic acids seem to be a common constituent of some young and immature crude oils.[1] Other high molecular weight components have been identified as porphyrin structures forming chelated complexes with metals such as nickel and vanadium.[16] These metal porphyrins exhibit strong interfacial activity and film-forming characteristics.

From the complexity of the chemistry of the process, it can be deduced that a great number of species are involved in the reactions with alkali and, consequently, an overall parameter such as the acid number is not a sufficient criterion to predict the behavior of a crude oil, since all the hydrolyzed acid species are not interfacially active. Moreover, as evidenced by Trujillo, and Sharma *et al.*, IFT lowering is a very complex mass transfer process of compounds between two immiscible fluid phases accompanied by chemical reactions. It follows that IFT behavior is time dependent.[17,18] Many authors have reported

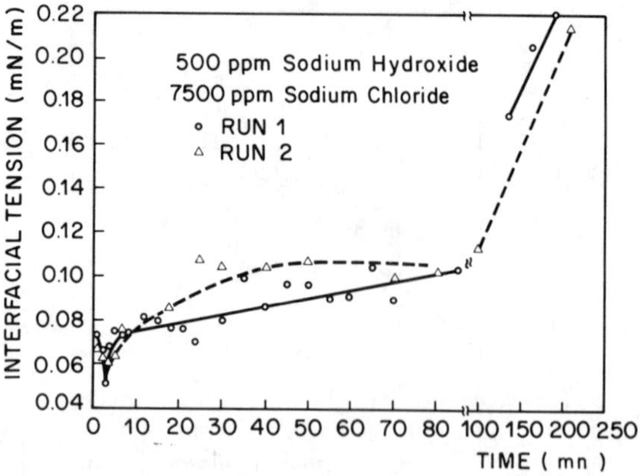

Fig. 4.5 Variation of the interfacial tension with time for long Beach crude oil at a low NaOH concentration. (From Ref. 18, reproduced by permission of The Society of Petroleum Engineers.)

transient IFT data for crude oil/caustic systems when measurements are made by a spinning drop apparatus. This technique gives readings in the first minute (Fig. 4.5). Experimental results showing IFT maxima and minima indicate that phenomena are controlled by convective diffusion. They depend on many parameters such as pH and ionic composition. The presence of calcium ions, even in small concentrations, destroys the capability of caustic soda to reduce IFT.

Chemical models taking into account the kinetics and the chemistry of the process have been developed to explain the phenomena observed.[18,19] Calculations indicate that IFT reduction seems to be caused by more than one chemical species. This is coherent with experimental observations, leading to the conclusion that the dynamic behavior of a crude oil is governed by the interfacial activity of the individual components.[20] In the most refined model,[18] a phenomenological surface phase approach is used to represent the interface, and surface active species are modeled by a two-component system. Each acid species diffuses out into the surface phase where it dissociates to an extent governed by the pH of the surface phase (Fig. 4.6).

$$HA_o \rightleftarrows HA_s \tag{4.1}$$

$$HA_s \rightleftarrows H^+ + A_s^- \tag{4.2}$$

where HA_o and HA_s are the acid species concentrations respectively in

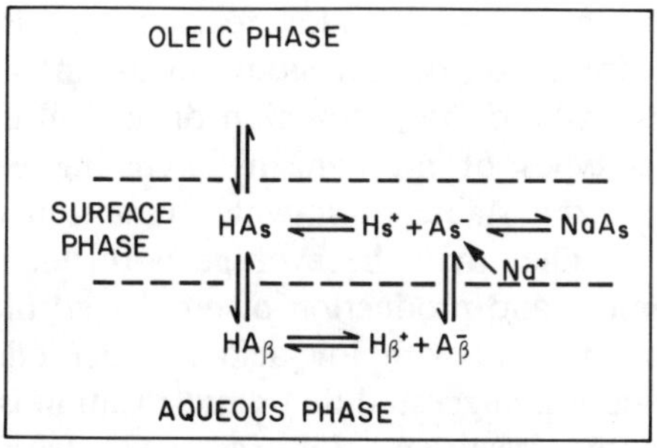

Fig. 4.6 Schematic representation of the kinetic steps involved in the reaction between a caustic solution and crude oil. (From Ref. 18, reproduced by permission of The Society of Petroleum Engineers.)

the oil phase and in the surface phase. In the two-component model, the existence is postulated of two kinds of species HA_1 and HA_2 with two different ionization constants KA_1 and KA_2 and two diffusion and reaction rates.

The IFT of the system will depend on the concentration of the surface active species A_s^- at the interface. Henry's law is used for dilute solutions. At high concentrations, a nonlinear dependence of IFT on surface concentration may be expected. Qualitative agreement is achieved between two-component model calculations and experimental observations.

4.3.2 *Emulsification*

The lowering of interfacial tension due to the reaction between alkalis and organic acids is probably not the primary mechanism of oil recovery. If we consider the concept of capillary number, as previously defined in Section 1.2.1, it is admitted that a conventional waterflood corresponds to capillary numbers in the range of 10^{-6} (see Chapter 1). From the relationship between capillary number and residual oil saturation, it can be deduced that to improve oil recovery, the capillary number must be increased to 10^{-4}–10^{-3}. This value can be achieved with some crude oil–caustic soda systems, where IFT values as low as 0·003 mN/m have been reached. However, in most of the cases described in the literature, interfacial tensions are higher than 0·01 mN/m, even if transient behavior is considered.

It has been shown that, at these low tensions, emulsification of fluids is favored. This formation of emulsions can be inferred from laboratory flooding tests where the production of fine oil droplets has been reported.[21] Two types of mechanisms involving emulsification are distinguished.[22] If the stable oil-in-water emulsion droplets are the same size as or smaller than the average pore diameter of the rock sample, entrainment and production of emulsified phases occur. This mechanism was observed with low-acid number oils in low salinity environments, and it is suggested that emulsification is induced by high shear-driven forces resulting, for instance, from high flooding velocities.

Emulsification with entrapment occurs if the average drop size of the emulsion phase is larger than the average pore throats of the porous medium. This mechanism is encountered with oils that have a moderate acid number and that become emulsified under limited shear. Emulsion droplets are entrapped in the porous medium, and high pressure gradients result. In this case, the improvement in oil recovery comes from improved sweep efficiency due to the reduction of the mobility of the aqueous phase.

The surface of oil-in-water emulsion droplets that are stabilized by surface active polyelectrolytes usually carries a negative charge, with the interfacial film consisting of a molecular complex with an oil-soluble component and an ionizable water-soluble component. Soaps from monovalent metals favor the formation of oil-in-water emulsions while soaps from polyvalent ions more usually give water-in-oil emulsions. The latter can also be formed at high NaCl concentrations. So, inversion may occur by the addition of a calcium salt to an oil-in-water emulsion. The two-stage demulsification process goes through the aggregation of small droplets (flocculation) and then the gathering of aggregates to form a single drop (coalescence). The rate of coalescence can be determined by measuring the ratio of the emulsified phase volume to the total volume as a function of time (Fig. 4.7).

Oil recovery experiments with hydroxides of divalent cations ($Ca(OH)_2$, $Ba(OH)_2$) have shown higher production and recovery efficiencies than alkaline floods with hydroxides of monovalent ions.[21] In this case, high recovery efficiencies resulted from a combined emulsification–coalescence process. Emulsification (and sometimes partial wettability alteration) gives rise to a reduction of the permeability to the aqueous phase as a result of flow restriction due to the

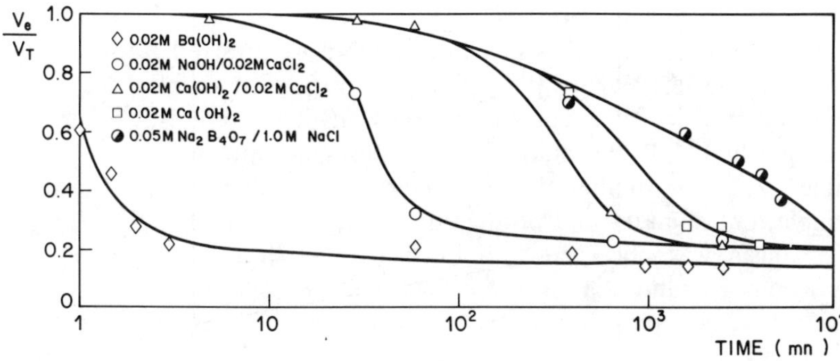

Fig. 4.7 Coalescence rate of emulsions as a function of time (From Ref. 21, reproduced by permission of Plenum Publishing Corporation.)

swelling of the residual oil phase. Coalescence of emulsified oil ganglia leads locally to high oil saturation and subsequently the formation of an oil bank with a corresponding increase in oil permeability. The two processes, emulsification and coalescence, are interrelated in the sense that coalescence requires unstable emulsions and that this instability depends on the emulsification process.

As discussed by Nelson *et al.*[9] surfactant flooding and alkaline flooding can be viewed as particular cases of the same process. In this respect, a test called '*Emulsion Screening Test*' can be used to determine the composition of the formulation corresponding to minimum interfacial tension. Glass tubes containing oil, brine and chemicals, namely surfactant and alkali, are equilibrated at reservoir temperature. For under-optimum systems, oil droplets are dispersed in the aqueous phase to form an oil-in-water emulsion while oil tends to be the continuous phase in over-optimum systems. Between these two regions, lies the optimum or active region, which corresponds to the Winsor III type, as defined in Section 3.2.3.

This methodology has been applied to investigate the effects of various alkaline chemicals on the phase behavior of brine, oil and petroleum sulfonate mixtures.[23] Tests have shown that the presence of any alkali lower the optimal salinity. As a matter of fact, alkaline chemicals affect phase behavior because they provide an additional source of electrolyte. For the surfactant under investigation, the alkali anion had little or no effect on the phase behavior, whereas cations were effective for decreasing the optimal salinity in the order: potassium > sodium ≫ ammonium (Fig. 4.8).

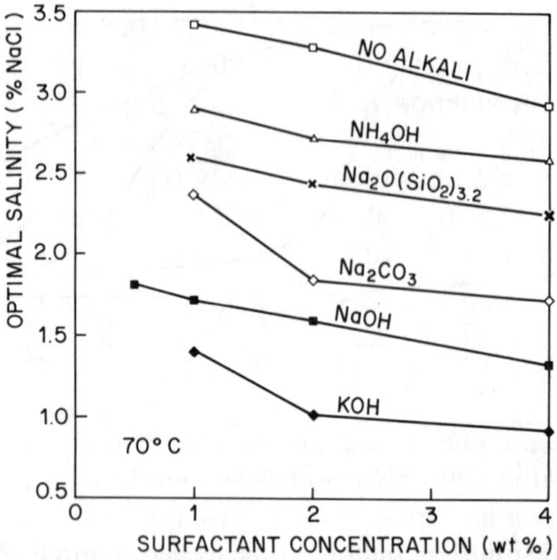

Fig. 4.8 Influence of combined cation content on optimal salinity for 1% wt alkaline chemicals, brine, *n*-tetradecane and surfactant. (From Ref. 23, reproduced by permission of The Society of Petroleum Engineers.)

4.3.3 *Precipitation of salts and hydroxides*

Reactions of alkalis with divalent ions such as calcium and magnesium lead to the formation of precipitates that can definitively impair the permeability of the reservoir.

Three origins can be attributed to the presence of divalent ions in high pH environment: the mixing of the alkaline solution with resident brines, the ion exchange process, and the dissolution process.

A factor that is fundamental in determining whether or not scales form is the pH value of the system. The pH value defines the composition and the amount of precipitates and provides qualitative estimates of the possible damage that could occur in the reservoir and at the injection and production wells. Accurate prediction of scale formation however is difficult because of the complexity of the mechanisms involved, which generate mixed precipitates associated with clay particles and emulsions.[24]

Another major factor is the type of alkali, caustic soda, carbonate or silicates, that is used, because the solubility of divalent salts and hydroxides varies considerably with the pH. This aspect must be taken into account when selecting chemicals, on the one hand by considering their efficacy in removing hardness ions and therefore the protection they offer with regard to the surfactant slug, and on the other hand, when looking for remedies capable of treating scale problems at the production wells.

Scales can originate from the reactions of alkalis with the carbonate minerals of the rock, i.e. dolomite CaMg $((CO_3)_2)$, siderite $(FeCO_3)$, magnesite $(MgCO_3)$ or calcite $(CaCO_3)$, through exchange type reactions that can be represented as follows:

$$FeCO_3(s) + SiO_3^{2-} \rightarrow FeSiO_3 \downarrow + CO_3^{2-} \tag{4.3}$$

$$MgCO_3(s) + 2OH^- \rightarrow Mg(OH)_2 \downarrow + CO_3^{2-} \tag{4.4}$$

$$CaCO_3(s) + SiO_3^{2-} \rightarrow CaSiO_3 \downarrow + CO_3^{2-} \tag{4.5}$$

Although solubilities of carbonate minerals could be lower than those of corresponding silicates or hydroxides, the continuous supply of fresh alkaline solution under the dynamic field conditions may be expected to result in a continuous release of carbonate ions from rock minerals into the solution. This effect can be prevented by using carbonates as alkali, because carbonate ions brought by the solution oppose calcite and magnesite dissolution.[25] Silicates do not profit from this kinetic effect. In another way, care must be taken when using silicates with rocks having high cation exchange capacity (CEC). Because of ion exchange processes, alkalinity loss is significant. This results in the formation of a region extending ahead of the pH front where the fluids are over-saturated with respect to silica. In contrast, for the carbonate injection, ion exchange results in the hydrolysis of the CO_3^{2-} ions to the highly soluble form HCO_3^-.

Silicate precipitates are generally hydrated, flocculent and highly plugging even at low concentrations. Carbonate precipitates are relatively granular and less adhering on solid surfaces.[26] Thus, under equivalent experimental conditions of porosity and flow rate, Na_2CO_3 shows the least degree of permeability damage in the presence of hard water (Fig. 4.9). Moreover, if calcium carbonate scales can be successfully removed at production wells by acidizing or by using inhibitors, no long-term treatment exists for the control of silicate-containing precipitates.

Other minerals, gypsum and anhydrite, can cause substantial alkali consumption. These sulfate-containing minerals react in alkaline solution to release calcium ions. As opposed to ion exchange, this process is time dependent and may affect the properties of the micellar–polymer fluids that are injected after an alkaline preflush.

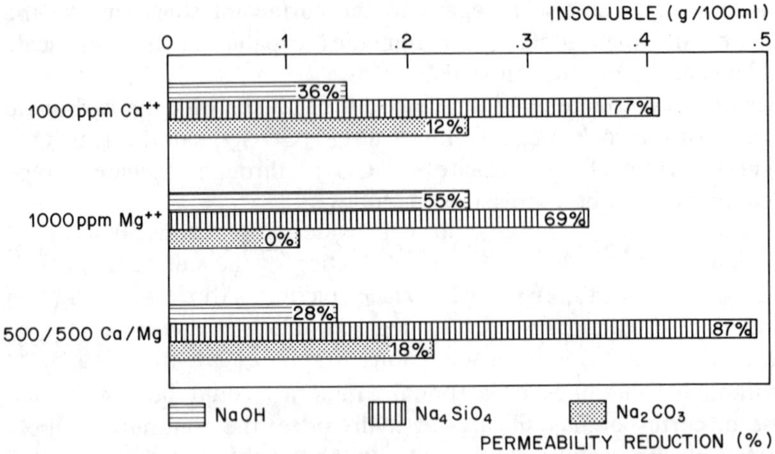

Fig. 4.9 Effects of alkali precipitation with different hardness solutions. Extent of alkali precipitation (bars) is correlated to permeability reduction (% within bars) in flow experiments. (From Ref. 26, reproduced by permission of The Society of Petroleum Engineers.)

4.3.4 *Effect of alkalis on the chemical stability of surfactants and polymers*

As a matter of fact, the chemical stability of surfactants is largely dependent on their molecular structure. In order to enlarge the range of application of chemical flooding processes, substitutes for conventional aromatic sulfonates have been proposed for high salinity and high temperature conditions.

The chemical stability of aromatic sulfonates is estimated to be satisfactory in the range of temperatures normally considered for a surfactant flooding process, i.e. from 30 to 120°C. On the other hand, other EOR compounds, such as alkylsulfates and alkylethoxy sulfates, alpha-olefin sulfonates or ethoxy sulfonates could suffer from some chemical degradation giving rise to a loss of activity. Several degradation mechanisms have been identified with some coupling effects that make the kinetic behavior very intricate.

Sulfate surfactants are found to be very sensitive to OH^- ions through a nucleophilic substitution mechanism. For instance, for alkylethoxy sulfates, this displacement reaction is described by the following scheme:

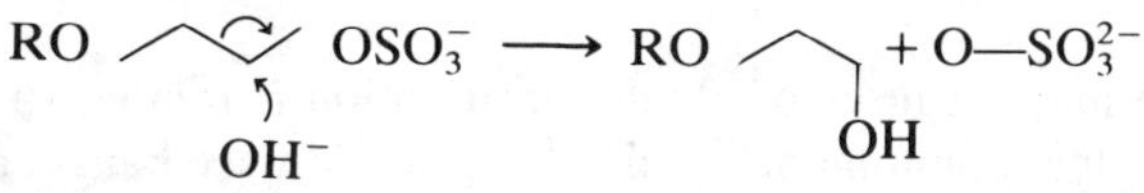

(4.6)

Calcium ions greatly accelerate hydroxide ion reactions. It has been postulated that the mechanism is a reduction of the sulfate–hydroxide repulsion energy in the micellar surface region.[27] Above pH 8, modest amounts of calcium cause severe degradation. It is therefore imperative to carefully remove divalent ions if sulfates are intended to be used.

Alpha-olefin sulfonates are specially sensitive to oxygen which affects the unsaturated species. The breaking of olefin double bonds results in the formation of carboxylic acid groups and consequently in a decrease of the pH. Interestingly, it has been found that basic pH inhibits oxidative degradation, even at high temperature.[28] For ethoxy sulfonates, although the oxidation mechanism is slowed down by the sulfonate group, it is not completely suppressed at high pH.

The question of compatibility of water soluble polymers with alkaline reagents has received little attention to date. Significant work has been done by Ryles who performed anaerobic tests to study the stability limits of some water-soluble polymers in alkaline media.[29] It was observed that, at 90°C, the viscosity of polyacrylamide solutions increased rapidly in the first days, then decreased. But, for the period investigated, i.e. more than one year, the final viscosity was generally greater than the initial one. This behavior reflects the hydrolysis of amide groups. This shows that, under anaerobic conditions, polyacrylamides remain stable for long periods at temperatures up to 90°C, irrespective of the nature of the alkaline reagent. However, their tolerance with respect to divalent ions, and specially calcium ions, is greatly reduced. By comparison, xanthan was totally degraded at all temperatures. It is believed that base-catalyzed fragmentation reactions play an important role in the degradation process.[30] The usefulness of xanthan in alkaline flooding is therefore questionable, particularly under high-temperature conditions.

4.4 Rock–fluid interactions

4.4.1 *Ion exchange*

One of the major causes of alkali consumption is related to clays and ionic exchange phenomena. Indeed, alkali ion exchange alone can decrease the pH of a caustic slug to a value not high enough to remove the residual oil and to increase recovery.[31]

The problem of describing cation distribution between solution and clay minerals has been widely investigated in the literature, but most of the published results refer to the description of cation exchanges that occur when an aqueous solution is displaced by another fluid with different salinity and hardness. The pH is fairly constant during the process, so it is legitimately assumed that the cation exchange capacity of the mineral is unvarying.

Several approaches have been used to quantify the exchange phenomena. One is the thermodynamic approach based on the mass action law. An exchange reaction of a divalent–monovalent cation such as the Ca–Na exchange is represented by the following equilibrium:

$$CaM_2 + 2Na^+ \rightleftarrows 2NaM + Ca^{2+} \tag{4.7}$$

where M is an exchange site on clay. The selectivity coefficient $K_{Ca\text{–}Na}$ is given by:

$$K_{Ca\text{–}Na} = \frac{(X_{Na}^2 \cdot a_{Ca})}{(X_{Ca} \cdot a_{Na}^2)} \tag{4.8}$$

where X denotes the mole fraction of the adsorbed ions and a denotes the solution activity.[32] Another equation has been proposed by Gapon:[33]

$$K_{Ca\text{–}Na} = \frac{E_{Na}}{E_{Ca}} \cdot \frac{(a_{Ca})^{1/2}}{a_{Na}} \tag{4.9}$$

where E_{Na} and E_{Ca} are the equivalent fractions of the total exchange capacity occupied by the sodium and the calcium ions, respectively. Other equations based on the diffuse double layer theory have also been proposed but the equations mentioned above are usually suitable for calculating ion exchange processes.

The problem is more complex when ion exchange occurs together with a large variation of the pH, which occurs when a clayey sandstone is contacted by alkali. Experimental observations indicate that new exchange sites are created. The total exchange capacity is then the CEC determined at neutral pH augmented by the hydrogen exchange capacity (HEC). The latter can be evaluated by potentiometric titration.

By analogy with divalent–monovalent cation exchange, the following equation has been proposed for describing the sodium–hydrogen exchange:

$$\mathrm{MOH + Na^+ \rightleftarrows MONa + H^+} \qquad (4.10)$$

where MOH denotes a negative lattice exchange site occupied by a hydronium ion.[31]

Combined with the water dissociation equilibrium, we obtain:

$$\mathrm{MOH + Na^+ + OH^- \rightleftarrows MONa + H_2O} \qquad (4.11)$$

Each sodium site MONa consumes one hydroxide ion. Hence, sodium-ion adsorption is equivalent to hydroxide-ion adsorption.

More refined theories represent the clay surface as being constituted of two types of sites, with one being totally dissociated at neutral pH. At any pH, the cation exchange capacity is given by:

$$Q_V = Q_1 + q_2 \qquad (4.12)$$

where Q_1 is the CEC at neutral pH and q_2 is the hydrogen exchange capacity HEC, calculated by a Langmuir type equation:[34]

$$q_2 = Q_2 \frac{\mathrm{k_e[OH]}}{1 + \mathrm{k_e[OH]}} \qquad (4.13)$$

where $\mathrm{k_e}$ is the ion exchange equilibrium constant.

An example of the evolution of the HEC with pH is given in Fig. 4.10. HEC may reach 40% of CEC, and thus significant quantities of alkali may be consumed through ion exchange processes.

4.4.2 *Wettability alteration*

It is widely recognized that oil recovery efficiency by waterflooding is dependent on the wettability of the oil–water–rock system, and many efforts have been made to alter the original wettability. Phenomenological investigations using reverse-wetting agents, such as oil soluble amines, demonstrate that transient adhesion–tension alterations result

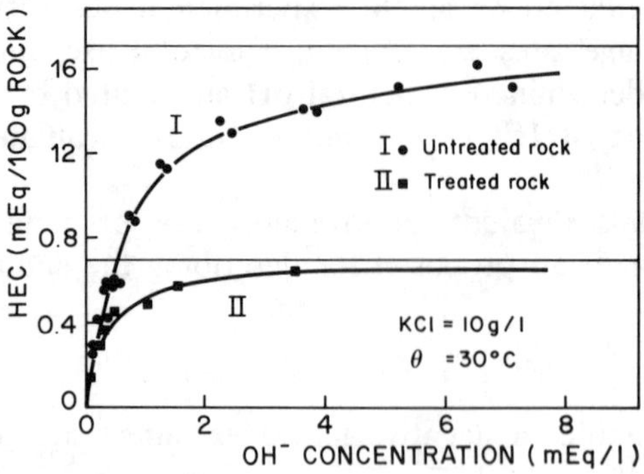

Fig. 4.10 Hydrogen exchange capacity (HEC) of a clayey sandstone as a function of OH^- concentration. (From Ref. 25.)

in the spontaneous accumulation of oil in large continuous masses that are subsequently mobilized.[35] It is deduced that, through a wettability reversal, increases in waterflood oil recovery may be obtainable from reservoirs that are naturally oil-wet because of adsorbed natural surface active substances.

An interesting approach to the question consists of considering the influence of electrical surface charges on the wetting properties of crude oils. In-situ wettability depends on the initial distribution of oil and interstitial water with respect to the rock surface, and the stability of water films between crude oil and solid is a key factor in determining wettability. If the film is unstable, polar compounds from the oil will be adsorbed on the rock surface, and wettability to the oil will occur.

The existence of stable water films has been shown to depend on repulsions between negative surface charges at the solid/water and water/oil interfaces.[36] Adhesion of oil on solid surfaces can be calculated by considering the interactive forces that come into play in the interfacial regions, i.e. electrical double layer repulsive forces, London-van der Waals attractive forces and short-range hydration forces. Adhesion behavior can also be determined experimentally by means of tests using crude oils and glass slides.[37] Results indicate that adhesion properties related to wettability alteration depend on the pH and ionic strength. With the oils considered, no adhesion occurs at alkaline pH and low salt content, indicating that the system is water-wet (Fig. 4.11). The same conclusion was drawn from a series of

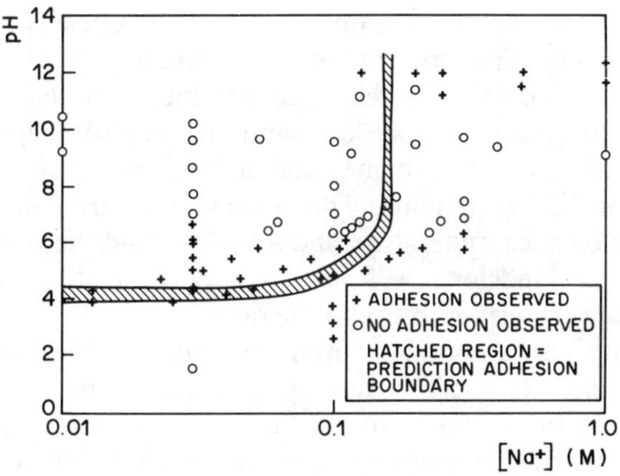

Fig. 4.11 Adhesion mapping for Moutray crude oil. (From Ref. 37, reproduced by permission of The Society of Petroleum Engineers.)

laboratory caustic waterfloods using crude oils originating from 19 oil reservoirs.[38] Relative permeability end-points before and after caustic flooding were used to estimate wettability and to infer the probable mechanism at the origin of increased recovery. Following caustic flooding, relative permeability to water ranged from 0·05 to 0·15 for all crude oils tested, indicating a strongly water-wet behavior.

At high ionic strength and high pH, experimental results are not so clear and adhesion can be observed at pH values above 10 (Fig. 4.11). Similar results were obtained by Brown and Neustadter,[39] who measured contact angles of crude oil/aqueous solutions on silica. The dependency of wettability on pH is explained in terms of a charge repulsion/attraction model. In the presence of a 36 000 ppm NaCl concentration, charge effects are suppressed and oil wettability is maintained through the 2–10 pH range. It may also be noted that multivalent ions (e.g. Ca^{2+}, Mg^{2+}) can promote oil wettability, even at low concentrations, because of specific ion interactions with the crude oil.

4.4.3 *Dissolution of minerals by alkalis*

Another cause of alkali consumption lies in the reaction of alkaline agents with rock minerals. Contrary to ion exchange, which is a fast reversible process, dissolution of reservoir silicate minerals is a long-term irreversible process limited by kinetics.

Sandstone reservoirs, which constitute the main class of candidates for alkali injections, are predominantly composed of quartz or silica, clays and feldspars. Carbonates and sulfates can also be present, but generally in lesser amounts. This makes alkalinity consumption very dependent on rock mineralogy and lithology, and the overall reactivity of a natural sandstone will be a complex function of elementary reaction rates referring to each mineral.

Alkalinity consumption is often evaluated by measuring what is called available or useful alkalinity. This parameter is determined by HCl titration to an end point pH that varies from 8 to 11. Of course, the meaning of this parameter is questionable on account of the buffering properties of dissolved species and the specificity of the relationship between crude oil/aqueous solution interfacial tension and pH.[40] The extent of dissolution is more conveniently estimated from the measurement of dissolved silicon and aluminum concentrations. In the range of pH where dissolution occurs, the dissolved species are anionic. For silicon and aluminum, the predominent species are $Si(OH)_3O^-$, $Si(OH)_2O_2^{2-}$, $Al(OH)_4^-$ and $Al(OH)_5^{2-}$, with the proportion of highly charged ions increasing with pH (Fig. 4.12). Consideration of silicon and aluminum effluent concentration behavior gives some insight into the mechanisms of attack. Short-time points can be used to reach initial reaction rates and to detect the inhibiting effect of clays on quartz dissolution.[41,42] Long-term behavior reveals that some new aluminosilicates are formed and precipitate as the reaction proceeds. Scanning electron microscopy and X-ray diffraction investigations indicate that many of the new precipitates belong to the zeolite family (Fig. 4.13).[43,44] In an Na_2CO_3 environment, from effluent composition behavior and electron spectroscopy chemical analysis, analcime with ideal formula $NaAlSi_2O_6 . H_2O$ was the only zeolite species unambiguously identified.[42] Based on thermodynamic data, mineral stability diagrams can be built. (Fig. 4.14). Of course, the type and composition of the zeolite formed depends on the nature and concentration of cations, such as Ca^{2+}, K^+, Na^+, Mg^{2+}.[40] In any case it appears that, under high pH conditions, transformations from clays to zeolite are expected, although these transformations may take a long time due to kinetic factors.

Because of the solution chemistry and the mineralogy complexity, a constant-order reaction is not anticipated. Even the silica dissolution reaction is not simple because of the interference of the reverse

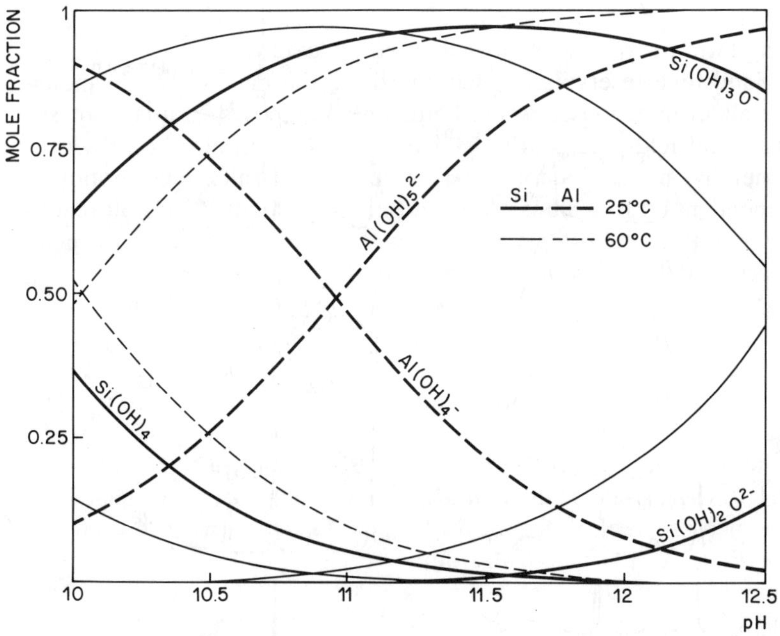

Fig. 4.12 Distribution of silicon and aluminum species as a function of pH.

condensation reaction. Different kinetic models have been proposed. In one of them, chemical reactions are written as a basis for defining equilibrium between the solution and representative minerals.[45] For each mineral, an equilibrium quotient Q is defined which is incorporated into a simple rate equation. The reaction for quartz is written as follows:

$$SiO_2(s) + OH^- + H_2O \rightleftarrows H_3SiO_4^- \tag{4.14}$$

with the corresponding equilibrium quotient:

$$Q_{QTZ} = \frac{[Si]}{[OH^-]} \tag{4.15}$$

Kaolinite would react according to the following equation:

$$[(Al_2O_3), (SiO_2)_2, (H_2O)_2](s) + 4\,OH^- + 3\,H_2O \rightleftarrows 2H_3SiO_4^- + 2Al(OH)_4^- \tag{4.16}$$

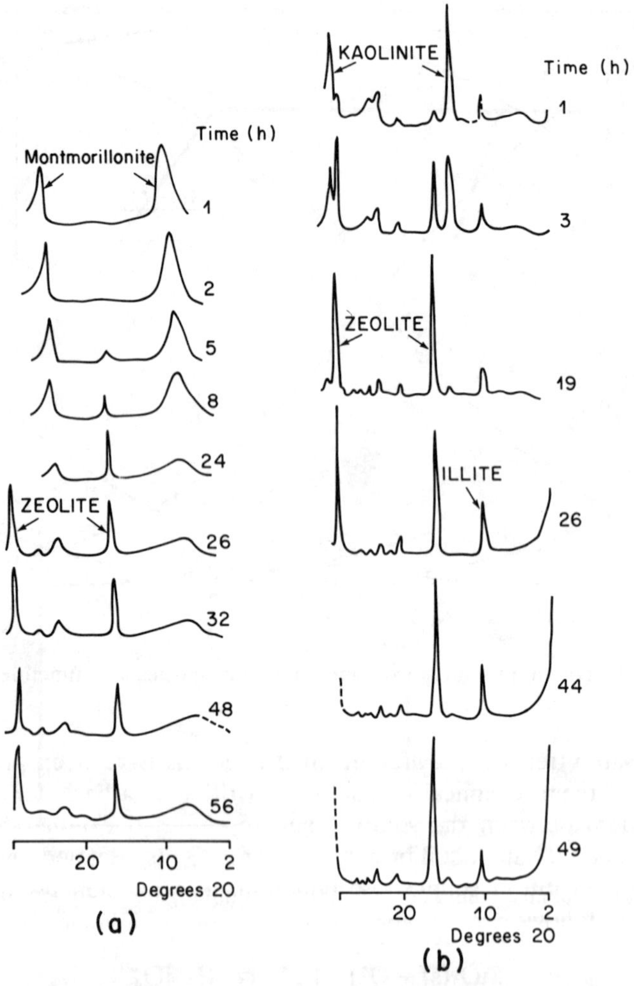

Fig. 4.13 Evolution of X-ray diffraction diagrams versus time for (a) montmorillonite and (b) kaolinite in 15% wt NaOH. (From Ref. 43, reproduced by permission of The Society of Petroleum Engineers.)

with the equilibrium quotient:

$$Q_{KAOL} = \frac{[Si][Al]}{[OH^-]^2} \qquad (4.17)$$

where [Si] and [Al] are the concentrations of the silicon and aluminum

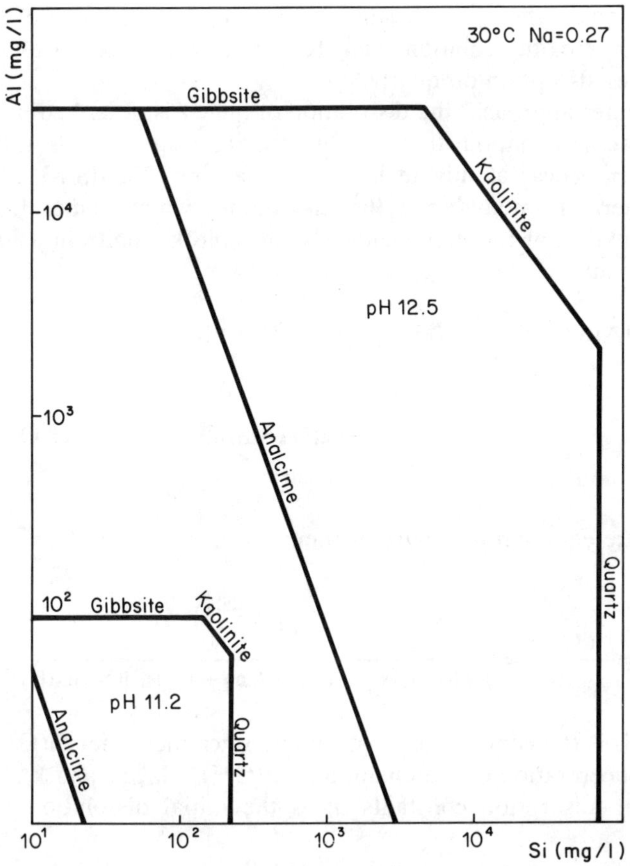

Fig. 4.14 Stability diagram of the kaolinite–gibbsite–quartz–analcime system. (From Ref. 42, reproduced by permission of Editions Technip.)

species in solution. The rate equations for quartz and kaolinite are respectively:

$$r_{QTZ} = \frac{\gamma_{QTZ} \cdot r^0_{QTZ}}{(1 + K_{Al} \cdot [Al])} \cdot \left(1 - \frac{[Si]}{Q_{QTZ} \cdot [OH^-]}\right) \tag{4.18}$$

$$r_{KAOL} = \gamma_{KAOL} \cdot r^0_{KAOL}\left(1 - \frac{[Si][Al]}{Q_{KAOL}[OH^-]^2}\right) \tag{4.19}$$

where r is the net dissolution rate of each mineral, γ is the solid-to-liquid ratio, r^0 is the dissolution rate of each mineral in pure sodium hydroxide solution and K_{Al} is a Langmuir constant for aluminate adsorption on quartz.

In another approach, the dissolution of quartz is described by means of a kinetic equation that reveals the competitive adsorption of monomeric silica, aluminate ions and analcime, $NaAlSi_2O_6 . H_2O$, a new mineral formed during the dissolution, which comes from the transformation of thermodynamically unstable kaolinite in a high pH environment:[42]

$$\underset{\text{quartz}}{2SiO_2} + \underset{\text{kaolinite}}{[Al_2O_3, (SiO_2)_2, (H_2O)_2]} + 2OH^- + 2Na^+ \rightarrow \underset{\text{analcime}}{2[NaAlSi_2O_6, H_2O]} + 2H_2O \quad (4.20)$$

The rate equation for quartz is then:

$$r_{QTZ} = -\frac{dQTZ}{dt} = r_0 \frac{\left(1 - \frac{[Si]}{k_0}\right)}{(1 + k_s[si] + k_a[al][Na] + k_{an}[al][Na][si]^2)} \quad (4.21)$$

where [si] is the concentration of soluble silica monomer $Si(OH)_4$, [al] is the concentration of aluminate ion $Al(OH)_4^-$; k_s, k_a and k_{an} are the Langmuir adsorption constants; r_0 is the initial dissolution rate for quartz.

The rate equation for kaolinite is as follows:

$$r_{KAOL} = -\frac{dKAOL}{dt} = r'_0 \frac{[OH^-]}{[OH^-] + k_k[si][al]} \quad (4.22)$$

which assumes, by analogy with quartz dissolution, an intermediate step of coadsorption of the species $Si(OH)_4$ and $Al(OH)_4^-$; k_k is the equilibrium constant, and r'_0 is the initial dissolution rate for kaolinite. These kinetic equations depict the coupling of elementary dissolution processes and the role played by the adsorbed species on sand and clay reactivities. The validity of the model has been tested in various operating conditions.[42] The model is able to predict the behavior observed during the alkaline attack of a clayey sand (Fig. 4.15).

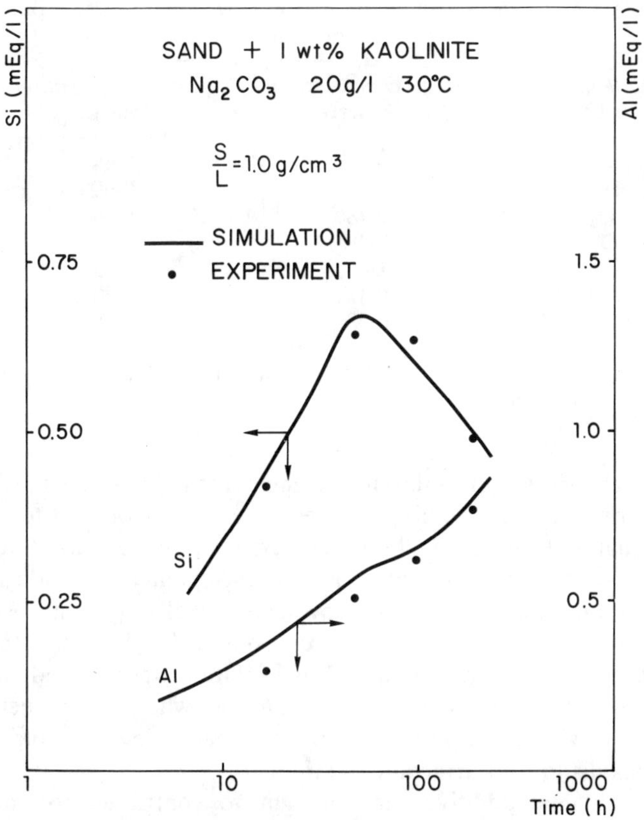

Fig. 4.15 Dissolution of a sand–kaolin mixture by Na_2CO_3. Comparison of numerical and experimental data. (From Ref. 42, reproduced by permission of Editions Technip.)

4.4.4 *Effect of alkalis on the retention of surfactants and polymers*

A major problem that affects the efficiency of oil recovery by surfactants and polymers is the loss of chemicals from interaction with rock minerals. If adsorption is excessive, the surfactant concentration is rapidly reduced as the micellar slug moves through the reservoir. In other respects, mobility control at the rear of the surfactant slug will not be ensured. Increasing the pH of the injected fluids is one of the most promising ways with regard to efficiency and economic considerations.

Table 4.2 Surfactant retention during flow experiments in Berea cores

Alkali type	*Alkali concentration (% wt)*	*Surfactant retention (mg/g)*
None	—	0·68
NaOH	0·38	0·65
Na_2CO_3	0·38	0·26
Na_3PO_4	0·38	0·28
Na_4SiO_4	1·0	0·25
Na_2O, $1{\cdot}6SiO_2$	0·43	0·20
Na_2O, $3{\cdot}2SiO_2$	0·38	0·15

(From Ref. 46, reproduced by permission of The Society of Petroleum Engineers.)

Although it is admitted that increasing pH is a way of decreasing the polymer retention, no detailed studies have been published to date on the mechanisms involved. In other respects, the effect of alkaline chemicals on surfactant retention is well documented. As indicative of the gains that can be expected by using alkalis, typical results of surfactant retention experiments are given in Table 4.2.[46] In these experiments, a 0·235 pore volume NaCl solution was injected ahead of the dilute surfactant alkaline slug. On the whole, a reduction of retention varying from 60 to 80% of the original level is obtained by adding alkalis to the surfactant solution.

These alkaline additives are thought to contribute to improved surfactant efficiency by a combination of different mechanisms. Among these are hardness ion removal and reduction in surfactant retention.

Sulfonates can be adsorbed on clay surfaces because of electrostatic interactions. At high pH values, the solid surfaces acquire a negative charge that gives rise to a large repulsion term. Thereby, a decrease in the adsorption of the anionic sulfonate species is expected. As discussed previously, a high ionic strength of the brine leads to a lowering of the electrostatic potential of the surfaces. The benefit of basic pH is therefore attenuated. In this connection, divalent ions such as Ca^{2+} and Mg^{2+} have a more pronounced effect than sodium ions.

To explain the results obtained under very alkaline pH conditions, other mechanisms have been suspected to play a role. But, up to now, they have not been fully established. It would seem that the type of inorganic species present in the system plays a major role in controlling the retention behavior of sulfonates.[47] In this way, aluminate ions resulting from clay dissolution could participate in an exchange reaction with sulfonate ions and then could be a contributing factor for determining the residual retention.

4.5 Process design

As shown above, oil recovery by alkalis is a complex function of oil–alkali interactions, alkali–rock interactions and chemical precipitation. Because of the coupling of these mechanisms, *alkaline flooding* even in its conventional form, is site-specific in terms of process design.

Data available concerning field applications indicate that field tests have been performed in various operating conditions (Table 4.3).

Alkali concentrations vary from 0·2 to 5% wt, with a majority of tests performed with high pH agents, such as caustic soda and sodium orthosilicate. Water for making up the alkaline solution may be softened either by an ion exchange treatment or by precipitation of divalent ions by an alkaline additive and filtration.

In some recent tests, orthosilicate has been used because it keeps calcium and magnesium out of solution better than sodium hydroxide.

Table 4.3 Characteristics of some alkali field tests

Field location	*Reservoir temperature (°C)*	*Acid number (mg KOH/g)*	*Type of alkali material*	*Alkali concentration (% wt)*	*Slug size (%PV)*
Southeast, TX	44	2·4	Na_2CO_3	3·2	—
Harrisburg, NE	93	low	NaOH	2·0	9
North Ward Estes, TX	46	0·2	NaOH	5·0	15
Singleton, NE	—	low	NaOH	2·0	8
Whittier, CA	49	—	NaOH	0·2	20
Brea-Olinda, CA	57	—	Na_4SiO_4	0·1	1·2
Orcutt Hill, CA	60	0·6	Na_4SiO_4	0·4	2
Dominguez, CA	—	—	NaOH	0·4	5
Van, TX	—	—	Na_4SiO_4	0·1	12
Wilmington, CA	52	2·0	Na_4SiO_4	0·4	60
Huntington Beach, CA	74	0·7	Na_4SiO_4	1·0	40

(From Ref. 6, reproduced by permission of The Society of Petroleum Engineers.)

But high pH sodium silicate and sodium hydroxide are consumed in the reservoir by reaction with rock minerals. In this respect, current trends are towards buffered alkali systems with moderate pH values, such as soda ash.[46]

The difficulty for selecting alkali in the conventional alkali process is to satisfy antagonist criteria, such as low interfacial tension and suitable transport properties. As observed by many investigators, the lowest interfacial tensions between a crude oil and alkali occur at low concentrations of the alkaline agent. However, alkali consumption by ion exchange and mineral dissolution requires a relatively high concentration to propagate the alkaline front at an acceptable rate. This dilemma can be solved by using alkalis in combination with surfactants and polymers in the same slug. It has been found that standard alkalis could be equally effective when systems are properly formulated.[9] Systems are then designed by adjusting the surfactant molecular weight. The choice of an alkali will depend on operating parameters such as temperature, rock mineralogy, crude oil composition, water composition, as well as the price and availability of chemicals.

The efficacy of *surfactant–alkali flooding* will depend on the simultaneous propagation of the primary surfactants or petroleum soaps, formed by reaction between alkali and crude oil components, and the added surfactant or cosurfactant. This can be achieved by carefully controlled salinity of the preflood ahead of the alkaline slug (Fig. 4.16).

In all these processes, mobility control is of prime importance, specially when the oil to be recovered is viscous. Generally speaking, adding polymer improves not only sweep efficiency, but also microscopic displacement efficiency because of an increased capillary number. Typical coreflood results concerning the *alkali–polymer process* show that the oil production is similar in nature to a micellar–polymer flood, with a well-formed oil bank (Fig. 4.17). Hardness decreased to near zero with the increase in pH, and, as a consequence, the interfacial tension dropped down to a few 10^{-3} mN/m.

At the present time, using water-soluble polymers are the only way to control mobility during a chemical flood. As mentioned earlier, polyacrylamides are more stable under alkaline conditions than xanthans, but neither polymer can tolerate much iron and oxygen, and it is recommended to include an oxygen scavenger in the formulation.

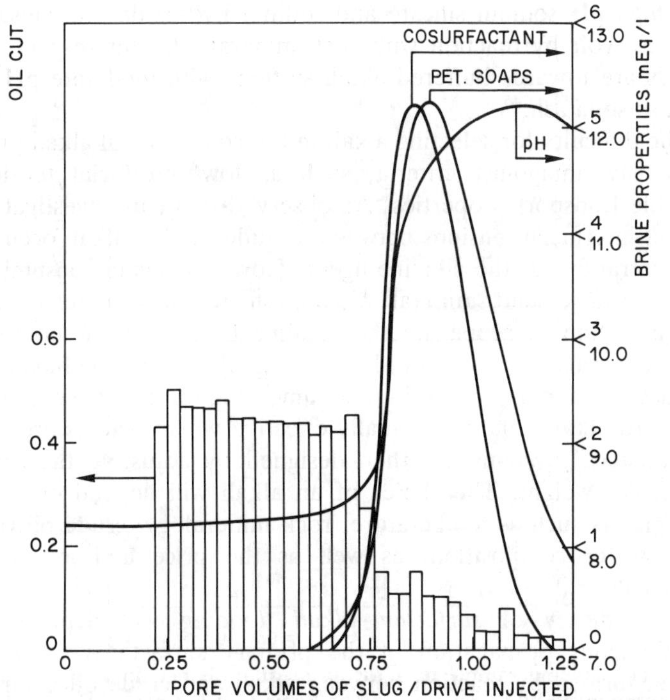

Fig. 4.16 Composition of liquids produced from a cosurfactant-enhanced alkaline flood. (From Ref. 9, reproduced by permission of The Society of Petroleum Engineers.)

For xanthans subject to bacterial degradation, biocides, such as formaldehyde, must be used.

The design and development of an *alkali–surfactant–polymer process* originate from the existing micellar–polymer process which has been proved technically successful. But the high cost of the injected chemicals is one of the reasons that field applications of this latter process have been disappointing. The idea was then to substitute cheap alkali for part of the more expensive surfactants and polymers.

Alkali can be used in a preflush in order to reduce surfactant loss and interactions with divalent ions.[7] Preflush requirements are determined by considering ion exchange properties and rock reactivity. It has been found that gypsum and anhydrite can cause serious alkali loss. In this case, larger volumes of preflush are needed to obtain substantial oil recovery improvement.

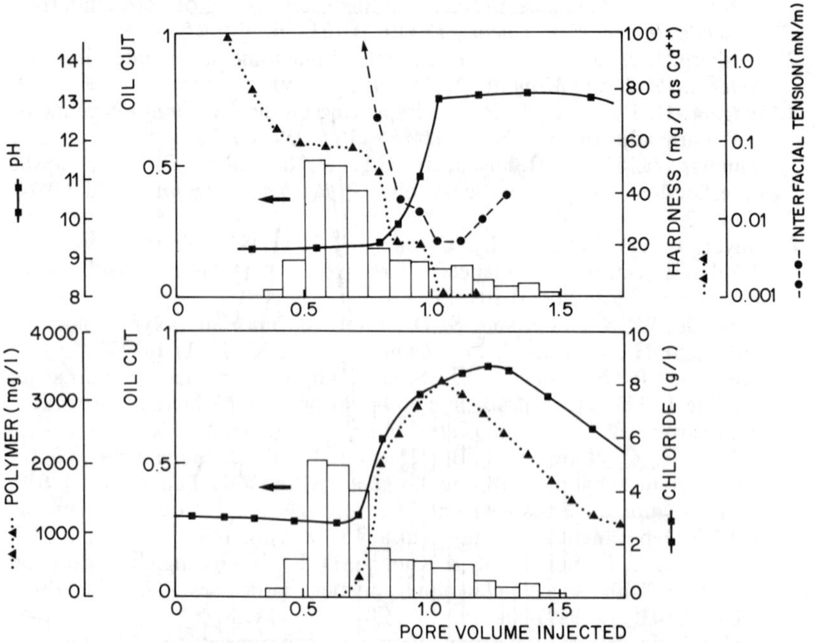

Fig. 4.17 Effluent sample data for an alkali–polymer flood. (From Ref. 10, reproduced by permission of The Society of Petroleum Engineers.)

Alkali can also be incorporated in the polymer drive immediately following the micellar slug.[10] Therefore, the effect of alkali in the rear mixing zone of the micellar slug must be taken into account in the process design, by phase behavior studies as in a conventional micellar–polymer process. As a matter of fact, alkali requirements for a field test are difficult to establish because of long-term consumption reactions. Models, such as those described in Section 4.4.3, can be used to calculate pH propagation at field scale. As far as alkali consumption is concerned, Na_2CO_3 may be preferred to NaOH in so far as it has been shown to be effective for displacing oil.

4.6 References

1. Squires, F., Method of recovering oil and gas. US Patent No. 1 238 355 (1917).

2. Nutting, P. G., Chemical problems in the water driving of petroleum from oil sands. *Ind. & Eng. Chem.*, **17** (1925) 1035–6.
3. Reisberg, J. & Doscher, T. M., Interfacial phenomena in crude oil–water systems. *Producers Monthly*, **21**(1) (1956) 43–50.
4. Wagner, O. R. & Leach, R. O., Improving oil displacement efficiency by wettability adjustment. *Trans. AIME*, **216** (1959) 65–72.
5. Jennings, Jr. H. Y., Johnson, Jr. C.E. & McAuliffe, C.D., A caustic waterflooding process for heavy oils. *J. Petroleum Technol.*, **26** (1974) 1344–52.
6. Mayer, E. H., Berg, R. L., Carmichael, J. D. & Weinbrandt, R. M., Alkaline injection for enhanced oil recovery—A status report. *J. Petroleum Technol.*, **35** (1983) 209–21.
7. Holm, L. W. & Robertson, S. D., Improved micellar/polymer flooding with high-pH chemicals. *J. Petroleum Technol.*, **33** (1981) 161–72.
8. Krumrine, P. H., Falcone, J. S. & Campbell, Jr. T. C., Surfactant flooding 1: The effect of alkaline additives on IFT, surfactant adsorption, and recovery efficiency. *Soc. Petroleum Engineers, J.*, **22**(4) (1982) 503–13.
9. Nelson, R. C., Lawson, J. B., Thigpen, D. R. & Stegemeier, G. L., Cosurfactant-enhanced alkaline flooding. SPE/DOE Paper No. 12672, Fourth Symp. on Enhanced Oil Recovery of Soc. Petroleum Engineers and US Department of Energy, Tulsa, 15–18 April 1984.
10. Schuler, P. J., Lerner, R. M. & Kuehne, D. L., Improving chemical flood efficiency with micellar/alkaline/polymer processes. *J. Petroleum Technol.*, **41**(1) (1989) 80–8.
11. Jennings, Jr. H. Y., A study of caustic solution–crude oil interfacial tensions. *Soc. Petroleum Engineers J.*, **15**(3) (1975) 197–202.
12. Minssieux, L., Intérêt de l'injection d'eau alcaline en récupération assistée. *Revue de l'Inst. Français du Pétrole*, **33** (1978) 47–57.
13. Seifert, W. K. & Howells, W. G., Interfacially active acids in a California crude oil. *Analyt. Chem.*, **41** (1969) 554.
14. Farmanian, P. A., Davis, N., Kwan, J. T., Weinbrandt, R. M. & Yen, T. F., Participation of selective native petroleum fractions in lowering interfacial tensions of aqueous alkaline systems. *ACS Symp. Series*, **91**, 103–14, ACS Washington DC, 1979.
15. Speight, J. G. & Moschopedis, S. E., On the molecular nature of petroleum asphaltenes. *Adv. in Chem. Series*, **195**, 1–14, ACS Washington DC, 1981.
16. Yen, T. F., *The Role of Trace Metals in Petroleum*, Ann Arbor Inc., Ann Arbor, 1975.
17. Trujillo, E. M., The static and dynamic interfacial tensions between crude oils and caustic solutions. *Soc. Petroleum Engineers J.*, **23**(4) (1983) 645–56.
18. Sharma, M. M., Jang, L. K. & Yen, T. F., Transient interfacial tension behavior of crude oil/caustic interfaces. *SPE Reservoir Engineering*, **4**(2) (1989) 228–36.
19. deZabala, E. F., Vislocky, J. M., Rubin, E. & Radke, C. J., A chemical theory for linear alkaline flooding. *Soc. Petroleum Engineers J.*, **22**(2) (1982) 245–58.

20. Pasquarelli, C. H. & Wasan, D. T., The effect of film forming materials on the dynamic interfacial properties in crude oil–aqueous systems. *Surface Phenomena in Enhanced Oil Recovery,* ed. D. O. Shah, Plenum Press, New York, 1981, pp. 237–248.
21. Castor, T. P., Somerton, W. H. & Kelly, J. F., Recovery mechanisms of alkaline flooding. In *Surface Phenomena in Enhanced Oil Recovery,* ed. D. O. Shah, Plenum Press, New York, 1981, pp. 249–91.
22. Johnson, Jr. C. E., Status of caustic and emulsion methods. *J. Petroleum Technol.,* **28** (1976) 85–92.
23. Martin, F. D., Oxley, J. C. & Lim, H., Enhanced recovery of a 'J' sand crude oil with a combination of surfactant and alkaline chemicals. SPE Paper No. 14293, 60th Annual Techn. Conf. and Exhibition of Soc. Petroleum Engineers, Las Vegas, 22–25 September 1985.
24. Krumrine, P. H., Mayer, E. H. & Brock, G. F., Scale formation during alkaline flooding. *J. Petroleum Technol.,* **37**(9) (1985) 1466–74.
25. Labrid, J. & Bazin, B., Alkaline preflush in a low-permeability clayey sandstone. *J. Petroleum Sci. Engineering,* **3** (1989) 111–20.
26. Cheng, K. H., Chemical consumption during alkaline flooding: a comparative evaluation. SPE/DOE Paper No. 14944, Fifth Symp. on Enhanced Oil Recovery, Tulsa, 20–23 April 1986.
27. Talley, L. D., Hydrolytic stability of alkylethoxy sulfates. *SPE Reservoir Engineering,* **3**(1) (1988) 235–42.
28. Bavière, M., Bazin, B. & Labrid, J., Chemical stability of surfactants for micellar flooding under North Sea conditions. *In Situ*, **13**(3) (1989) 101–20.
29. Ryles, R. G., Chemical stability limits of water-soluble polymers used in oil recovery processes. *SPE Reservoir Engineering,* **3**(1) (1988) 23–34.
30. Seright, R. S. & Henrici, B. J., Xanthan stability at elevated temperatures. *SPE Reservoir Engineering,* **5**(1) (1990) 52–60.
31. Bunge, A. L. & Radke, C. J., The origin of alkali ion exchange with reservoir rock. SPE Paper No. 11978, Int. Symp. on Oilfield and Geothermal Chemistry of Soc. Petroleum Engineers, Denver, 1–3 June 1983.
32. Sposito, G., The Gapon and the Vanselow selectivity coefficients. *Soil Sci. Soc. American J.,* **41** (1977) 1205–06.
33. Gapon, E. N., On the theory of exchange adsorption in soils. *J. Gen. Chem. USSR,* **3** (1933) 144.
34. Bazin, B. & Labrid, J., Ion exchange and dissolution/precipitation modeling—Application to the injection of aqueous fluids into a reservoir sandstone. *SPE Reservoir Engineering,* **6**(2) (May 1991) 233–8.
35. Michaels, A. S. & Porter, M. C., Water–oil displacements from porous media utilizing transient adhesion–tension alterations. *AIChE J.,* **11** (1965) 617–24.
36. Takamura, K. & Chow, R. S., A mechanism for initiation of bitumen displacement from oil sand. *J. Canadian Petroleum Technol.,* **22**(6) (1983) 22–30.
37. Buckley, J. S., Takamura, K. & Morrow, N. R., Influence of electrical

surface charges on the wetting properties of crude oils. *SPE Reservoir Engineering,* **4**(3) (1989) 332–40.
38. Ehrlich, R. & Wygal, R. J., Interrelation of crude oil and rock properties with the recovery of oil by caustic waterflooding. *Soc. Petroleum Engineers J.,* **17**(4) (1977) 263–70.
39. Brown, C. E. & Neustadter, E. L., The wettability of oil/water/silica systems with reference to oil recovery. *J. Canadian Petroleum Technol.,* **19**(3) (1980) 100–10.
40. Mohnot, S. M., Bae, J. H. & Foley, W. L., A study of mineral–alkali reactions. *SPE Reservoir Engineering,* **2**(4) (1987) 653–63.
41. Diallo, M. S., Jenkins-Smith, N. L. & Bunge, A. L., Dissolution rates for quartz, aluminum-bearing minerals, and their mixtures in sodium and potassium hydroxide. SPE Paper No. 16276, Int. Symp. on Oilfield Chemistry of Soc. Petroleum Engineers, San Antonio, 4–6 February 1987.
42. Labrid, J. & Duquerroix, J. P., Thermodynamic and kinetic aspects of the dissolution of quartz–kaolinite mixtures by alkalis. *Revue de l'Institut Français du Pétrole* **46**(1) (1991) 41–58.
43. Bertaux, J. & Lemanczyk, Z. R., Importance of dissolution/precipitation mechanisms in sandstone–alkali interactions. SPE Paper No. 16278, Int. Symp. on Oilfield Chemistry of Soc. Petroleum Engineers, San Antonio, 4–6 February 1987.
44. Huang, L., Yang, P. & Qin, T., A study of caustic consumption by clays. SPE/DOE Paper No. 14945, Fifth Symp. on Enhanced Oil Recovery of Soc. Petroleum Engineers and US Department of Energy, Tulsa, 20–23 April 1986.
45. Thornton, S. D., Reaction of sodium hydroxide with silicate minerals. US Department of Energy Report No. Niper 129, 1986.
46. Krumine, P. H. & Falcone, J. S., Beyond alkaline flooding: design of complete chemical systems. SPE Paper No. 16280, Int. Symp. on Oilfield Chemistry of Soc. Petroleum Engineers, San Antonio, 4–6 February 1987.
47. Somasundaran, P. & Hanna, H. S., Adsorption of sulfonates on reservoir rocks. *Soc. Petroleum Engineers J.,* **19**(4) (1979) 221–32.

MISCIBLE GAS DISPLACEMENT

5 Thermodynamics

Franklin M. Orr, Jr
Stanford University, California, USA

5.1	**Introduction**	159
5.2	**Thermodynamic background**	160
5.3	**Equations of state**	162
5.3.1	Pure component parameters—law of corresponding states	163
5.3.2	Mixing rules	166
5.3.3	Examples of equations of state	167
5.4	**Volume translation**	172
5.5	**Characterization of crude oils**	176
5.5.1	Hydrocarbon analysis	176
5.5.2	Choice of pseudocomponents	177
5.5.3	Pseudocomponent properties	177
5.5.4	Tuning of EOS parameters	179
5.6	**Other models**	179
5.7	**Conclusions**	181
5.8	**References**	181

5.1 Introduction

In miscible flood processes, some combination of transfer of components from the oil displaced to the injected fluid and from the injected fluid to the oil takes place as the phases flow through the porous medium. When the required transfer is efficient enough, local displacement efficiency can approach 100%. At the microscopic level, therefore, displacement efficiency is controlled by the phase behavior of the fluid mixtures that form as the flow evolves.

At the pore level it is reasonable to assume that fluids present are well mixed and in chemical equilibrium. Consider, for example, flow in a capillary tube. If the flow is slow enough, then radial diffusion can eliminate concentration variations induced by the fact that flow is faster near the center of the tube than it is near the tube wall. Taylor[1] found that radial diffusion is fast enough to mix fluid adequately in the direction transverse to flow as long as

$$\frac{14{\cdot}4LD}{ur_t^2} \gg 1 \tag{5.1}$$

where L is the pore length, u the flow velocity, r_t the radius of the tube, and D the diffusion coefficient. For flow in a pore of length 100 μm and diameter 100 μm at a velocity of $3{\cdot}5 \times 10^{-4}$ cm/s with diffusion coefficient 10^{-5} cm^2/s, for example, Taylor's parameter estimate yields a value of 165. Thus, the local equilibrium assumption is a good one.

The fundamental problem, therefore, is to represent the equilibrium phase behavior of mixtures of injected fluid with the crude oil present in the formation. In the remainder of this chapter, we mainly consider the representation of phase behavior using an equation of state. Additional models are cited in Section 5.6. Representation of phase behavior on ternary diagrams is detailed in Chapter 6.

5.2 Thermodynamic background

The calculation of equilibrium phase compositions is based on the assumptions that (1) energy is conserved and (2) heat and work are equivalent. These statements are reflected in the fundamental property equation

$$\mathrm{d}U = T\,\mathrm{d}S - P\,\mathrm{d}V \tag{5.2}$$

where U is the internal energy per mole of a mixture of fixed composition, T is the absolute temperature, S is the entropy per mole, P is the pressure, and V is the molar volume. In what follows, it will be convenient to make use also of the definition of the Helmholtz free energy,

$$A = U - TS \tag{5.3}$$

Differentiation of eqn (5.3) along with substitution of eqn (5.2) gives

$$\mathrm{d}A = -S\,\mathrm{d}T - P\,\mathrm{d}V \tag{5.4}$$

Consider a mixture of fixed composition at fixed temperature.

Under such conditions, eqn (5.4) reduces to

$$dA = -P\,dV \tag{5.5}$$

The change in total Helmholtz energy is given by

$$d(nA) = -P\,d(nV) \tag{5.6}$$

where n is the total number of moles. With addition and subtraction of terms involving RT/V on the right side, eqn (5.6) can be integrated to give

$$n(A - A^\circ) = \int_V^\infty \left[P - \frac{RT}{V}\right] dV + nRT \ln\frac{V^\circ}{V} \tag{5.7}$$

In eqn (5.7), A° and V° refer to values of the Helmholtz function and molar volume at a reference state at which the pressure is low enough for the mixture to behave as an ideal gas. R is the gas constant. Equation (5.7) simply states that the Helmholtz energy of a mixture can be calculated in a straightforward way if the volumetric behavior of the mixture is known over the full range of volume.

The utility of eqn (5.7) for phase equilibrium calculations comes from the fact that the chemical potential μ_i of a component in the mixture can be calculated easily if the Helmholtz energy is known. The required expression is

$$\mu_i = \frac{\partial}{\partial n_i}(nA - A^\circ)_{T,n,V,n_{j\neq i}} \tag{5.8}$$

Chemical potential of a component is of fundamental importance because it is used to state the requirements of chemical equilibrium as derived by Gibbs in the 1870s. At equilibrium, the chemical potential for each component must be the same in every phase present. Thus, at equilibrium for phases j and k,

$$\mu_{ij} = \mu_{ik} \qquad i = 1, n_c \tag{5.9}$$

where n_c is the number of components. For most applications, it is more convenient to use the partial fugacity of a component in a mixture, $\hat{f}_i$, to represent the equilibrium condition. Fugacity, defined by Lewis in 1900, has units of pressure, and is equal to pressure for a perfect gas. The relationship between partial fugacity and chemical potential is

$$RT \ln\left[\frac{\hat{f}_i}{x_i P^\circ}\right] = \mu_i \tag{5.10}$$

In eqn (5.10), the reference partial fugacity has been chosen to be $x_i P^\circ$ where x_i is the mole fraction of component i in the mixture, and P° is the reference pressure corresponding to V° in eqn (5.7). Substitution of eqn (5.10) into the condition for equilibrium, eqn (5.9) yields a similar expression for equality of component partial fugacities,

$$\hat{f}_{ij} = \hat{f}_{ik} \qquad i = 1, n_c \tag{5.11}$$

Substitution of eqns (5.8) and (5.7) into eqn (5.10) gives, after some rearrangement,

$$RT \ln \hat{f}_i = \int_{V^t}^{\infty} \left[\left[\frac{\partial P}{\partial n_i} \right]_{T, V^t, n_{j \neq i}} - \frac{RT}{V^t} \right] \mathrm{d}V^t - RT \ln \left[\frac{V}{x_i RT} \right] \tag{5.12}$$

where V^t is the total volume, $V^t = nV$. According to eqn (5.12), the partial fugacity of a component in a mixture can be calculated directly if the relationship between pressure, volume, temperature and composition is known for the full range of volumes from V^t.

5.3 Equations of state

If the PVT relationship required by eqn (5.12) were determined solely from experimental data, an impractically large number of experiments would be required. Fortunately, another approach is available, the use of an equation of state (EOS), i.e. an analytical relation between pressure, volume and temperature. In the early 1870s, at around the same time that Gibbs was working on the statement of chemical equilibrium, a Dutch physical chemist, J. D. van der Waals, developed an EOS that represented qualitatively, at least, the behavior of both liquid and vapor phases,

$$P = \frac{RT}{V - b} - \frac{a}{V^2} \tag{5.13}$$

Parameter b, known as the repulsion parameter, forces the pressure to climb rapidly as V approaches b; behavior similar to that found in liquids. Parameter a is a simple representation of the fact that molecules experience attractions in addition to repulsions. Without them, liquefaction would not be possible. In the limit of a large volume, eqn (5.13) reduces to the ideal gas law.

Since 1870 much work has been devoted to the improvement of the van der Waals equation of state. The repulsion term has remained unchanged, while many corrections have been made to the attractive term.

5.3.1 *Pure component parameters—law of corresponding states*

Once an analytical relationship between *P*, *V* and *T* has been defined, the various coefficients involved in that relation have to be evaluated.

First, consider the calculation of pure component coefficients. In the case of van der Waals EOS, the two coefficients *a* and *b* of eqn (5.13) were derived directly from the fact that at the critical point, where $P = P_c$, $V = V_c$, and $T = T_c$,

$$\left[\frac{\partial P}{\partial V}\right]_T = \left[\frac{\partial^2 P}{\partial V^2}\right]_T = 0 \tag{5.14}$$

Differentiation of eqn (5.13) and substitution of P_c, V_c and T_c gives the following expressions for *a* and *b*,

$$a = \frac{27R^2T_c^2}{64P_c} \tag{5.15}$$

$$b = \frac{RT_c}{8P_c} \tag{5.16}$$

Van der Waals EOS is one expression of the idea of corresponding states. The corresponding states assumption is that, whatever the type of component, the *PVT* analytical relation can be reduced to a dimensionless equation where coefficients no longer depend on the component studied.

The appropriate dimensionless variables are generally reduced pressure P_r, reduced temperature T_r and reduced volume V_r. They are defined as the ratio of actual pressure, temperature and volume to critical pressure, temperature and volume, respectively.

Van der Waals equation of state is an example of an EOS based on the corresponding states principle. If *a* and *b* coefficients, as defined in eqns (5.15) and (5.16), are inserted in eqn (5.13), this relation can be written as:

$$\left(P_r + \frac{3}{V_r^2}\right)(3V_r - 1) = T_r \tag{5.17}$$

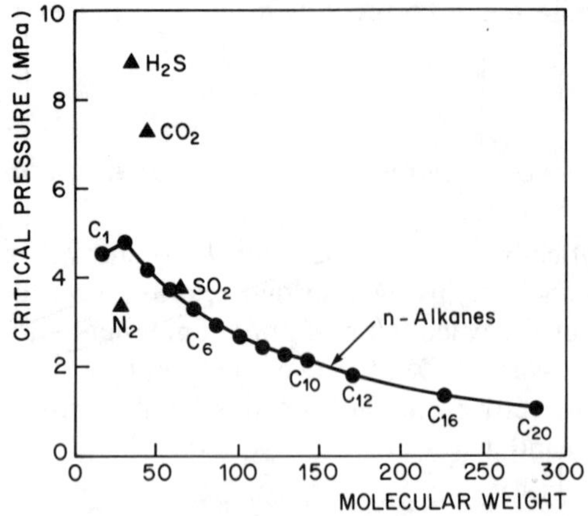

Fig. 5.1 Component critical pressures.

A complete description of the *PVT* behavior of a component is available, given P_r and T_r, that is given data of P_c and T_c, (V_c is deduced from P_c and T_c values), such as those shown in Figs 5.1 and 5.2 for normal alkanes and other selected components.[2] For hydrocarbons, critical pressures decrease as the size of the molecule increases (Fig. 5.1), and critical temperature increases (Fig. 5.2).

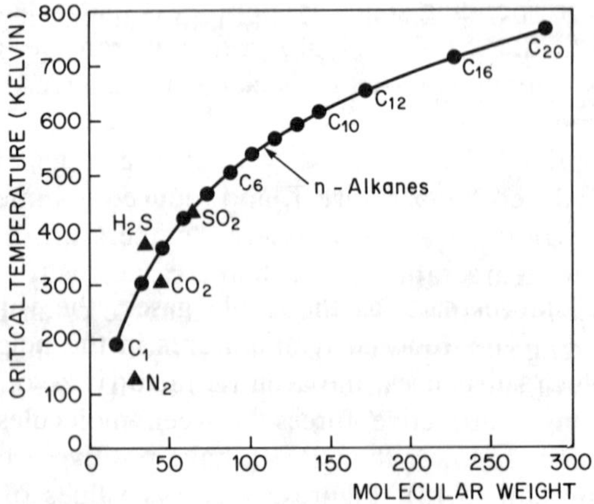

Fig. 5.2 Component critical temperatures.

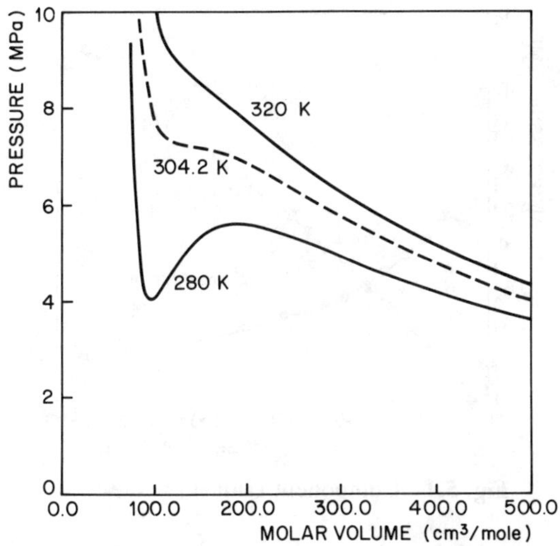

Fig. 5.3 Molar volume of CO_2 according to van der Waals EOS.

Figure 5.3 shows the volumetric behavior of CO_2 calculated with eqns 5.13–5.16. For temperatures below T_c, three molar volumes are observed for some pressures. The lowest value is taken to be the molar volume of the liquid phase, and the largest value that of the vapor. The intermediate value is unstable because $\partial P/\partial V > 0$, and has no physical significance.

A third parameter may also be used in the corresponding state law—the Pitzer's acentric factor, which is defined as:

$$\omega = \log_{10}\left[\frac{P_c}{10P^{sat}}\right]_{T_r=0{\cdot}7} \tag{5.18}$$

For simple substances such as the noble gases, the vapor pressure, P^{sat}, is about $P_c/10$ for reduced temperatures in the neighborhood of $T_r = 0{\cdot}7$. For such substances, the acentric factor is zero. Components that show stronger attractive forces between molecules, and hence have vapor pressures lower than $P_c/10$, have positive acentric factors, as indicated by eqn (5.18). Figure 5.4 shows values of the acentric factor for some *n*-alkanes and other selected components.

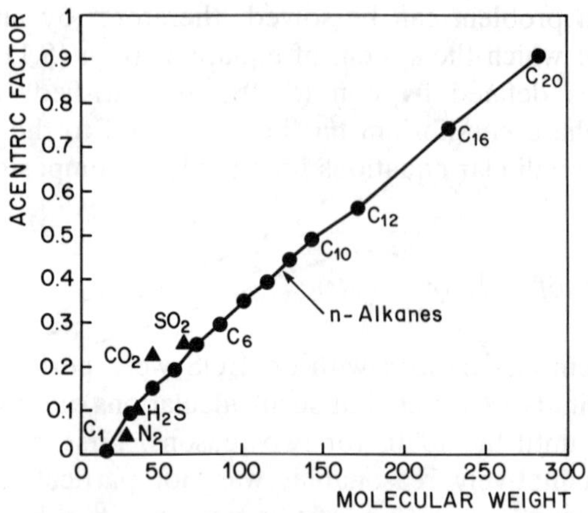

Fig. 5.4 Component acentric factors.

5.3.2 *Mixing rules*

Evaluation of pure component properties is very important, but the main problem in the oil industry is the representation of *PVT* behavior of mixtures.

Suitable equations must be provided which allow the calculation of the coefficients of equation of state (a_m and b_m, for example, in the van der Waals EOS) for the mixture. These equations are called mixing rules. The ones proposed for the van der Waals EOS are:

$$\sqrt{a_m} = \sum_i x_i \sqrt{a_i} \tag{5.19}$$

and

$$b_m = \sum_i x_i b_i \tag{5.20}$$

Given the mixing rules, eqns (5.19) and (5.20), and the form of the EOS, eqn (5.13), the derivative of pressure needed in eqn (5.12) can be evaluated, and the resulting expression can be integrated to yield an equation for the partial fugacity of a component in the mixture. The resulting expression given by Walas[3] is

$$\ln\left[\frac{\hat{f}_i}{x_i P}\right] = \frac{b_i}{V - b_m} - \ln\left[\frac{PV}{RT}\left[1 - \frac{b_m}{V}\right]\right] - \frac{2\sqrt{a_m a_i}}{RTV} \tag{5.21}$$

The equilibrium problem can be solved, therefore, by finding phase compositions for which the system of equations (eqn (5.11)), with the partial fugacities defined by eqn (5.20), are satisfied. Hence, the calculation of phase equilibrium has been reduced to the problem of solving a set of nonlinear equations for the phase compositions.

5.3.3 *Examples of equations of state*

Phase equilibrium calculations with an EOS were performed by van der Waals for binary mixtures, but such calculations did not come into widespread use until the 1970s for two reasons. First, van der Waals EOS, while qualitatively reasonable, was not particularly accurate, and hence other methods such as excess free energy models were used to represent the behavior of the liquid phase. Second, the solution of the nonlinear equilibrium equations was quite difficult without the use of a computer. Both difficulties were eventually resolved, though nearly a century had passed before the full value of the combined work of van der Waals and Gibbs was apparent. Much improved accuracy was obtained by a series of investigators who proposed changes in the form of the attraction term (a/V^2) in eqn (5.13).

Most of those modifications left the cubic form of the EOS unchanged, while some others proposed higher order polynominal equations, mainly to improve the representation of volumetric properties.

(a) *Redlich–Kwong EOS.* In 1949 Redlich and Kwong[4] offered the form

$$P = \frac{RT}{V - b} - \frac{a}{T^{1/2}V(V + b)} \tag{5.22}$$

with

$$a = \frac{0{\cdot}42748R^2T_c^{5/2}}{P_c} \tag{5.23}$$

and

$$b = \frac{0{\cdot}08664RT_c}{P_c} \tag{5.24}$$

The Redlich–Kwong EOS was used by Wilson[5] to describe both the liquid and vapor phases in calculations of equilibrium phase compositions for cryogenic systems. Thereafter, the application of cubic EOSs to phase equilibrium expanded rapidly.

(b) *Soave EOS*. An additional modification suggested by Soave[6] is now in widespread use.

$$P = \frac{RT}{V-b} - \frac{a\alpha}{V(V+b)} \tag{5.25}$$

where a and b are given by eqns (5.23) and (5.24), and α is given by:

$$\sqrt{\alpha} = 1 + (0{\cdot}48508 + 1{\cdot}55171\omega - 0{\cdot}15613\omega^2)(1 - \sqrt{T/T_c}) \tag{5.26}$$

and the following mixing rules apply

$$(a\alpha)_m = \sum_i \sum_j x_i x_j (a\alpha)_{ij} \tag{5.27}$$

$$(a\alpha)_{ij} = (1 - k_{ij})\sqrt{(a\alpha)_i (a\alpha)_j} \tag{5.28}$$

$$b_m = \sum_i x_i b_i \tag{5.29}$$

The attraction term used by Soave incorporates Pitzer's acentric factor ω.

The mixing rule for the attraction parameter (eqn (5.28)) also makes use of a binary interaction parameter, k_{ij}.[7] The binary interaction parameter adjusts the magnitude of the attraction term in an attempt to reflect the attractions between pairs of unlike molecules. Values of k_{ij} are chosen to improve the agreement between measured and calculated phase equilibrium data for binary mixtures. The resulting values, however, are applied to multicomponent mixtures as well.

The approach is based on the assumption that pairwise interactions have a greater effect than those involving more molecules, and on the practical consideration that experimental data required for the determination of interactions of more than two components are rarely available.

(c) *Peng–Robinson EOS*. Another widely used EOS is that of Peng and Robinson.[8]

$$P = \frac{RT}{V-b} - \frac{a\alpha}{V^2 + 2bV - b^2} \tag{5.30}$$

with

$$a = 0{\cdot}45724 \frac{R^2 T_c^2}{P_c} \tag{5.31}$$

$$b = 0{\cdot}07780 \frac{R T_c}{P_c} \tag{5.32}$$

and

$$\alpha = [1 + (0{\cdot}37464 + 1{\cdot}54226\omega - 0{\cdot}2699\omega^2)(1 - \sqrt{T_r})]^2 \tag{5.33}$$

Mixing rules for the Peng–Robinson EOS are identical in form to those of the Soave–Redlich–Kwong EOS, although, of course, the values of the binary interaction parameters differ.

(d) *Simonet–Behar–Rauzy EOS* (*SBR*). The SBR equation of state is a noncubic equation. It is of the fifth degree for the volume.

$$P = \frac{RT}{V-b} - \frac{a}{V(V+b)}\left(1 - \gamma\frac{b}{V} + \delta\left(\frac{b}{V}\right)^2\right) \tag{5.34}$$

The main advantage of this equation is that, without any correction factor (see Section 5.4) volumetric properties of pure components and mixtures are better represented. Values a, b, γ and δ, are the four coefficients of the equation. They are evaluated according to the principle of the corresponding state law through a set of equations given in detail in Ref. 9.

An indication of the improvement in the accuracy of the representation of phase behavior as a result of the modifications to the van der Waals (VDW) EOS made in the Redlich–Kwong (RK), the Soave–Redlich–Kwong (SRK), and the Peng–Robinson (PR) equations is given in Fig. 5.5, which compares pressure–composition phase diagrams calculated with each EOS for CO_2-n decane mixtures. The curves shown in Fig. 5.5 give the compositions of the saturated phases, liquid and vapor, at a given pressure. Vapor phase compositions are given by the nearly vertical curves at high CO_2 compositions, and liquid phase compositions by the curves on the left side of the figure. Two phases form for mixtures with compositions between the liquid and vapor loci (see Ref. 10 for a review of the properties of these and other phase diagrams). Data points shown are those reported by Reamer and Sage.[11] As Fig. 5.5 indicates, both the RK and VDW equations give two-phase regions that are substantially smaller than

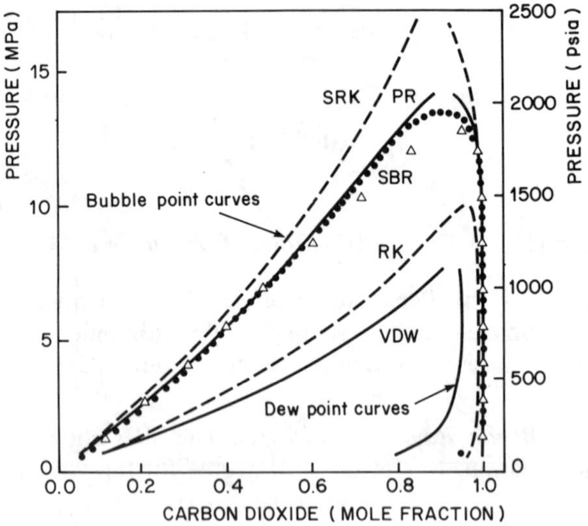

Fig. 5.5 Comparison of EOS predictions with measured phase behavior of CO_2–decane mixtures at 70°C. (From Ref. 10.) Δ, Experimental data.

that observed experimentally. The calculated two-phase region for the SRK EOS is larger than that observed experimentally when the binary interaction parameter used is $k_{ij} = 0{\cdot}1377$, the value recommended by Graboski and Daubert,[12] although the agreement could undoubtedly be improved by adjustment of that value. The interaction parameter used in the PR EOS is $k_{ij} = 0{\cdot}0942$, a value reported by Deo *et al.*[13]

The sensitivity of the calculated phase equilibrium to the value of the binary interaction parameter is shown in Fig. 5.6 for the PR EOS. The two-phase region is much larger when the interaction parameter is $k_{ij} = 0{\cdot}15$, the value suggested by Katz and Firoozabadi,[14] than it is when a smaller value is used. As Fig. 5.5 suggests, the accuracy of the calculated phase compositions can be improved substantially by a suitable choice of the interaction parameter, though the fact that such empirical parameters are required is an indication that the simple form of the representation of attractions and repulsions is far from perfect. The poorer agreement in the neighborhood of the critical point is another manifestation of the limitations of current EOS implementations. Nevertheless, with suitable adjustment of binary interaction parameters to match available experimental data, reasonable repre-

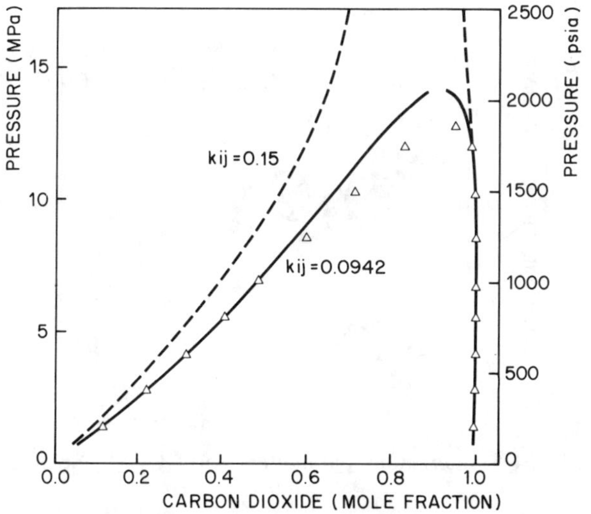

Fig. 5.6 Sensitivity of phase compositions calculated with PR EOS to the binary interaction parameter for CO_2–C_{10} mixtures at 70°C.

sentations of phase behavior can be obtained for use in calculations of miscible flood performance.

The range of variation in interaction parameters for CO_2–alkane mixtures is given in Fig. 5.7. Values shown for CO_2–alkane interaction parameters are those reported by Deo *et al.*[13] for the PR EOS. Several values are shown for some carbon numbers. They represent best fit values to different experimental data sets. Part of the variation is due to the fact that best fit values vary slightly with temperature and part may be due to normal variations in experimental data. Clearly, the interaction parameters themselves can be determined with a precision no greater than that of the experimental data. The values shown for C_1–alkane mixtures are those recommended by Katz and Firoozabadi.[14] The values for C_1 increase with the carbon number of the alkane, while those for CO_2 decrease. Peng and Robinson[8] suggested that a value of 0·01 be used for ethane and propane interactions with heavier hydrocarbons. Interaction parameters for other hydrocarbon–hydrocarbon pairs are usually taken to be zero, though Slot–Peterson[15] found that accurate representation of the behavior of near-critical fluids required adjustment of interaction parameters for C_2, C_3 and C_4 with heavier hydrocarbons.

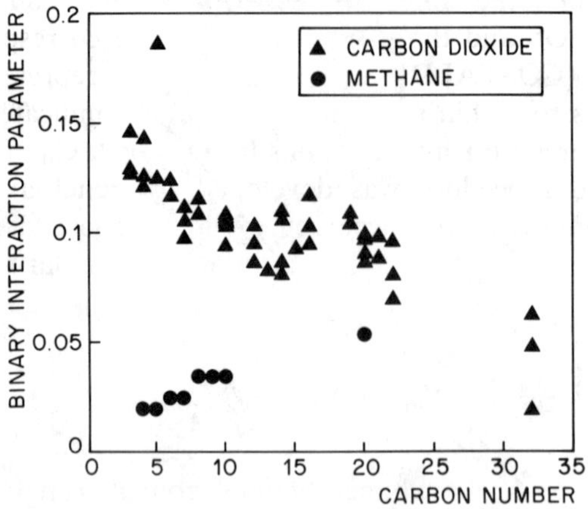

Fig. 5.7 PR EOS interaction parameters for CO_2–alkane and C_1–alkane pairs.

5.4 Volume translation

The modern equations of state generally give accurate predictions of vapor phase densities, but calculated liquid phase densities are often less accurate, particularly if heavy hydrocarbon components are present. The difficulty arises even for single components, as Fig. 5.8

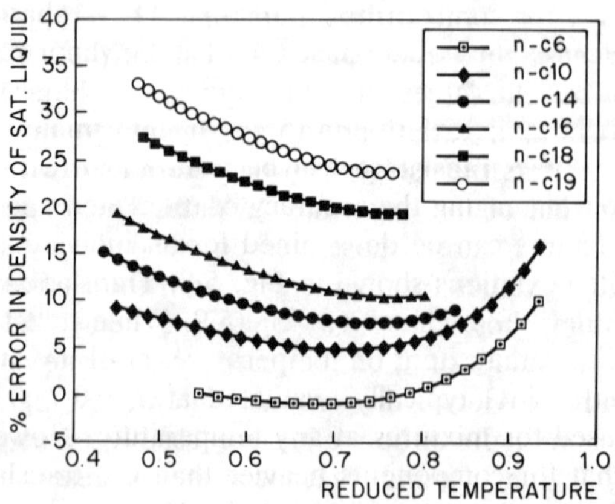

Fig. 5.8 Agreement of PR EOS estimates of saturated liquid density with the COSTALD correlation. (From Ref. 13, reproduced by permission of The Society of Petroleum Engineers.)

shows. It reports the difference between molar volumes calculated with the PR EOS and those obtained from the corresponding states liquid density (COSTALD) correlation,[16] which represents saturated liquid densities to within about 2%. As Fig. 5.8 indicates, substantial errors are observed for hydrocarbons heavier than C_{10}. Fortunately, a straightforward procedure was developed by Peneloux *et al.*[17] that substantially improves the accuracy of liquid phase density calculations. They showed that a more accurate mixture molar volume can be calculated from the equation

$$V_{\text{trans}} = V_{\text{EOS}} - \sum_{j=1}^{n_c} x_i c_i \tag{5.35}$$

where V_{EOS} is the molar volume obtained from any original cubic EOS and c_i is a constant translation factor for component *i*. If the volume given by eqn (5.33) is used in the calculation of component partial fugacities, the translated partial fugacity differs from the original value only by a multiplicative constant. If the same translation is used for both phases, therefore, both partial fugacities in eqn (5.11) are multiplied by the same constant, which then divides out of the equation. Hence, equilibrium phase compositions calculated with the translated molar volume are identical to those calculated with the original EOS; a useful property because considerable progress has been made toward the goal of accurate calculation of phase compositions.

The effect of volume translation on calculated phase volume is generally greater for the liquid phase than for the vapor phase. At low pressure, at least, vapor molar volume is much larger than liquid volume, and the corrections of eqn (5.34) are too small to have much effect. Thus, volume translation can be used to correct liquid phase densities without damaging the accuracy of calculated vapor densities.

Translation factors can be determined for an individual component to eliminate the deviations shown in Fig. 5.8. Translation factors that result for the PR EOS are shown in Fig. 5.9. While translation factors clearly depend to some extent on temperature, most investigators have used a constant value, typically evaluated at $T_r = 0{\cdot}7$. This constant value is then used for mixtures at any temperature. However, Deo *et al.*[13] suggest that for components heavier than C_{16} translation factors evaluated at the temperature of interest give better results. Translation factors for the SRK EOS have been reported by Peneloux *et al.*[17] and for the PR EOS by Jhaveri and Youngren.[18]

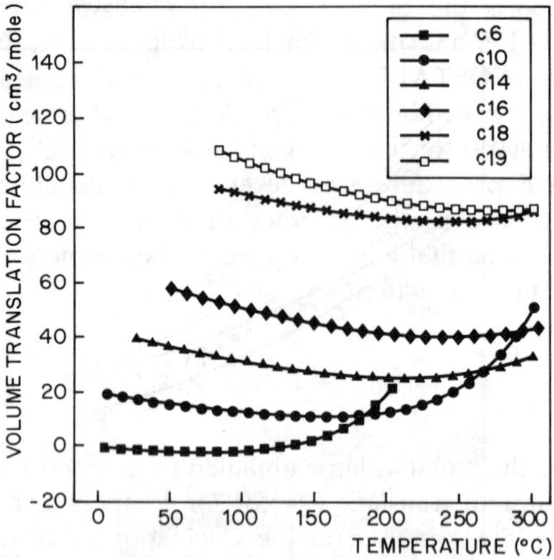

Fig. 5.9 Volume translation factors for the PR EOS estimated from the COSTALD correlation for saturated liquid densities. (From Ref. 13, reproduced by permission of The Society of Petroleum Engineers.)

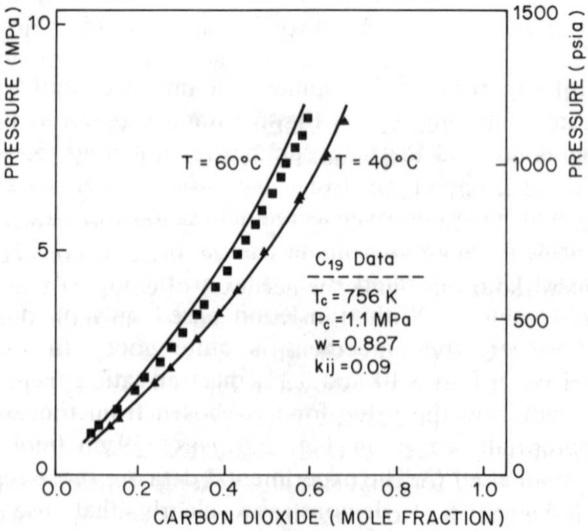

Fig. 5.10 Comparison of PR EOS bubblepoint pressures with measured data for CO_2–C_{19} mixtures. (From Ref. 19.)

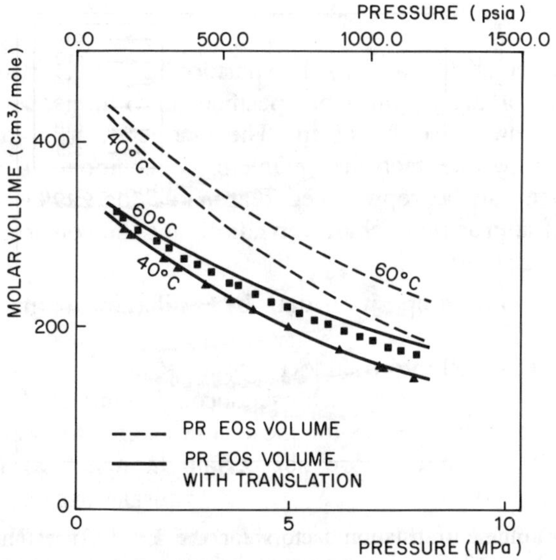

Fig. 5.11 Effect of volume translation on calculated molar volume (PR EOS) for CO_2–C_{19} mixtures at the bubble point pressure. Data points shown are from Fall *et al.* (From Ref. 19.)

Use of the translated volumes can improve liquid phase density calculations substantially, as Figs 5.10 and 5.11 show. For the bubble point pressure calculations summarized in Fig. 5.10, the binary interaction parameter was fixed at $k_{ij} = 0{\cdot}09$, though a slightly smaller value would have improved agreement at 60°C at the expense of the match at 40°C. Without volume translation, the PR EOS gives molar volumes of CO_2-saturated n-C_{19} that are about 30% higher than the measured values.[18] With translation based only on pure component data, however, the agreement is quite good. In the calculations summarized in Figs 5.10 and 5.11, the translation factor for CO_2 was 3·7 cm³/mol, and the value for C_{19} chosen from the extrapolation of the appropriate curve in Fig. 5.9 was 119 cm³/mol at 40°C and 114 cm³/mol at 60°C. The experimental data are those reported by Fall *et al.*[19] Figure 5.11 demonstrates clearly that use of the pure component translation factors also improves calculated liquid phase molar volumes for mixtures.

5.5 Characterization of crude oils

If EOS calculations are to be performed for a crude oil, the composition of the oil must be specified, and the properties of each component must also be given. The characterization of the oil is complicated by the fact that many more components are actually present than can be represented reasonably for EOS computations. Instead, an approximate characterization must be constructed through the following steps:

(1) Obtain oil composition data by distillation or gas chromatography (GC).
(2) Divide the oil into pseudocomponents.
(3) Assign properties to each pseudocomponent: T_c, P_c, ω, and k_{ij}.
(4) Adjust the properties for those components for which values of T_c, P_c, ω, and k_{ij} are not certain to match available phase behavior data.

In the remainder of this section, we consider each of these steps in turn.

5.5.1 *Hydrocarbon analysis*

Measurement of reservoir fluid composition may be undertaken by gas chromatography, distillation, or some combination of the two. Gas compositions are routinely determined by chromatography.[20] The traditional method for liquid analysis is distillation. This technique produces samples for which density, molecular weight, and other properties can be measured directly. Chromatography can also be used to analyze the hydrocarbons present in the oil. Several options are available depending on the detail required of the analysis. Simulated distillation[21,22] with packed columns can be used to obtain a weight fraction distribution similar to that obtained by true boiling point (TBP) distillation, though the GC analysis does not produce the liquid samples for measurement of density and molecular weight. In simulated distillation, hydrocarbons are assigned to carbon number cuts on the assumption that those that elute between C_n and C_{n+1} all have carbon number $n+1$. The molecular weight of the cut can then be estimated as that of the normal alkane. Chromatographic analysis performed on capillary columns can provide additional resolution over that available with packed columns.[20]

5.5.2 *Choice of pseudocomponents*

Any crude oil mixture contains hundreds of hydrocarbon components; far more than can be represented explicitly in EOS calculations. Hence, pseudocomponents that are themselves mixtures are generally used for components heavier than C_4, and even the C_4 fractions are frequently lumped. The goal of any lumping scheme is to represent as a single component a collection of actual components that exhibit approximately equal partition coefficients (K-values) at equilibrium. The idea is that components that partition equivalently need not be represented separately. Some investigators have suggested that the C_{7+} fraction be divided into pseudocomponents with equal mole fractions or weight fractions. Others have proposed more complex schemes to divide the fractions based on molecular weight intervals[23] or component K-values.[24,25]

The resolution required depends on the phase behavior being represented. For example, bubble point pressure calculations for oils with a low content of light fractions are most sensitive to the properties of methane and other light gases present. For such systems, therefore, representation of the C_{7+} fraction by two to three pseudocomponents is often adequate. Dew point pressure and liquid dropout calculations are most sensitive to the interactions of the light and heavy components. Adequate resolution of heavy components is required, therefore, if good agreement between experiment and EOS predictions is to be obtained. Newley and Merrill[25] have shown, however, that good agreement is still possible when the total number of pseudocomponents is held to six or seven, when one very heavy component with a small mole fraction is present. Adequate representation of the behavior in the neighborhood of the critical point generally requires the use of more pseudocomponents.[14]

5.5.3 *Pseudocomponent properties*

Pseudocomponent properties needed for EOS calculations are T_c, P_c, ω, and k_{ij}. In addition, component molecular weights must be known if molar densities are to be converted to mass densities. For hydrocarbons up to C_6 and common nonhydrocarbon constituents, the values needed are given in Figs 5.1, 5.2 and 5.4. For heavier pseudocomponents, some sort of estimation scheme is required. The approach that is appropriate depends on the oil characteristic data available. If the analysis has been obtained by distillation, for example, then critical properties and acentric factors for each component can be estimated for each boiling range cut from the measured boiling point temperature and the specific gravity. In addition, molecular weights can be measured for each cut. An example of the many correlations proposed is that of Riazi and Daubert.[26]

$$T_c = 24{\cdot}2787 T_b^{0{\cdot}58848} \rho^{0{\cdot}3596} \tag{5.36}$$

$$P_c = 3{\cdot}12281 \times 10^9 T_b^{-2{\cdot}3125} \rho^{2{\cdot}3201} \tag{5.37}$$

where ρ is the specific gravity of the cut at 15°C/15°C, (about 60°F/60°F) T_b is the boiling point. T_c and T_b are both in °R (1 Rankine = 1·8 Kelvin), and P_c is in psia (1 psia = $6{\cdot}895 \times 10^{-3}$ MPa). The acentric factor can be estimated from Edmister's correlation as[27]

$$\omega = \frac{3}{7}\left[\frac{\log_{10} P_c}{T_c/T_b - 1}\right] - 1 \tag{5.38}$$

Many other correlations are also available (see Refs 20 and 28 for additional correlations and for examples of crude oil characterizations).

If chromatography data are used as the extended analysis, or if an approximation to the hydrocarbon distribution[23] is used, boiling point and specific gravity data will not be available. In that case, molecular weights for single carbon number cuts can be estimated from curves given by Katz and Firoozabadi[14] or Pedersen *et al.*[20] Both authors note that molecular weights at a given carbon number do not differ greatly from oil to oil. Specific gravities can be estimated by Whitson's procedure.[23] In that approach, the Watson characterization factor, defined as

$$K_W = \frac{T_b^{1/3}}{\rho} \tag{5.39}$$

is determined for the C_{7+} fraction. The value of K_W is chosen to match the specific gravity of the C_{7+} fraction[28]

$$K_W = \left[\frac{0{\cdot}16637 \rho_{7+}}{x_{7+} M_{7+}} \sum_{j=7}^{N} x_i M_i^{0{\cdot}82053}\right]^{-0{\cdot}84573} \tag{5.40}$$

The specific gravity for each carbon number can then be calculated from

$$\rho_i = 6{\cdot}0108 M_i^{0{\cdot}17947} K_W^{-1{\cdot}18241} \tag{5.41}$$

Critical properties and acentric factors can then be calculated from eqns (5.36)–(5.38).

While Whitson's procedure generates estimated critical properties and acentric factors, additional correlations must be used for binary interaction parameters. The correlation of Katz and Firoozabadi[14] can be used for methane interaction parameters, and the values reported by Deo *et al.*[13] can be used to estimate values for CO_2. Interaction parameters for hydrocarbon pairs are generally taken to be zero, although Slot-Peterson[15] found that nonzero values for hydrocarbon pairs improved agreement between calculation and experiment for near-critical mixtures.

5.5.4 *Tuning of EOS parameters*

The estimates of EOS parameters discussed above are inevitably uncertain, given all the assumptions made. Furthermore, the uncertainty about parameters is greater for heavy hydrocarbons. Hence, many investigators choose to adjust some or all of the parameters for the heaviest component(s) to improve the match between calculation and experiment. Saturation pressures, particularly dewpoints, are frequently quite sensitive to binary interaction parameters, and all calculated values depend on the choice of critical parameters and acentric factors. There are dangers lurking in any curve-fitting exercise, however, and this one is no exception. If many parameters are adjusted simultaneously, the minimum-error match is not likely to be unique because the parameters are not strictly independent. Furthermore, adjustments of parameters to match data of one type (say saturation pressure) may worsen other calculated values (say liquid volume fractions), as Pedersen *et al.*[20] have shown. Thus, parameter tuning, though often necessary, should be performed with care, and the range of adjustment permitted should be monitored carefully to prevent the use of nonphysical values. Finally, it should be noted that the resulting EOS representation of phase behavior will be most accurate for the range of experimental conditions matched. Extrapolation outside that range is inherently more dangerous.

5.6 Other models

Equations of state are currently the most widely used models for the representation of multicomponent phase equilibria in oil recovery calculations, but other models, such as group contribution models and continuous thermodynamics, are also used in some applications.

In group contribution methods, each molecule or component is decomposed into characteristic groups. For example, an *n*-alkane is made up of CH_3— and —CH_2— groups. Component partial fugacities (usually represented in terms of activity coefficients) are then estimated as a sum, according to specific rules, of contributions from the constituent groups. The idea behind this approach is that there are many fewer functional groups than there are molecules. Hence, if representations of the thermodynamic behavior of the groups can be obtained, the calculation of the behavior of complex molecules is simplified. Typically, excess free energies or activity coefficients are calculated directly. Examples include the widely used UNIFAC and UNIQUAC models.[29]

The advantage of group contribution methods lies in the fact that predictions can be made once a limited number of parameters has been determined for typical constituent groups. Thus, detailed analytical information, if available, can be used directly to construct the thermodynamic representation. For crude oil systems, however, the large number of components and wide range of molecular structures present make application of such methods difficult.

Continuous thermodynamic methods were developed to deal with the problems posed by the presence of large numbers of components. In that approach, the representation of the phase equilibrium is based on a description of the oil composition as a continuous distribution function in place of discrete mole fractions of individual components.[30] Any thermodynamic function, then, can be defined as an integral function of the molar distribution:

$$I = \int_{x_1}^{x_2} f(x)\,\mathrm{d}x \tag{5.42}$$

where x_1 and x_2 define the interval of continuous composition, and $f(x)$ is some thermodynamic function (molar free energy, entropy, volume, etc.).

The advantage of the continuous thermodynamics approach is that the choice of pseudocomponents is reduced to the specification of the number of intervals in the distribution. Additional development is needed for consistent distribution functions appropriate to both liquid and vapor phases, but successful applications to natural gas systems[31] suggest that the techniques of continuous thermodynamics hold considerable promise for the future.

5.7 Conclusions

Calculations of phase equilibrium are a fundamental part of enhanced oil recovery process design, particularly for miscible flooding, where substantial material exchanges occur between oil and injected solvent. In the 1990s, equations of state will remain the most widely used tool to perform these calculations.

The biggest difficulty of the modeling is to characterize the oil by a limited number of pseudocomponents and to evaluate the thermodynamic parameters of these pseudocomponents. Given the inevitable uncertainties inherent in the characterization of crude oils in terms of a small number of pseudocomponents, some experimental phase behavior data are still required, and in most cases some empirical adjustment of binary interaction parameters and volume translation factors will be necessary, especially for heavy pseudocomponents. Even so, the use of an EOS provides a thermodynamic framework for rational interpolation and (careful) extrapolation of limited experimental data.

5.8 References

1. Taylor, G. I., Dispersion of soluble matter in solvent flowing slowly through a tube. *Proc. Royal Soc.*, **A219** (1953) 186–203.
2. Reid, R. C., Prausnitz, J. M. & Sherwood, T. K., *The Properties of Gases and Liquids,* McGraw-Hill, New York, 1987.
3. Walas, S. M., *Phase Equilibria in Chemical Engineering,* Butterworths, Stoneham, 1985.
4. Redlich, O. & Kwong, J. N. S., On the thermodynamics of solutions: V—An equation of state. Fugacities of gaseous solutions. *Chem. Rev.*, **44** (1949) 233–44.
5. Wilson, G. M., Vapor–liquid equilibria correlation by means of a modified Redlich–Kwong equation of state. *Advances in Cryogenic Engineering,* **9** (1964) 168–76.
6. Soave, G., Equilibrium constants from a modified Redlich–Kwong equation of state. *Chem. Engineering Sci.*, **27**(6) (1972) 1197–203.
7. Zudkevitch, D. & Joffe, J., Correlation and prediction of vapor–liquid equilibria with the Redlich–Kwong equation of state. *AIChE J.*, **16** (1970) 112–19.

8. Peng, D. & Robinson, D. B., A new two-constant equation of state. *Ind. Engineering Chem. Fundamentals,* **15**(1) (1976) 59–64.
9. Behar, E., Simonet, R. & Rauzy, E., A new non-cubic equation of state. *Fluid Phase Equilibria,* **21** (1985) 237–55.
10. Orr, Jr, F. M., & Taber, J. J., Displacement of oil by carbon dioxide. Final Report to the US Department of Energy, Report No. DOE/BC/10331 (1984).
11. Reamer, H. H. & Sage, B. H., Phase equilibria in hydrocarbon systems. Volumetric and phase behavior in the *n*-decane–CO_2 system. *Ind. Engineering Chem.,* **8** (1963) 508.
12. Graboski, M. S. & Daubert, T. E., A modified Soave equation of state for phase equilibrium calculations. 2: Systems containing CO_2, H_2S, N_2 and CO. *Ind. Engineering Chem. Process Des. Dev.,* **17** (1978) 448–54.
13. Deo, M. D., Nutakki, R. & Orr, Jr, F. M., Schmidt–Wenzel and Peng–Robinson equations of state for CO_2–hydrocarbon mixtures: Binary interaction parameters and volume translation factors. Paper SPE No. 18796, 59th Annual California Regional Meeting of Soc. Petroleum Engineers, Bakersfield, 5–7 April 1989.
14. Katz, D. L. & Firoozabadi, A., Predicting phase behavior of condensate/crude-oil systems using methane interaction coefficients. *J. Petroleum Technol.,* **20**(11) (1978) 1649–55.
15. Slot-Petersen, C., A systematic and consistent approach to determine binary interaction coefficients for the Peng–Robinson equation of state. *SPE Reservoir Engineering,* **4**(4) (1989) 488–94.
16. Hankinson, R. W. & Thomson, G. H. A new correlation for saturated densities of liquids and their mixtures. *AIChE J.,* **25** (1979) 653–63.
17. Peneloux, A., Rauzy, E. & Freze, R., A consistent correction for Redlich–Kwong–Soave volumes. *Fluid Phase Equilibria,* **8** (1982) 7–23.
18. Jhaveri, B. S. & Youngren, G. K., Three-parameter modification of the Peng–Robinson equation of state to improve volumetric predictions. *SPE Reservoir Engineering,* **3**(3) (1988) 1033–44.
19. Fall, D. J., Fall, J. L. & Luks, K. D., Liquid–liquid–vapor immiscibility limits in carbon dioxide + *n*-paraffin mixtures. *J. Chem. Engineering Data,* **30**(1) (1985) 82–90.
20. Pedersen, K. S., Fredenslund, Aa. & Thomassen, P., *Properties of Oils and Natural Gases,* Gulf Publishing Company, Houston, 1989.
21. Chorn, L. G., Simulated distillation of petroleum crude oil by gas chromatography. Characterizing the heptanes-plus fraction. *J. Chromatographic Sci.,* **22**(1) (1984) 17–21.
22. Green, L. E., Schmauch, L. J. & Worman, J. C., Simulated distillation by gas chromatography. *Anal. Chem.,* **36**(8) (1964) 1512–16.
23. Whitson, C. H., Characterizing hydrocarbon plus fractions. *Soc. Petroleum Engineers J.,* **23**(4) (1983) 683–94.
24. Li, Y.-K., Ngheim, L. X. & Siu, A., Phase behavior calculations for reservoir fluids: effect of pseudocomponents on phase diagrams and simulation results. *J. Canadian Petroleum Technol.,* (1985) 29–36.
25. Newley, T. M. J. & Merrill, Jr, R. C., Pseudocomponent selection for compositional simulation. Paper SPE No. 19638, 64th Annual Technical

Conference and Exhibition of Soc. Petroleum Engineers, San Antonio, 8–11 October 1989.

26. Riazi, M. R. & Daubert, T. E., Simplified property predictions. *Hydrocarbon Proc.*, **59** (1980) 115–16.
27. Edmister, W. C., Applied hydrocarbon thermodynamics. Part 4. Compressibility factors and equations of state. *Petroleum Refinery,* **37** (1958) 173–9.
28. Whitson, C. H., Effect of C_{7+} properties on equation-of-state predictions. *Soc. Petroleum Engineers J.*, **24**(6) (1984) 685–96.
29. Fredenslund, Aa., Gmehling, J. & Rasmussen, P. *Vapor–Liquid Equilibria Using UNIFAC. A Group Contribution Method.* Elsevier, Amsterdam, 1977.
30. Kehlen, H., Rätzsch, M. T. & Bergmann, J., Continuous thermodynamics of multicomponent systems. *AIChE J.*, **31**(7) (1985) 1136–48.
31. Cotterman, R. L. & Prausnitz, J. M., Application of continuous thermodynamics to natural gas. *Revue de l'Institut Français du Pétrole,* **45**(5) (1990) 633–43.

6 Miscible gas flooding

Danielle Morel
Institut Français du Pétrole, Rueil-Malmaison, France

6.1	**Introduction**	185
6.2	**Vaporizing gas drive**	186
6.2.1	Mechanism and representation by a ternary diagram	186
6.2.2	Minimum miscibility pressure	190
6.3	**Condensing gas drive**	196
6.3.1	Mechanism and representation by a ternary diagram	197
6.3.2	Miscibility determination	199
6.3.3	Combined condensing–vaporizing mechanism	203
6.4	**First contact miscible process**	205
6.4.1	Mechanism and representation by a ternary diagram	205
6.4.2	Method for determining miscibility conditions	206
6.5	**Miscible flooding parameters**	207
6.5.1	Mobility ratio	207
6.5.2	Diffusion–dispersion	208
6.6	**Field applications**	211
6.6.1	Field equipment	211
6.6.2	Injection scheme	212
6.6.3	Field operations	212
6.7	**References**	212

6.1 Introduction

When a fluid is injected into a field to displace oil towards the production well, oil recovery is not complete (see Chapter 1). One way to improve recovery is to reduce or suppress the interfacial tension σ between the oil and injected fluid. This happens when injected fluid, gas for example, is miscible with oil ($\sigma = 0$).

In this chapter, the term 'solvent' will often be used to describe whatever injected fluid is able to exchange components with the oil (it may be a gas or a liquid).

Three processes are involved in miscible gas drive, each of which is detailed in this chapter, and some additional information is given about factors affecting miscible flooding. A separate chapter (Chapter 7) deals with the CO_2 process, due to the particular properties of this gas and its increasing use in EOR applications.

Some injected fluids mix directly with reservoir oil, without any multiphase behavior, developing a 'first contact miscibility' process (see Section 6.4). It is the simplest way to achieve miscibility, but also the least used, because the solvents that give this type of miscibility are expensive.

In other cases, the injected fluid undergoes phase separation from the oil. Nevertheless, intensive mass transfer occurs. Light to intermediate components are exchanged between oil and injected fluid. A transition zone spreads out in which both fluids are miscible. This type of miscibility is called 'multiple-contact miscibility' or dynamic miscibility, and develops the following two processes:

- vaporizing gas drive (see Section 6.2)
- condensing gas drive (see Section 6.3)

Basic information and extensive details on those processes may be found in the monograph on miscible displacement by Stalkup.[1]

6.2 Vaporizing gas drive

Vaporizing gas drive is a particular case of multiple contact miscibility. It is based on the vaporization of the intermediate components from the reservoir oil to the injected gas creating a miscible transition zone.

The C_2–C_5 fraction is preferentially extracted. This mainly occurs at high pressure, by injecting nitrogen, flue gas or natural gas. It is also partly true for CO_2, though higher molecular weight hydrocarbons may be extracted by this gas.

6.2.1 *Mechanism and representation by a ternary diagram*

(a) *Hydrocarbon gas.* A practical way to visualize the development of vaporizing gas drive miscibility is to draw a ternary diagram. The three poles of that ternary diagram are defined as:

- the light component, methane C_1,
- the intermediate components, generally C_2 to C_6 (all the paraffins from ethane to hexane, with both limits included),
- the heavy components, for example C_{7+} (heptane and heavier fractions).

Under determined conditions of pressure and temperature (P and T), the dew and bubble point curves are plotted inside the triangle, thus defining the two-phase zone (Fig. 6.1). Point C is representative of the mixture of the three pseudocomponents, which has T and P as critical temperature and pressure. The high-pressure injected gas is represented by the point G, the original oil by point O.

As will be detailed later, it is absolutely necessary that, if miscibility is to develop, the representative point of the oil should be on the right side of the critical tie line (i.e. the tangent to the phase envelope at point C). This implies that oil is relatively rich in intermediate components.

The various exchanges involved are schematically represented inside the diagram (Fig. 6.1), and in Fig. 6.2.[2]

Initially, virgin oil and injected gas are immiscible and a representative line GO passes through the two-phase zone. This implies that

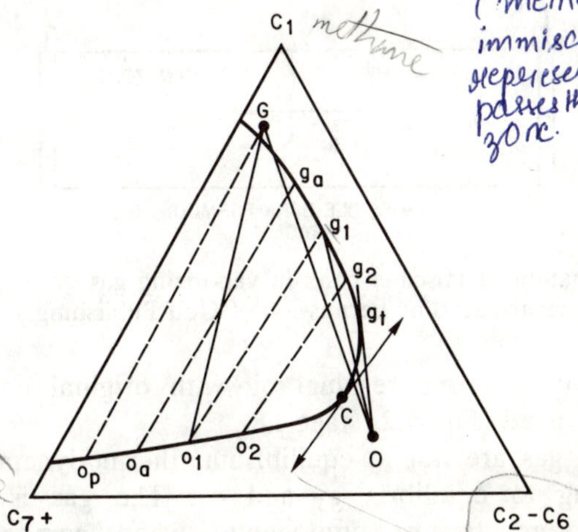

Fig. 6.1 Ternary diagram: vaporizing gas drive process.

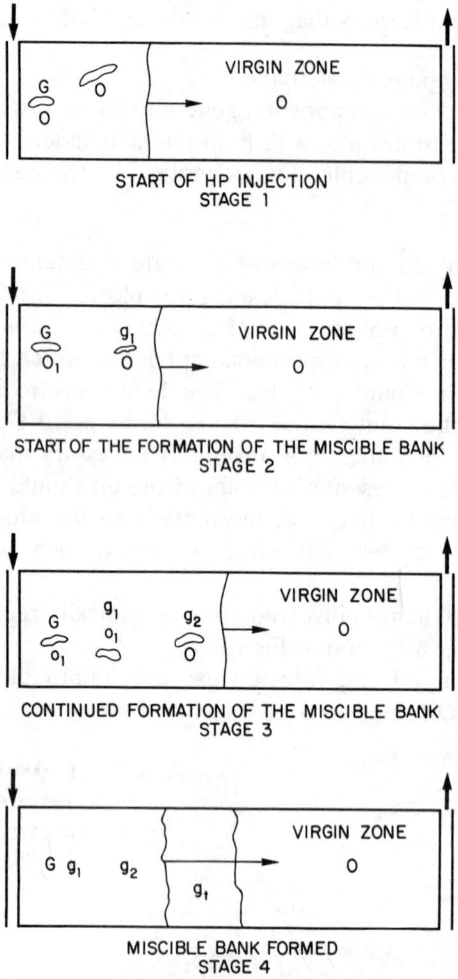

Fig. 6.2 Formation of miscible bank in vaporizing gas drive. (From Ref. 2, reproduced by permission of Gulf Publishing.)

near the wellbore some residual oil with original composition O remains unchanged (Fig. 6.2, stage 1).

As oil and gas are not in equilibrium, thermodynamic exchanges occur, leading to equilibria g_1 and o_1. The gas is enriched in intermediate and heavy components, when compared to G's composition.

Oil o_1 generally occupies a smaller volume than oil O, so oil saturation remains underneath critical mobility value behind the front. Gas g_1 moves ahead, chased by the fresh injected gas G, while the oil remains in place.

At that step of the process (Fig. 6.2, stage 2), g_1 contacts virgin oil O. Since they are not in equilibrium, they divide into two phases g_2 and o_2, with g_2 being in contact with the front of the injected gas. On the other hand, oil o_1 in contact with gas G gives oil o_a which is even poorer in intermediate components (Fig. 6.2, stage 3). This whole process will go on until the gas in contact with virgin oil reaches point g_t, defined as the intercept of the tangent to two-phase envelope from the oil representative point O. There, full miscibility is achieved and no residual oil remains.

Behind the miscible bank (Fig. 6.2, stage 4), residual oils o_1, o_2, etc., previously formed, continue becoming poorer in light fractions while in contact with fresh G. The extreme composition of these residual oils is o_p, placed on the tie line that passes through G gas composition. This o_p oil does not exchange any intermediate component with gas G, and will remain trapped in the reservoir.

It appears from field experience that the miscible bank is perfectly formed, a few dozen meters away from the injector well.[1] It implies that residual oil losses are totally negligible.

(b) *Nitrogen or flue gas.* Nitrogen may be used as a chase gas, once miscibility has been achieved with hydrocarbon gas, or directly as a miscible solvent for high pressure reservoir oils.

Flue gas, a mixture of approximately 12% CO_2 and 88% N_2, may also be injected. It is produced during methane combustion (practically, hydrocarbon gas is burned with air):

$$CH_4 + 8N_2 + 2O_2 \rightarrow CO_2 + 2H_2O + 8N_2 \quad (6.1)$$

Nine volumes of combustion gas are obtained per volume of methane, which is a very interesting ratio.

Furthermore, compressibilities of flue gas and nitrogen are much more favorable than that of hydrocarbon gas (i.e. a given standard volume of flue gas or nitrogen will give a higher volume under reservoir pressure and temperature than the same standard volume of hydrocarbon gas).

For flue gas and nitrogen, a ternary diagram representation of the equilibria involved is possible but the choice of the poles is difficult.

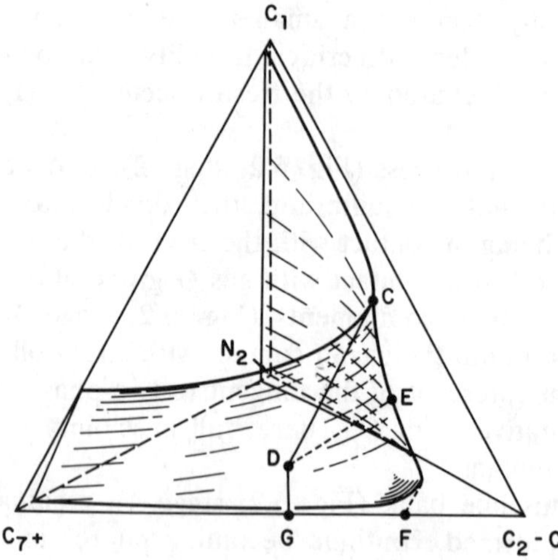

Fig. 6.3 Example of a quaternary diagram. (From Ref. 3, reproduced by permission of The Society of Petroleum Engineers.)

The best solution is to choose nitrogen (or flue gas) as a pole, with the two others being C_1–C_6 and C_{7+}. Oil is then represented by a pseudobinary mixture (N_2 amount in the original live oil is low), but the complex phase exchanges are poorly represented.

Some authors[1,3] prefer to plot a tetrahedron or pseudoquaternary diagram like that of Fig. 6.3. The front triangular face represents the classical ternary diagram with the phase envelope of the oil, and the bottom face represents the various equilibria of nitrogen with intermediate and heavy fractions of the oil. The nitrogen envelope is greater than that of the oil due to the lower solubility of nitrogen in hydrocarbons, when compared to that of methane. The phase boundary is the surface of a volume whose important areas are shaded in Fig. 6.3.

No data have been published on that type of diagram, but some experimental results are detailed in Koch's previously cited publication.[3]

6.2.2 *Minimum miscibility pressure*

(a) *Definition.* The minimum miscibility pressure (MMP) is the minimum pressure required to achieve multiple contact miscibility

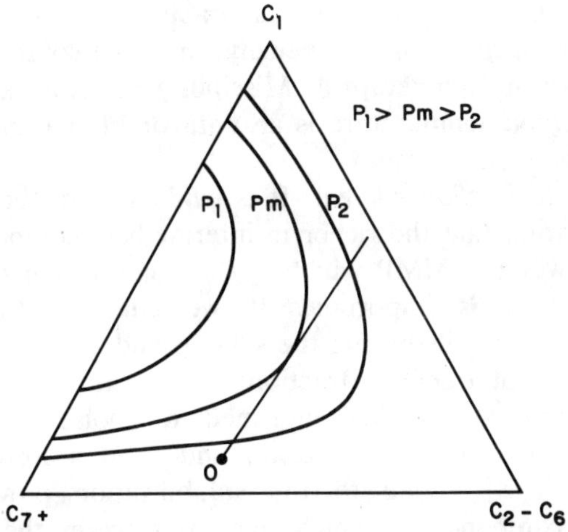

Fig. 6.4 Influence of pressure on diphasic envelope.

between injected fluid and oil. A pseudoternary diagram has to be used to conceptually define what is the minimum miscibility pressure.

Two-phase envelopes are plotted in Fig. 6.4, under various pressures at reservoir temperature. The oil composition must lie to the right of the critical tie line so that a vaporizing gas drive miscibility process may develop.

This condition is not satisfied at pressure P_2 where the phase envelope is so large that a tie line passes through the oil composition point O. Under Pm conditions, the critical tie line passes exactly through the oil composition; Pm is called the minimum miscibility pressure.

This pressure is perfectly defined by the thermodynamic data of the process, i.e. oil and gas compositions, and temperature. Variations of those parameters will affect the value of the MMP. Nevertheless, it is independent of the porous medium where the injection occurs.

(b) *Sensitive parameters*

- *Temperature.* Miscibility pressure is highly sensitive to the temperature level. Its influence will depend on the injected gas. Since the solubility of nitrogen in hydrocarbons increases with temperature under common reservoir temperatures (above 40°C),[4,5] the two-phase zone has a lower extent at higher temperature. Thus high reservoir temperature is a favorable parameter for miscibility pressure with nitrogen. Conversely, methane solubility decreases with increasing temperature. Miscibility pressure is thus higher. High reservoir temperature is an unfavorable parameter for lean hydrocarbon gas injection.

- *Intermediate fraction content.* It is evident from ternary diagram representation that the richer in intermediate components the oil is, the lower the MMP will be. The composition of this C_2–C_6 fraction itself is important. If we compare the two-phase envelope of two oils having the same pseudoternary composition, but with an intermediate fraction of different molecular weights, the oil with the higher intermediate molecular weight will have a smaller two-phase zone, and have a lower miscibility pressure. This is due to the solubilization power of intermediate components, which increases with their molecular weight.
- *Heavy fraction content.* Both the relative quantity of the heavier fraction and its molecular weight will influence the level of the MMP. It is obvious that an oil with a relatively low heavy fraction content will have a lower MMP. But if we consider two oils having the same overall composition of heavy and intermediate fractions, the miscibility pressure will be lower for the one that has the lighter heavy-fraction molecular weight, because this oil is more volatile.

(c) *Correlations.* Little work has been devoted to the predictive evaluation of minimum miscibility pressures when injecting lean hydrocarbon gas or nitrogen.

Past studies have mainly been devoted to rich hydrocarbon gas or CO_2 injection. As economical conditions change, rich hydrocarbon gas is now being upgraded and more rarely injected. Leaner hydrocarbon gas and nitrogen are becoming good candidates for injection. Since 1980 studies concerning vaporizing gas drive miscibility pressure correlations have been appearing.

Most correlations evaluating the miscibility pressure take into account: (a) the concentration of the intermediate fraction, (b) the molecular weight of the heavy fraction, and (c), of course, the temperature of the reservoir.

The ability of the gas to vaporize oil components is generally accounted for through its C_{2+} fraction content (CO_2 is included in that fraction, and nitrogen is grouped with methane).

The first correlation to be published was that of Stalkup[1] in 1983, but Stalkup himself recommends its use with caution. The basic problem is that very few experimental data are included in that work.

Two years later, Firoozabadi and Aziz[6] published a correlation based on an appreciable set of experimental data, but their work could not fully explore the temperature dependence of minimum miscibility pressure.

At the same time, Glaso[7] proposed a first set of MMP correlations for hydrocarbon gas, CO_2 and nitrogen. His work is based on Benham *et al.*[8] graphic correlations and gives MMP as a function of temperature, molecular weight of the C_{7+} fraction of the oil, mole per cent of methane in the injected gas, and molecular weight of the intermediate C_2–C_6 fraction of the gas.

CO_2 and N_2 are treated as pseudomixtures of methane–propane and methane–ethane, respectively. The paraffinic content of the oil is taken into account via a corrective factor, K.

1988 was a flourishing year for vaporizing gas drive miscibility prediction: Glaso[9] and Hudgins *et al.*[10] for nitrogen injection, Nouar and Flock[11] for lean hydrocarbon gas, and Eakin and Mitch[12] for all types of gas and miscibility processes (CO_2, N_2, hydrocarbon gas).

Hudgins *et al.* demonstrated the role of methane (and/or nitrogen) and intermediate oil cut in efficient prediction of MMP, but his work falls completely out of the range of that of Firoozabadi and Aziz's, as shown in Fig. 6.5, and does not line up with it. No explanation is given for this observation.

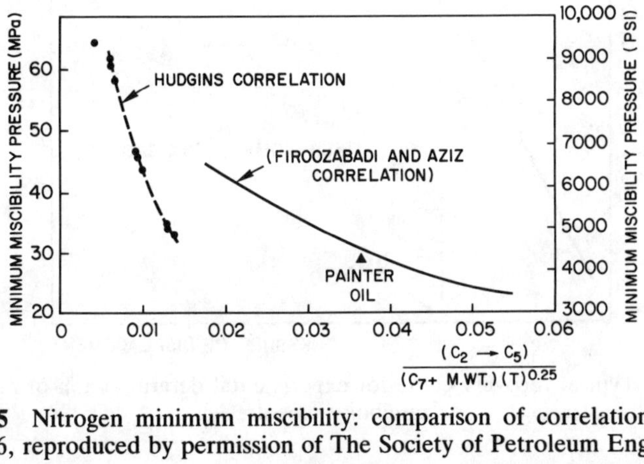

Fig. 6.5 Nitrogen minimum miscibility: comparison of correlations. (From Ref. 6, reproduced by permission of The Society of Petroleum Engineers.)

Nouar and Flock's work is somewhat different from the others because it is based directly on ternary diagram interpretation. Their correlation applies to vaporizing gas drive by a very lean hydrocarbon gas, mainly composed of methane. An equation of state evaluates the slope of the limiting tie line and the composition of the critical point for various pressures and temperatures. The parameters of the equation of state are fitted to data concerning various true ternary mixtures. For each system and temperature, the slope of the limiting tie line and critical point composition are correlated. Any more complex mixture is reduced to a pseudoternary mixture, and its MMP is interpolated between that of the nearest true ternary mixtures. It is relatively complicated to use.

Some work has also been done to evaluate the MMP by equilibrium cell calculations.[13,14] The principle of that type of calculation is to represent the dynamic mixing of solvent and oil by a succession of equilibrium cells. Initially, all the cells are filled with virgin oil. Solvent is injected in the first cell. A fraction of the resulting mixture is injected in the second cell together with fresh solvent, and so on. The recovery results are within 10–15% according to the nature of the gas (the accuracy is better for hydrocarbon gas than for CO_2 or N_2). These simulators will provide only a priori estimations of the MMP.

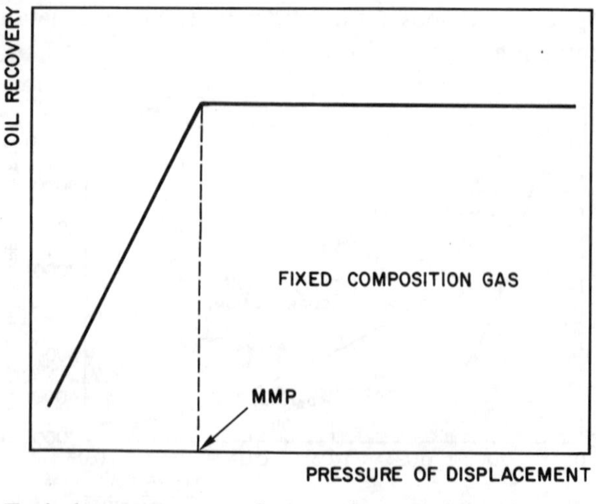

Fig. 6.6 Typical recovery curve for experimental determination of minimum miscibility pressure.

In conclusion to those correlations, much work remains to be done to predict the MMP between reservoir oils and lean gas or N_2. Extensive experimental data are not yet available to confidently recommend the use of a given correlation rather than another one. Experimental determinations are thus absolutely necessary, especially when nitrogen or flue gas has to be injected.

(d) *Experimental determination.* A standard way to determine the MMP is to perform several displacement tests with the gas to be injected and the reservoir oil, under various values of pressure, whatever the type of dynamic miscibility may be (vaporizing or condensing gas drive) (see Section 6.3).

According to the increasing level of pressure, recovery first increases, then becomes stabilized. A typical recovery curve is plotted in Fig. 6.6, where recovery after an injection of 1·2 pore volume (PV) of gas is plotted versus the pressure of displacement. The breakpoint of this curve is the minimum miscibility pressure. When there is no evidence of a precise definition for the breakpoint, the pressure level to which recovery reaches 90% of original oil in place is assumed to be the MMP.

A special experimental device has to be designed. Most proposed methods have to do with displacement tests conducted in a long stainless-steel tube (typically 0·6 cm in diameter) packed with sand of a specific particle size, called a 'slim tube'. A standard schematic diagram of the equipment is shown in Fig. 6.7.

The conventional slim tube is first saturated with oil. Gas is then injected at a specific flow rate while the outlet pressure is maintained constant. The length of the slim tube is still a subject of discussion. It may vary from 6 to 20 m. The recommended values of injection flow rates also fluctuate greatly (12 to 60 m/day). Some details on those procedures may be found in the literature.[15–17] However, these procedures can be challenged on two points:

- the model is in a horizontal position or has a very low dip angle, and a gravitational tongue may develop due to the high permeability of the model.
- thermodynamic equilibria may not be reached at each step, due to high velocity.

For those reasons, some authors recommend the use of a sand-pack tube of 1·5 cm in diameter ('fat tube'), in a vertical position.[18]

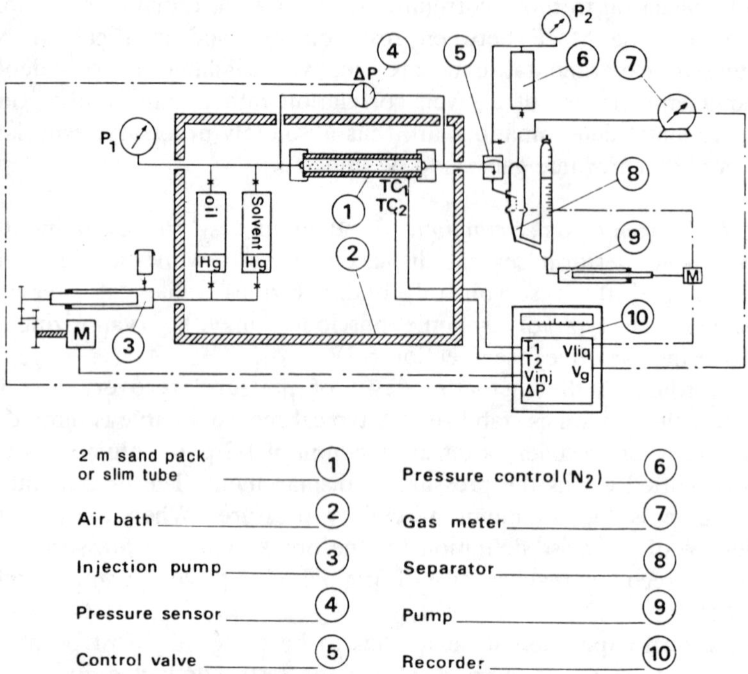

Fig. 6.7 Experimental device for MMP determination.

Relatively low flow rates may be used—6 m/day is recommended; 2 m length is sufficient for a miscible oil bank to mobilize.

Another method of experimental determination has arisen in the 1980s, called the 'rising bubble method'.[19] The principle is to let in a bubble of gas at the bottom of a visual tube cell filled with oil under reservoir temperature and pressure conditions. The tube is in a vertical position. The rising of the bubble in the liquid is observed. Three typical behaviors are observed: below, near and above the MMP where the bubble rapidly disappears. This method is still rarely used.

6.3 Condensing gas drive

Until 1980, condensing gas drive was considered to be a well-known phenomenon which arises when rich gas is injected in medium heavy oils. During the 1980s, much experimental and simulation work has shown that pure condensing gas drive as classically described in the literature does not often occur, whereas the process actually taking place is probably a dual vaporizing/condensing gas drive.

In this section 'pure' condensing gas drive mechanisms will be described, and then some papers establishing that both vaporization and condensation occur will be reviewed.

6.3.1 *Mechanism and representation by a ternary diagram*

When a rich gas is injected into a relatively heavy oil, oil and gas are initially immiscible, but a miscible bank forms through condensation of intermediate components from gas into oil.

This implies that the injected gas is rich in intermediate hydrocarbons. On the other hand, the oil to be recovered is relatively heavy (for a light oil, a lighter gas, i.e. poorer in intermediate components, is sufficient and cheaper for achieving miscibility).

In the ordinary ternary representation in Fig. 6.8, the representative point for gas, *G,* is nearer to the C_2–C_6 pole than for lean gas injection, and the representative point for oil, *O*, is close to the C_{7+}

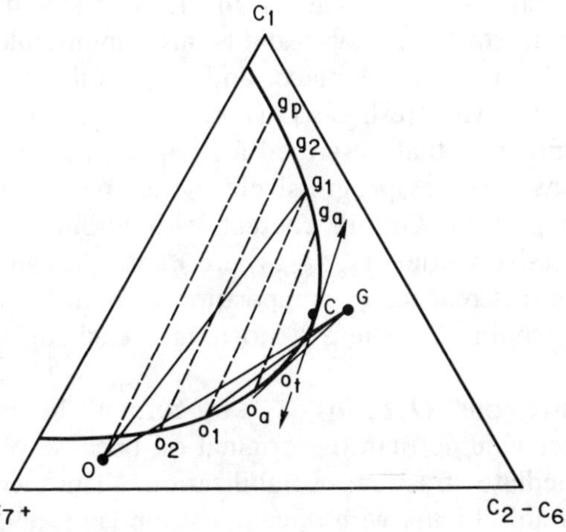

Fig. 6.8 Ternary diagram: condensing gas drive process.

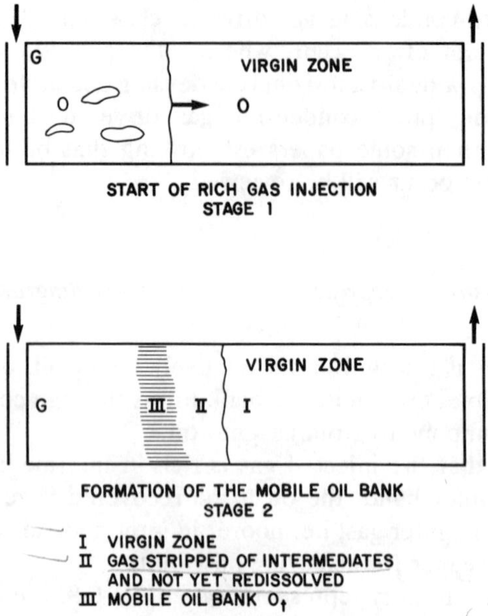

Fig. 6.9 Formation of miscible bank in condensing gas drive. (From Ref. 2, reproduced by permission of Gulf Publishing.)

pole. It is absolutely necessary for this type of miscibility to develop that the representative point of the oil will be on the left side of the critical tie line and that of the solvent on the right side of this line.

Once injection starts, because fluids are immiscible, a classical immiscible displacement takes place, and virgin oil, O, is left behind the front in contact with fresh gas G (Fig. 6.9, stage 1).

A process similar to that described for vaporizing will develop. Oil behind the front becomes progressively lighter until it reaches o_t oil composition (Fig. 6.8). Gas in contact with virgin oil at the front becomes progressively drier (G, g_1, g_2, etc.) and loses all its intermediate fractions until it reaches g_p composition as defined by the tie line passing through point O, where it no longer exchanges components with the oil.

The successive oils $O, o_1, o_2, \ldots, o_t$ formed behind the front occupy a greater volume than the original oil because of swelling due to the intermediate fractions solubilization. This will cause the mobilization of an oil bank with o_t composition (Fig. 6.9, stage 2).

If the phase equilibria between oil and gas are true two-phase liquid–vapor equilibria, i.e. no solid phase is precipitated such as, for example, asphaltenes and heavy fraction, no residual oil will be left in the reservoir if condensing gas drive occurs. Conversely, in vaporizing gas drive the resulting o_p oil is unrecoverable. The dry gas produced ahead of the bank is generally continuously dissolved in the virgin oil that it contacts.

6.3.2 *Miscibility determination*

(a) *Definition*

- *Minimum miscibility pressure* (*MMP*). Phase envelopes of oil and rich gas are drawn at various pressures in a classical ternary diagram in Fig. 6.10. Oil and gas compositions must lie on opposite sides from the tangent to the critical point so that condensing gas drive miscibility may develop. The pressure at which the critical tie line passes exactly through the gas representative point G_r, to be influenced as Fig. 6.10, is the minimum miscibility pressure. Conceptually, it is evident that the richer the gas in intermediate fraction, the lower the miscibility pressure.

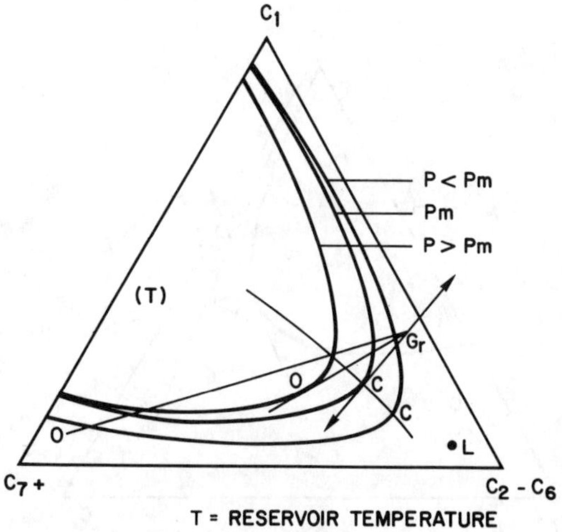

Fig. 6.10 Definition of minimum miscibility pressure with a ternary diagram: condensing gas drive.

- *Minimum miscibility composition for a given pressure level* (*MMC*). Rich gases are costly and very often they are obtained by the mixing of liquid solvents, such as liquefied petroleum gas (LPG), with less valuable gas, such as pipe gas. Even if there is an industrial or natural source of rich gas, the operator will be tempted to inject the minimum quantity of solvent and to dilute it with less expensive gas. In this case, one main parameter of operating conditions is the minimum solvent concentration required to achieve miscibility, under reservoir temperature and pressure.

  In Fig. 6.11, the composition of the rich gas is represented by point G_r, and that of lean gas by point g. Mixtures of the two are represented by the line g–G_r. The intersection of the line g–G_r with the critical tie line is the minimum miscibility composition. Cm

Practically, MMP and MMC are linked, and depending on the actual pressure of the reservoir and the economic conditions of solvent

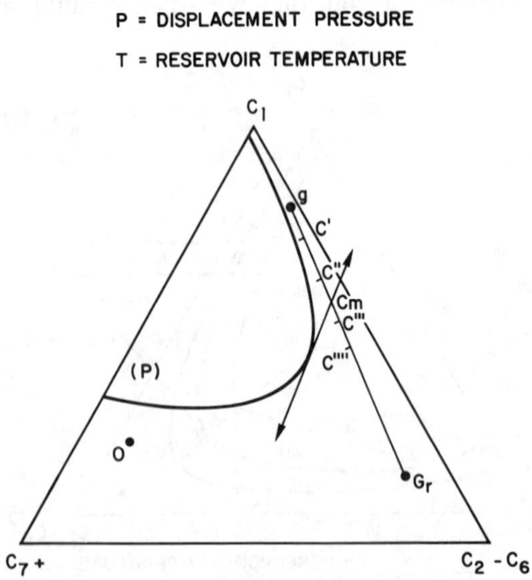

Fig. 6.11 Definition of minimum miscibility composition with a ternary diagram: condensing gas drive.

supplying, an economically optimized composition of injected fluid will be chosen.

(b) *Correlations.* Much work has been devoted to the predictive evaluation of minimum miscibility pressures, because experimental determinations represent long and tedious work. Some of the published correlations have met with great success, both for their predictive qualities and their ease of use. That of Benham *et al.*[8] was the pioneering work in MMP correlations, and it is still the most widely used for rich gas injection. The oil composition is accounted for by the molecular weight of the heavier fraction C_{5+}, and by the molecular weight of the intermediate fraction C_2–C_4. Gas composition is characterized by its content in intermediate fractions. At three levels of temperature, the minimum miscibility concentration of the intermediate fraction in the injected gas can be found at a given level of pressure, or, conversely, for a given gas composition, minimum miscibility pressure can be estimated. Benham's results are extensively given in Figs 6.12–6.14.

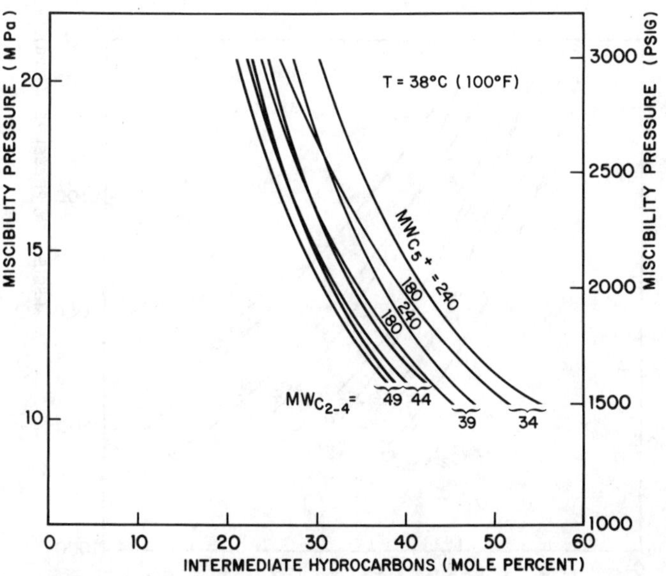

Fig. 6.12 Benham's correlation for condensing gas drive MMP (temperature = 38°C). (From Ref. 1, reproduced by permission of The Society of Petroleum Engineers.)

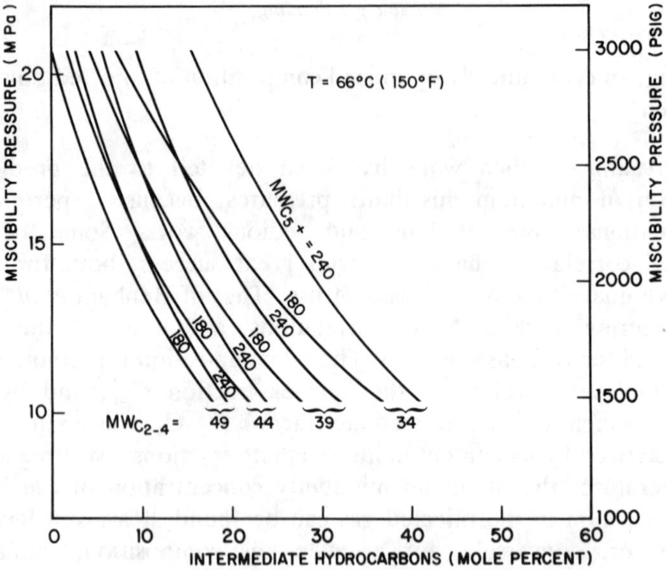

Fig. 6.13 Benham's correlation for condensing gas drive MMP (temperature = 66°C). (From Ref. 1, reproduced by permission of The Society of Petroleum Engineers.)

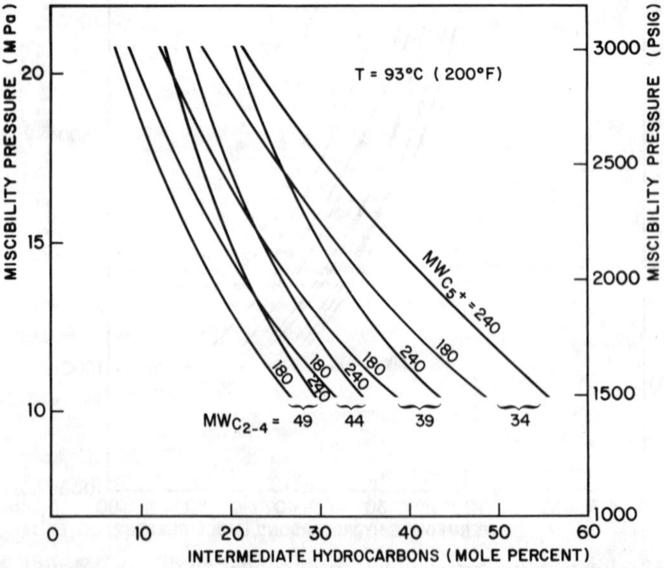

Fig. 6.14 Benham's correlation for condensing gas drive MMP (temperature = 93°C). (From Ref. 1, reproduced by permission of The Society of Petroleum Engineers.)

If the temperature is expected to be higher than 93°C no other correlation is available and extrapolation is difficult. This correlation gives only an estimate of MMP or MMC and is a good screening tool for the process. Nevertheless experimental determination will always be necessary for a precise evaluation.

(c) *Experimental determination.* The apparatus described in Section 6.2.2 for vaporizing gas drive is used. Two typical types of procedure are employed:

- *MMP determination.* As for vaporizing gas drive, several displacement tests are performed at various levels of pressure. Recovery first increases when pressure increases, then it becomes stabilized. The recovery levels are plotted versus pressure in Fig. 6.7. The breakpoint of the curve is the minimum miscibility pressure (same criteria as for vaporizing gas drive).
- *MMC determination.* Several displacement tests are performed at a single level of pressure, which is that of the reservoir, with gases of different compositions. Recovery increases while the mixture concentration in intermediate fractions increases, and then becomes stabilized. A curve similar to that of MMP is plotted, except that recovery levels are functions of concentration (Fig. 6.15). Here again the breakpoint of the recovery curve, or the 90% recovery at 1·2 PV of injected fluid criteria (cf. Section 6.2.2(d)), gives the minimum miscibility composition.

6.3.3 *Combined condensing–vaporizing mechanism*

As stated before, experimental observations and calculations with equations of state have shown that miscible displacement by rich gas injection seems to be due to a combined condensing–vaporizing gas drive mechanism rather than a classical condensing gas drive mechanism. According to Zick,[20] although minimum miscibility pressure and minimum miscibility concentration of solvent are observed in rich gas injection, true miscibility is probably not developed. A nearly miscible (but two-phase) transition zone develops and is propagated during the laboratory displacements. It is preceded by a zone of original oil in equilibrium with gas that has lost its intermediate fractions (typical of condensing gas drive), but a heavy residual oil remains trapped

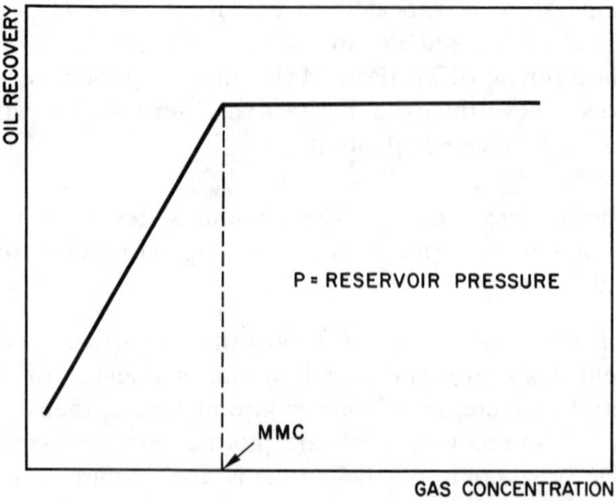

Fig. 6.15 Typical recovery curve for experimental determination of minimum miscibility composition.

backward (typical of vaporizing gas drive). Zick concludes that ternary diagrams cannot be used to elucidate the miscibility mechanism of reservoir oil and solvent when combined condensing–vaporizing occurs.

Stalkup[21] has performed a series of compositional simulations of slim-tube displacements. He demonstrates that basically, although a condensing–vaporizing mechanism is found with crude oils during rich gas injection, displacement behavior in the transition zone does not depart from traditional ternary concepts, except that a residual oil saturation is left.

Novosad and Costain[22] obtained similar results for rich gas drive at low temperature. They even found that at higher temperatures liquid extraction occurred in hot reservoirs. Simulation and experiment results show that the mechanism is controlled by solvent phase behavior under reservoir conditions.

The main conclusions of these articles are as follows:

- a condensing–vaporizing gas drive mechanism is more likely than a pure condensing gas drive when rich gas is injected into a reservoir oil;
- a pseudomiscible zone develops quite similar to that in a condensing gas drive;
- some residual oil remains trapped behind the displacement as in a vaporizing gas drive.

6.4 First contact miscible process

A first contact miscibility is obtained when both fluids are completely miscible, in all proportions, without any multiphase behavior. This is usually the case for solvents such as propane, butane or mixtures of LPG. Other solvents could be good candidates, such as organic solvents (toluene), but their high cost definitely precludes their use.

6.4.1 *Mechanism and representation by a ternary diagram*

In Figure 6.16, oil is represented by point *O*, and solvent by point *S*. The line which passes through points *O* and *S* represents the various mixtures of solvent and oil. The limiting conditions to achieve first contact miscibility between oil and solvent is that the line *O–S* does not cross the two-phase zone.

Since first contact miscible solvents are expensive, their injection is generally reduced to a slug followed by a worthless gas, mainly composed of methane.

Mixtures of solvent and chase gas are totally miscible. If the composition of chase gas is *G* (Fig. 6.16), mixtures of oil and solvent (line *O–S*) will remain miscible with *G* as long as their composition is between S_1 and *S* points.

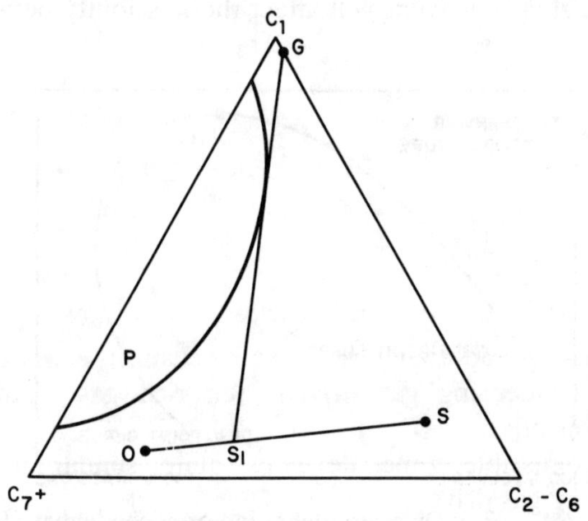

Fig. 6.16 Ternary diagram: first contact miscibility process.

The solvent content of the slug may generally be reduced to 30 or 40% and miscibility with methane is preserved. In many practical cases it has been reduced to as little as 2% of the displaceable volume without reducing the recovery efficiency.

6.4.2 *Method for determining miscibility conditions*

The pressure required for first contact miscibility between solvent and oil can be experimentally determined by establishing the P, x diagram of the oil and solvent mixtures; P being the saturation pressure of the mixture of oil and solvent with a molar content of solvent dissolved in oil equal to x. The cricondenbar, i.e. the maximum pressure of the phase envelope, must be lower than the reservoir pressure to achieve first contact miscibility (Fig. 6.17).

Another important parameter to be determined is the maximum permissible concentration of drive gas.

Pseudoternary phase behavior calculations are generally performed by using an equation of state. The main sensitive parameters of the process are the following:

- *Solvent miscibility*. Miscibility pressure between oil and solvent will decrease with increasing temperature, and increase with the molecular weight of the solvent.
- *Drive gas composition*. Composition of drive gas to be injected after the slug of solvent will affect the miscibility between solvent

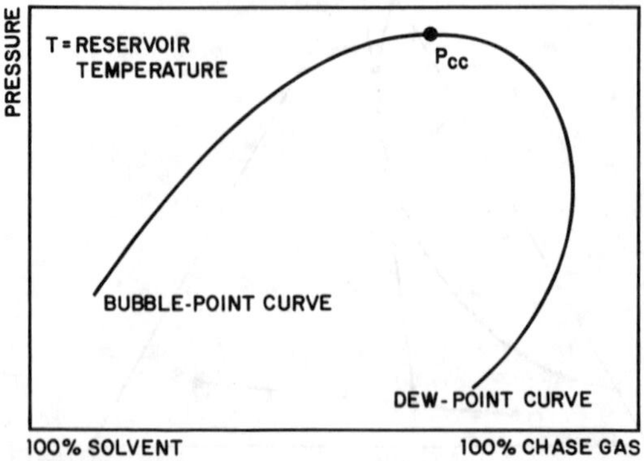

Fig. 6.17 P, x Diagram: definition of cricondenbar (P_{cc}).

and gas. Schematically, a higher content in nitrogen and/or methane will induce a higher miscibility pressure. On the contrary, a higher content of ethane and/or CO_2 will have a beneficial effect, and reduce the miscibility pressure.

- *Asphaltene content.* Propane is the typical antisolvent for asphaltenes, and in most first contact miscible determinations, the solvent to be studied will have a high propane content. The effect of that precipitation is complex. The resulting lightened oil phase is perfectly miscible with the solvent and with the original oil, so a pseudomiscible process goes on in the reservoir. However, asphaltene deposits may have a detrimental plugging effect on the rock material and in the wells. Experimental determinations should be performed to evaluate these plugging effects. For an asphaltenic oil, special care should be taken with miscibility pressure measurements, and a visual cell has to be used to determine whether precipitation occurs or not.

6.5 Miscible flooding parameters

Many flooding displacement aspects are similar to those of immiscible displacements, encountered in water injection for example (viscous fingering, volumetric efficiency, etc.). Nevertheless some specific points need to be mentioned. More complete information can be found in the literature cited.

6.5.1 *Mobility ratio*

This point is developed in Chapter 1. As far as immiscible fluids are concerned, the properties of each fluid are perfectly defined, and so is their mobility ratio. If any mixing occurs, the effective ratio of viscosities will always be smaller than that of pure fluids, and some ambiguity will remain as to its true value. The ratio will also fluctuate according to the position where it is evaluated inside the field.

Nevertheless the mobility ratio will always be higher than one. The consequences are widely developed in Section 1.2.3.

6.5.2 *Diffusion–dispersion*

As soon as two miscible fluids are in contact, the concentration of each component tends to be equalized in both fluids due to molecular diffusion. On the other hand, when fluids are in motion the presence of a porous medium influences the mixing process. Tortuosity induces a forced curvature of flooding lines and leads to a faster homogenization of the mixture.

This diffusion–dispersion phenomenon may be important in gas injection for two main reasons:

- it may influence the propagation of the miscible front;
- it is a promising way to produce oil from fractured reservoirs.

This latter point enlarges the application of gas injection, and may lead to a process widely used in the future.

Some valuable information can be found in the literature. Perkins and Johnston[23] have made a relatively complete review of the calculation of diffusion and dispersion coefficients in 1963. Stalkup[1] examines, in a whole chapter of his book, the factors affecting displacement behavior. He details the influence of dispersion on the propagation of the front. Saidi[24] explains how diffusion may participate in the extraction of oil from the matrix of a fractured reservoir.

(a) *Diffusion.* Fick's law for the equation of diffusion describes the movement of one component i through a mixture of several components, because of the concentration gradient of species i. For example, for a concentration gradient in the x direction, it may be written as follows:

$$\frac{dn_i}{dt} = -D_{0i} A \frac{dC_i}{dx} \qquad (6.2)$$

where

n_i = number of diffusing moles of species i,
t = time,
D_{0i} = molecular diffusion coefficient of species i,
A = cross-sectional area for diffusion,
C_i = molar concentration of species i,
x = distance.

Rigorously, the diffusion coefficients depend on mixture composition, but generally an average value is chosen, evaluated at a concentration of equimolar quantities of the two fluids.

Correlations for estimating molecular diffusion are available in the literature,[25,26] though high pressure evaluations are scarce.

Molecular diffusion coefficients have to be corrected for the presence of porous medium. Some authors have established a dependence via porosity and formation volume factor of the rock where the fluids are mixed:

$$D_e = \frac{D}{F\Phi} \tag{6.3}$$

D_e effective diffusion coefficient in porous medium
D molecular diffusion coefficient
F formation electrical resistivity factor
Φ fractional porosity

Equation (6.3) is based on the analogy between diffusion and electrical conductivity.[27]

Saidi[24] performed a series of experiments using rock samples with various permeabilities (from $0{\cdot}5 \times 10^{-3}$ μm^2 to a few μm^2), and found that this correction systematically overestimates the effective diffusion coefficient as demonstrated in Fig. 6.18.

(b) *Dispersion.* A dispersion mechanism may occur parallel to the gross motion, i.e. longitudinal dispersion, or perpendicular to it, i.e. transverse dispersion.

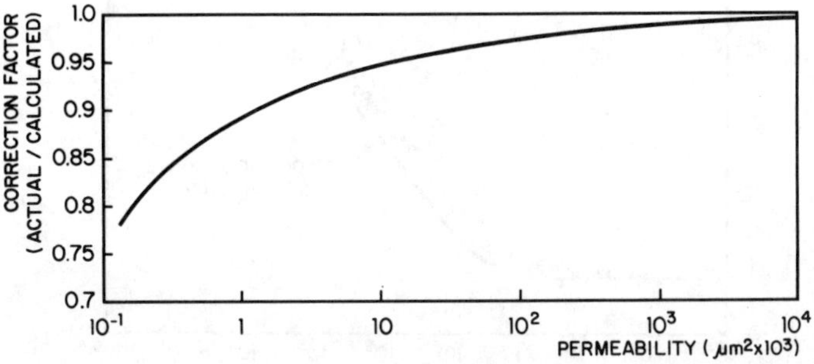

Fig. 6.18 Correction factor *FE* for calculating effective diffusion coefficient in porous media ($D_e = D/FE$) versus permeability. (From Ref. 24, reproduced by permission of Total Edition Press.)

- *Dispersion coefficient evaluation.* A satisfactory evaluation of these parameters has not yet been performed although it has been studied for a long time.[28] Perkins and Johnston[23] recommended the following equations:

—longitudinal dispersion:

$$\frac{K_l}{D_0} = \frac{1}{F\Phi} + \frac{0{\cdot}5V_i\delta d_p}{D_0}, \qquad \text{with} \qquad \frac{0{\cdot}5V_i\delta d_p}{D_0} < 50 \tag{6.4}$$

where

K_l = coefficient of longitudinal dispersion
D_0 = molecular diffusion coefficient
V_i = interstitial velocity
δ = heterogeneity factor
d_p = average particle diameter

—transverse dispersion:

$$\frac{K_t}{D_0} = \frac{1}{F\Phi} + \frac{0{\cdot}0157V_i\delta d_p}{D_0}, \qquad \text{with} \qquad \frac{0{\cdot}0157V_i\delta d_p}{D_0} < 10^{-4} \tag{6.5}$$

where

K_t = coefficient of transverse dispersion
V_i = interstitial velocity

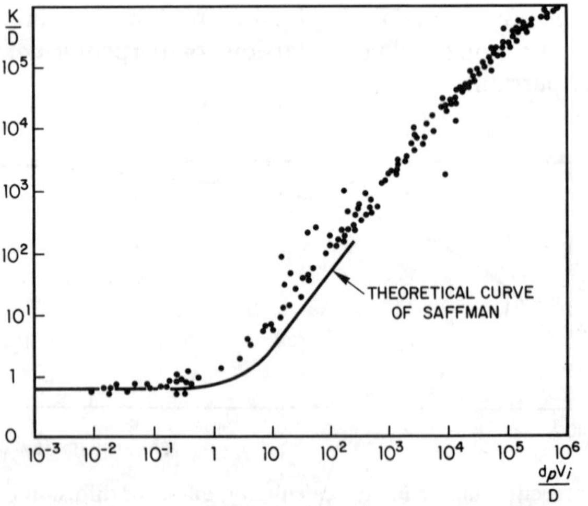

Fig. 6.19 Effect of Peclet number on longitudinal dispersion coefficient in porous media. (From Ref. 29, reproduced by permission of Editions Technip.)

Some experimental work has been done to measure longitudinal dispersion. It is reviewed in Fig. 6.19.[29,30] For a small Peclet number (which is the convection to diffusion force ratio: $P_e = d_p V_i / D$), dispersion is equal to molecular diffusion. Equations (6.4) and (6.5) have been established for fluids having the same density and viscosity. They give only approximate values of true dispersion coefficients. In particular, mobility and density ratios, which are never equal to one, will influence those coefficients.

- *Dispersion coefficient effects on miscible displacements.* Since oil-to-solvent mobility ratios are generally higher than one, viscous instabilities (fingers) develop between solvent and oil. Both transverse and longitudinal dispersion coefficients will affect those instabilities. Transverse dispersion will have a strong effect, mainly because it applies to a large surface area, i.e. that of the sides of the fingers of instabilities. Longitudinal dispersion will affect only the smaller surface of finger tips of instabilities.

  Transverse dispersion has been proved to play an important role in the growth of fingers. It is responsible for the spreading of fingers and the coalescence of small fingers to large ones. It has a beneficial effect on sweepout efficiency. Conversely, when dynamic miscibility is found, it can dilute the solvent front concentration below miscibility requirements. The effect is then detrimental to the process.

6.6 Field applications

6.6.1 *Field equipment*

The main equipment for hydrocarbon or nitrogen flooding process is a compression plant for the injected gas and a separation plant for the produced gas. This separation plant has two objectives; first to prevent any pollution of the ambient atmosphere, and second to allow the recycling of the produced gas into the injection wells, with a possible valorization of the byproducts.

For flue gas (then CO_2-containing gas) injection, specific anticorrosive equipment (Fiberglas tubings, for example) is needed due to corrosion problems which arise with CO_2 in the presence of water.

Gas delivery is in two modes, either made on the spot, or piped.

6.6.2 *Injection scheme*

Depending on the characteristics of the field to be flooded, gas is injected according to one of the following processes:

- *Continuous injection.* Continuous injection is the simplest way to proceed. The gas is injected into the reservoir until the production gas/oil ratio makes it uneconomical to continue. This mode of injection is applied only when large quantities of gas are available at a very low price.
- *Slug injection.* A slug of miscible gas is injected and a less expensive chase gas follows. The chase gas must remain miscible with the slug to achieve the best final recovery.
- *Water alternate gas (WAG) injection.* Slugs of water and gas are alternately injected in the reservoir. The main objective is to improve the sweep efficiency by controlling gas mobility (see also Chapter 7).

6.6.3 *Field operations*

Gas injection under miscible or near-miscible conditions has been implemented since the early 1960s with successful technical and economical results.

One of the first pilots was that of Hassi Messaoud,[31] which started in 1964 and in which miscibility was observed. Gas injection was still going on at field scale in the late 1980s. Another pioneering project was that of Intisar (Libya) in 1969, and it is still going on in 1990 with a daily production of 6000 m^3.[32]

However, gas injection is not a 'past' process: new projects are beginning every year. Reference 32 indicates that the biggest increase in EOR production deals with hydrocarbon injection with an additional recovery that doubled between 1988 and 1990. This clearly demonstrates the industrial importance and efficiency of the process in the future of EOR.

6.7 References

1. Stalkup, Jr, F. I., Miscible Displacement, Monograph vol. 8, L. Henry, Doherty Memorial Fund of AIME, Soc. of Petroleum Engineers, Richardson, Texas 1983.

2. Latil, M., *Enhanced Oil Recovery,* Gulf Publishing, Houston, 1980, pp. 99–121.
3. Koch, Jr, H. A. & Hutchinson, Jr, C. A., Miscible displacements of reservoir oil using flue gas. *Petroleum Trans. AIME,* **213** (1958) 7–10.
4. Rushing, M. D., Thomasson, B. C., Reynolds, B. & Crawford, P. B., Miscible displacement with nitrogen. *Petroleum Engineer Int.,* (November 1977) 26–30.
5. Rushing, M. D., Thomasson, B. C., Reynolds, B. & Crawford, P. B., Nitrogen may be used for miscible displacement in oil reservoirs. *J. Petroleum Technol.,* **30**(12) (1978) 1715–16.
6. Firoozabadi, A. & Aziz, K., Analysis and correlation of nitrogen and lean gas miscibility pressure. SPE Paper No. 13669, California Regional Meeting of Soc. Petroleum Engineers, Bakersfield, 27–29 March 1985.
7. Glaso, O., Generalized minimum miscibility pressure correlation. *Soc. Petroleum Engineers J.,* **25**(6) (1985) 927–34.
8. Benham, A. L., Dowden, W. E. & Kunzman, W. J., Miscible fluid displacement—prediction of miscibility. *Petroleum Trans. AiME,* **219** (1960) 219–37.
9. Glaso, O., Miscible displacement: recovery tests with nitrogen. SPE Paper No. 17378, SPE/DOE Sixth Symposium on Enhanced Oil Recovery of Soc. Petroleum Engineers and US Department of Energy, Tulsa, 17–20 April 1988.
10. Hudgins, D. A., Llave, F. M. & Chung, F. T. H., Nitrogen miscible displacement of light crude oil: a laboratory study. *SPE Reservoir Engineering,* **5**(1) (1990) 100–6.
11. Nouar, A. & Flock, D. L., Prediction of the minimum miscibility pressure of a vaporizing gas drive. *SPE Reservoir Engineering,* **3**(1) (1988) 182–8.
12. Eakin, B. E. & Mitch, F. J., Measurements and correlation of miscibility pressures of reservoir oils. SPE Paper No. 18065, 63rd Ann. Techn. Conf. and Exhibition of Soc. Petroleum Engineers, Houston, 2–5 October 1988.
13. Van Quy, N., Simandoux, P. & Corteville, J., A numerical study of diphasic multicomponent flow. *Soc. Petroleum Engineers J.,* **12**(2) (1972) 171–84.
14. Emanuel, A. S., Behrens, R. A. & McMillen, T. J., A generalized method for predicting gas/oil miscibility. *SPE Reservoir engineering,* **1**(5) (1986) 463–72.
15. Rathmell, J. J., Stalkup, Jr, F. I. & Hassinger, R. C., A laboratory investigation of miscible displacement by carbon dioxide. SPE Paper No. 3483, 46th Annual Techn. Conf. and Exhibition of Soc. Petroleum Engineers, New Orleans, 3–6 October 1971.
16. Holm, L. W. & Josendal, U. A., Discussion of determination and prediction of CO_2 minimum miscibility pressure. *J. Petroleum Technol.,* **32**(5) (1980) 870–1.
17. Yellig, W. F. & Metcalfe, R. S., Determination and prediction of CO_2 minimum miscibility pressures. *J. Petroleum Technol.,* **32**(1) (1980) 160–8.
18. Bardon, C. and Denoyelle, L., *Miscible Process.* Master Lecture, Second INTEVEP Int. Seminar on Advances in Oil Recovery Techniques, Los

Teques, Venezuela, Dec. 2–4, 1981. Proceedings edited and published by INTEVEP S. A., Caracas, Venezuela, **1**, 193–261.

19. Christiansen, R. L. & Kim, H., Rapid measurement of minimum miscibility pressure using the rising bubble apparatus. SPE Paper No. 13114, 59th Annual Techn. Conf. and Exhibition of Soc. Petroleum Engineers, Houston, 16–19 September 1984.
20. Zick, A. A., A combined condensing/vaporizing mechanism in the displacement of oil by enriched gases. SPE Paper No. 15493, 61st Annual Techn. Conf. and Exhibition of Soc. Petroleum Engineers, New Orleans, 5–8 October 1986.
21. Stalkup, Jr, F. I., Displacement behavior of the condensing/vaporizing gas drive process. SPE Paper No. 16715, 62nd Annual Techn. Conf. and Exhibition of Soc. Petroleum Engineers, Dallas, 27–30 September 1987.
22. Novosad, Z. & Costain, T. G., Mechanisms of miscibility development in hydrocarbon gas drives: new interpretation. SPE Paper No. 16717, 62nd Annual Techn. Conf. and Exhibition of Soc. Petroleum Engineers, Dallas, 27–30 September 1987.
23. Perkins, T. K. & Johnston, O. C., A review of diffusion and dispersion in porous media. *Soc. Petroleum Engineers J.*, **63**(1) (1963) 70–84.
24. Saidi, A. M., *Reservoir Engineering of Fractured Reservoirs,* Total Edition Press, 1987.
25. Crank, J., *Mathematics of Diffusion,* 1st edn, Oxford University Press, 1957.
26. Reid, R. C. & Sherwood, T. K., 2nd edn, The properties of gases and liquids, McGraw Hill Book Company—Chemical Engineering Series, 1958.
27. Klinkenberg, L. J., Analogy between diffusion and electrical conductivity in porous rocks. *Bull. GSA*, **62** (1951) 559.
28. Taylor, G. I., Dispersion of soluble matter in solvent flowing slowly through a tube. *Proc. Royal Soc.*, **A219** (1953) 186–203.
29. Pfannkuch, H. O., Contribution à l'étude des déplacements de fluides miscibles dans un milieu poreux. *Revue de l'Institut Français du Pétrole,* **18**(2) (1963) 215–70.
30. Saffman, P. G., A theory of dispersion in a porous medium. *J. Fluid Mechanics,* **6**(3) (1959) 321–49.
31. Pottier, J., Delclaud, C., Leduc, J., d'Herbes, J. & Thomère, R., Injection de gaz miscible à haute pression à Hassi Messaoud. World Petroleum Conference, Mexico City, April 1967.
32. Anon. CO_2 and HC Injection Lead EOR Production Increase. *Annual Production Report, Oil and Gas J.*, **88**(17) (1990) 49–82.

7 Carbon dioxide flooding

Mark A. Klins
Chevron USA Inc., Hobbes, New Mexico, USA

and

Charles P. Bardon
Institut Français du Pétrole, Rueil-Malmaison, France

7.1	**Introduction**	216
7.1.1	Physical properties of CO_2	216
7.1.2	CO_2 solubility in crude oil	220
7.1.3	CO_2 solubility in water	221
7.1.4	Rock–fluid interactions	223
7.2	**Mechanisms of miscibility**	225
7.2.1	Phase behavior	225
7.2.2	Minimum miscibility pressure (MMP)	228
7.2.3	Binary screening guides	231
7.3	**Partially miscible CO_2 displacements**	233
7.3.1	'Near miscibility' process with light-to-medium oils	233
7.3.2	Heavy oils	233
7.3.3	Role of diffusion	234
7.4	**Sweep efficiency**	234
7.4.1	Alternate CO_2–water injections	235
7.4.2	Foams	235
7.5	**Field implementation**	236
7.5.1	Injection schemes	236
7.5.2	Response	236
7.5.3	Operating problems	237
7.5.4	Asphaltene precipitation	237
7.5.5	Concluding remarks	238
7.6	**References**	239

7.1 Introduction

Because of its widespread availability and positive oil displacement characteristics, research scientists have been attracted to CO_2 for enhancing crude oil recovery.

Work during the 1950s helped to quantify those leading mechanisms that favor CO_2 flooding, namely: oil swelling, viscosity reduction and crude vaporization (potentially developing miscible displacement). However, in the 1960s interest in carbon dioxide flooding still remained limited in comparison to the industry-favored miscible hydrocarbon gas processes. Additional research done from 1960 to 1970 helped define the main displacement characteristics of the process, identify amenable reservoir characteristics and determine the influence of major design parameters. Then, in the 1970s widespread field testing occurred.

A recent article reviews this large number of CO_2 projects.[1] Worldwide there are currently 64 projects of which 90% are located in the United States. Three projects are underway in Canada, two in Hungary and one each in Trinidad and Turkey. Given a current injection rate of approximately 125 000 tonnes of gas per day, an incremental oil response of 20 000 tonnes of oil per day and a capital investment of approximately $5 billion, development costs have averaged about $3 per barrel of incremental oil.

Given the present level of activity and investment, CO_2 flooding seems to be a real success story; even though considerable progress remains to be made so as to reduce the technical risks and to facilitate the preparation and implementation of a specific project.

7.1.1 *Physical properties of CO_2*

Under ambient conditions, CO_2 is a colorless almost odorless gas. It becomes toxic when its content in the atmosphere reaches 10%. Its properties under standard conditions (1·01 MPa, 0°C) are:

- Molecular weight 44·010 g/mol
- Specific gravity with respect to air 1·529
- Density 1·95 kg/m^3
- Viscosity 0·0137 mPa/s

The phase diagram of CO_2 is shown in Fig. 7.1. Its critical properties

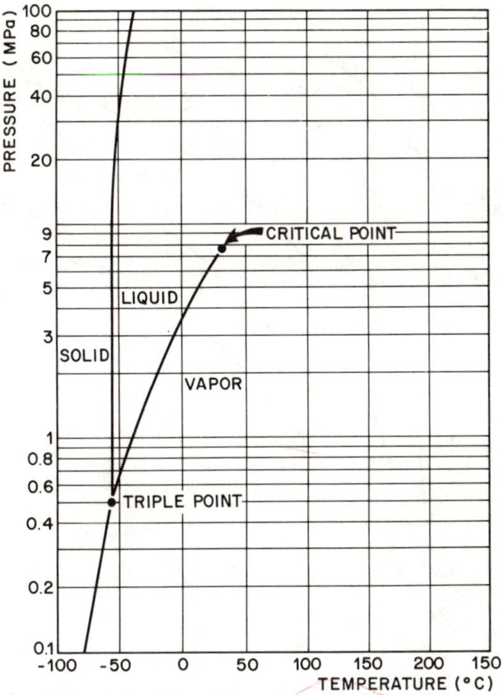

Fig. 7.1 Phase diagram for carbon dioxide. (From Ref. 3, reproduced by permission of the US Department of Energy.)

are:

- $P_c = 7{\cdot}39$ MPa
- $T_c = 304$ K (31·1°C)
- $V_c = 94\ \text{cm}^3/\text{mol}$

with a triple point pressure of 0·51 MPa and temperature of −56·6°C.

Having a critical temperature of 31·1°C (304·2 K), CO_2 behaves as a vapor under most reservoir conditions.

Figure 7.2[3] shows that fluid density increases with pressure at temperatures above critical conditions, but abrupt discontinuities appear at temperatures below the critical region. The dotted line gives the average density of CO_2 under 'average' reservoir conditions considering hydrostatic pressure and a conventional temperature gradient of 3°C per hundred meters of increasing depth. Note that its behavior is that of a very dense gas with densities, in many cases,

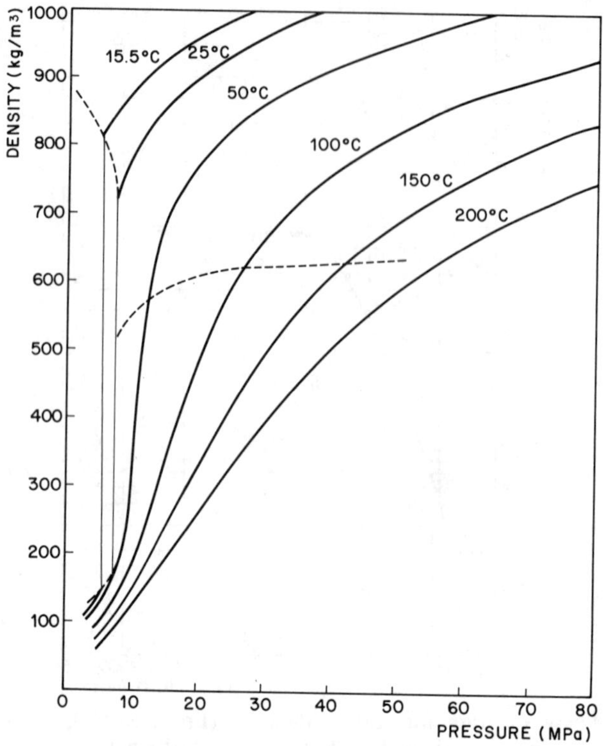

Fig. 7.2 CO_2 density versus pressure. (From Ref. 3, reproduced by permission of the US Department of Energy.)

similar to the resident crude oil. A 19 constant polynomial representation of this property has been presented by Reynolds.[4]

In addition, gas density can be calculated from the traditional real gas law such that:

$$\rho = \frac{PM}{ZRT} \tag{7.1}$$

where P is the pressure, M is the molecular weight, T is the absolute temperature, R is the universal gas constant and Z is the gas supercompressibility factor[5–7] whose variations with pressure and temperature is given in Fig. 7.3.

The viscosity of CO_2 strongly depends on pressure and temperature,[6] as shown in Fig. 7.4. Note that gas viscosity increases

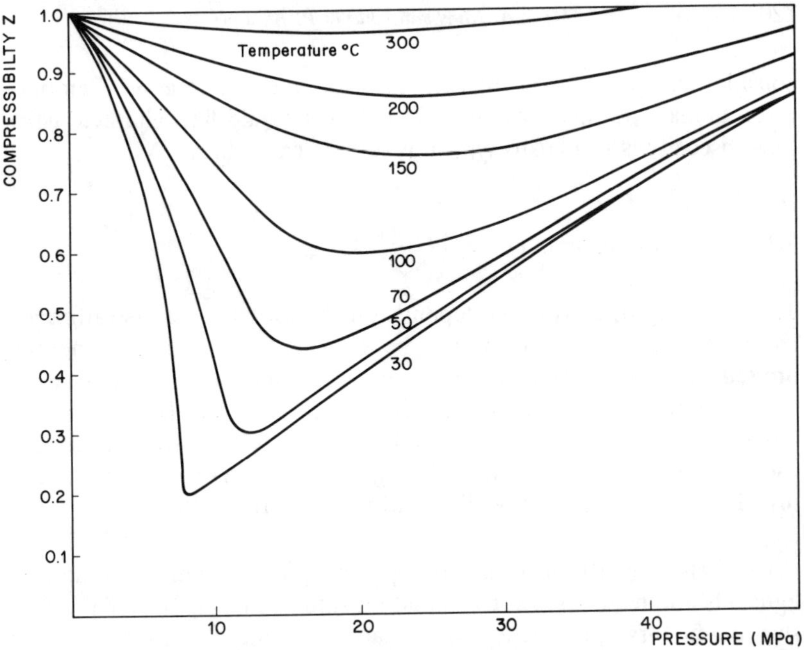

Fig. 7.3 Compressibility factor. (From Ref. 7.)

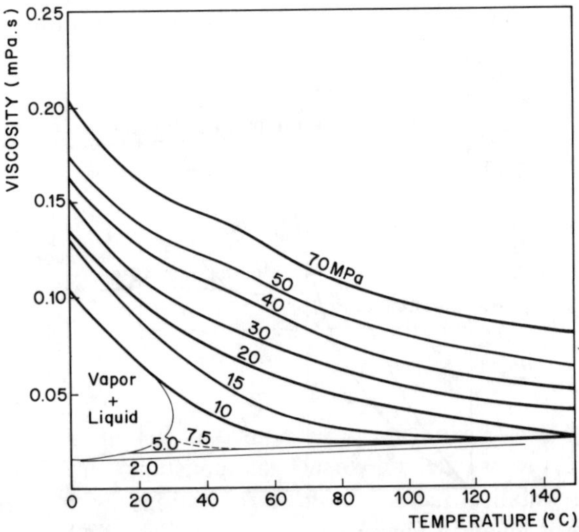

Fig. 7.4 Viscosity of CO_2 versus temperature. (From Ref. 6, reproduced by permission of The Society of Petroleum Engineers.)

considerably when pressure increases for a given reservoir temperature. Unlike density, where CO_2 exhibits liquid-like characteristics, CO_2 viscosity is similar to typical reservoir gas values.

7.1.2 CO_2 *solubility in crude oil*

The solubility of CO_2 in oil depends on the pressure, temperature and characteristics of the oil. As can be seen in Fig. 7.5, it is sometimes limited to a saturation value as pressure increases.[8–10] Figure 7.6 expresses the CO_2 solubility in terms of gas/oil ratio (GOR) for two oils with a specific gravity of 881·6 kg/m³ (29° API, in units commonly used in oil industry) and 982·0 kg/m³ (12.2° API) at 65°C. CO_2 dissolution results in oil swelling and viscosity reduction.[11]

(a) *Oil swelling (from 10 to 100%).* The phenomenon is greater for light oils and leads to lower residual oil saturation. Figure 7.7 gives the swelling factors as a function of the saturation pressure for the two oils mentioned above.

(b) *Viscosity reduction.* Viscosity reduction becomes greater as saturation pressure increases (Fig. 7.8). For any given saturation pressure

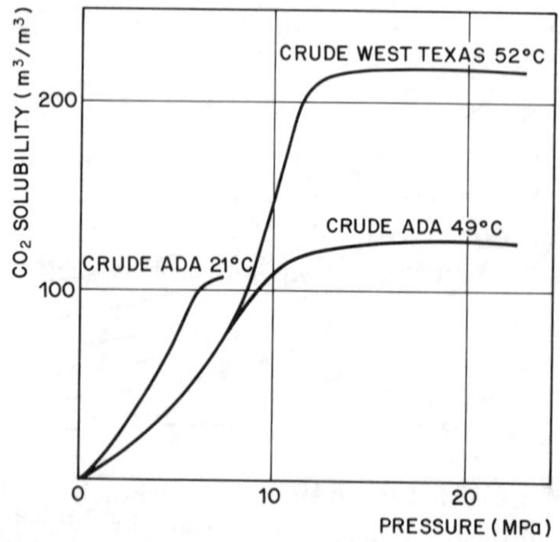

Fig. 7.5 CO_2 solubility in crude oil. (From Ref. 9, reproduced by permission of The Society of Petroleum Engineers.)

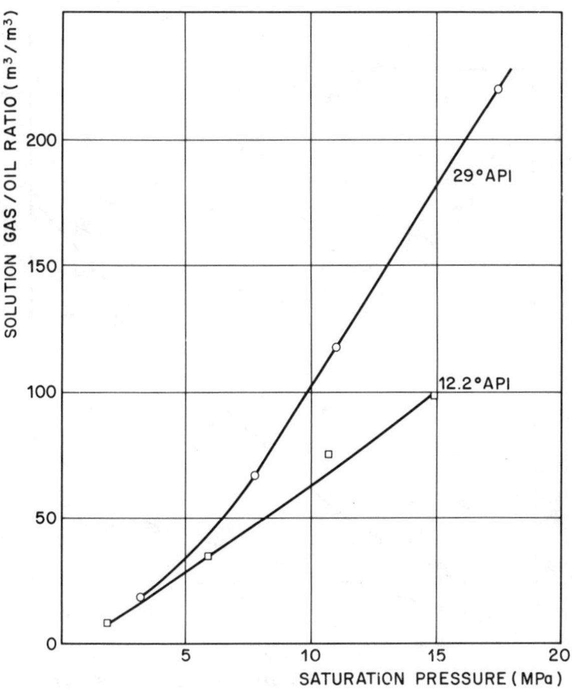

Fig. 7.6 GOR versus saturation pressure. (From Ref. 11, reproduced by permission of Editions Technip.)

the viscosity reduction becomes greater when the original oil viscosity is higher. As an example, the 12·2° API oil viscosity decreases from 900 mPa/s to 40 mPa/s, as saturation pressure increases to 15 MPa, whereas for the 29° API crude it decreases from 6 to 0·5 mPa/s. Thus, the viscosity reduction and its effect on the mobility ratio is more significant in medium and heavy oils and not as great in low viscosity oils.

7.1.3 *CO_2 solubility in water*

The solubility of CO_2 in formation water is much greater and hence more important in design considerations than the solubility of natural gas. This is an important factor that cannot be neglected in the simulation process. It is particularly true when CO_2 is injected after a

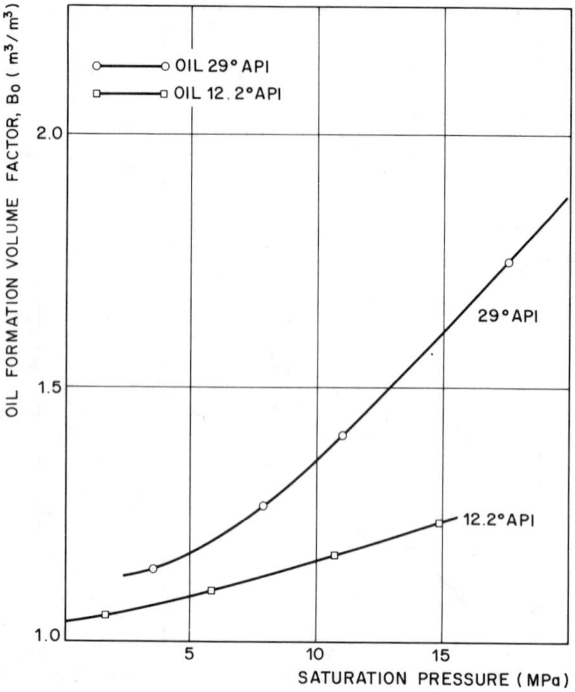

Fig. 7.7 Swelling factor versus saturation pressure. (From Ref. 11, reproduced by permission of Editions Technip.)

waterflooding or when CO_2 is being injected alternately with water (water alternate gas process, currently called the WAG process).

Unlike what happens with CO_2–oil solubilities, the pressure has a much weaker effect. The solubility increases significantly from atmospheric pressure up to approximately 7 MPa. Above 7 MPa, practically no more dissolution occurs. The solubility is also sensitive to temperature, and decreases with rising temperature. A thorough study of CO_2 dissolved in distilled water was presented by Dodds *et al.*[12] and is shown in Fig. 7.9. An increase in salinity of the water decreases the solubility as demonstrated by McRee. (Fig. 7.9).[13]

(a) *Water viscosity*. The effect of carbonation on water viscosity was investigated by Tumasyan *et al.*[14] Their report showed that CO_2 solutions prepared in filtered tap water at 10 MPa and 15 MPa, and 20°C, increased the water viscosity by 18·9% and 27·3%, respectively. This effect is shown in Fig. 7.10.

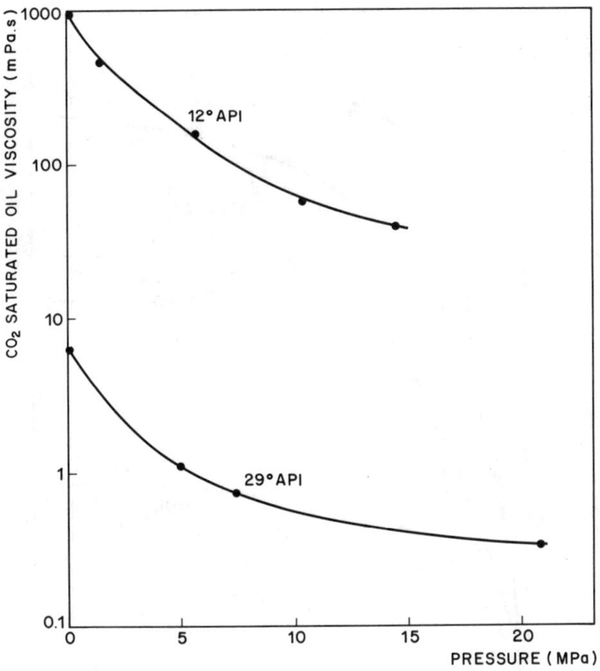

Fig. 7.8 Viscosity of CO_2–crude oil mixtures versus saturation pressure. (From Ref. 11, reproduced by permission of Editions Technip.)

7.1.4 *Rock–fluid interactions*

CO_2 reacts with wellbore hardware and reservoir mineral mainly by the carbonic acid it forms with water, according to the reversible reaction

$$CO_2 + H_2O \rightleftharpoons H_2CO_3 \tag{7.2}$$

In clay-laden rocks it stabilizes the clays by lowering the pH, thus preventing their swelling.

In carbonates, it can improve injectivity by partial dissolution of the rock, mainly in the vicinity of the injection well according to the following reactions:

$$H_2CO_3 + CaCO_3 \rightleftharpoons Ca(HCO_3)_2 \tag{7.3}$$

$$H_2CO_3 + MgCO_2 \rightleftharpoons Mg(HCO_3)_2 \tag{7.4}$$

The bicarbonates formed are quite soluble in water.

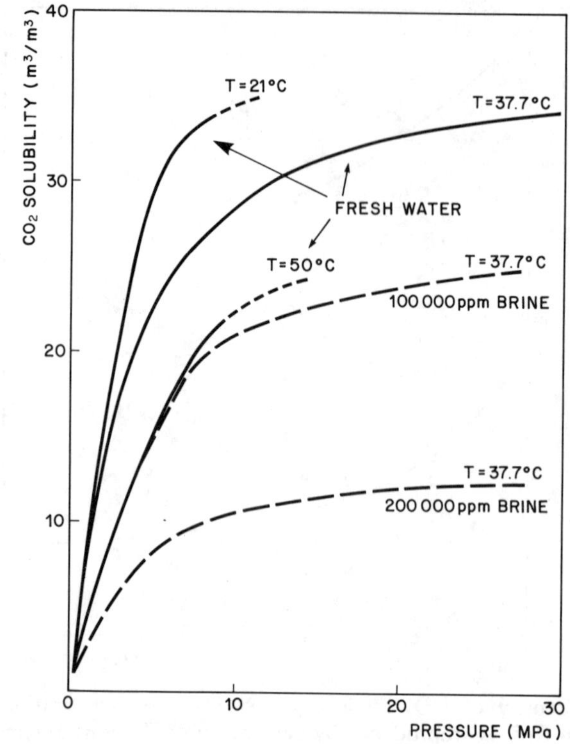

Fig. 7.9 CO_2 solubility in water versus pressure and temperature. (From Ref. 13.)

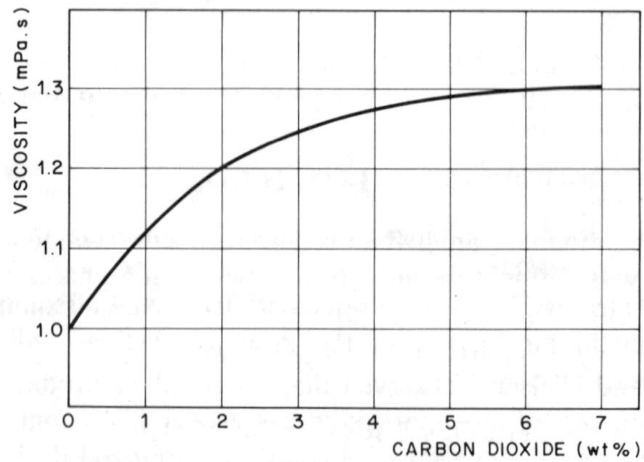

Fig. 7.10 Viscosity of CO_2–water mixtures. (From Ref. 14.)

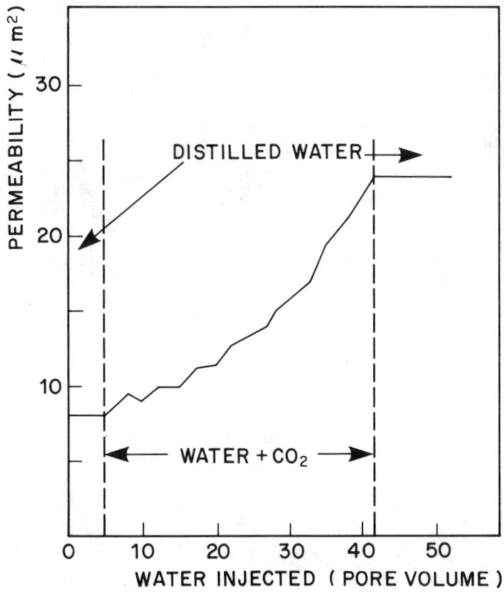

Fig. 7.11 Increase in rock permeability due to CO_2 injection. (From Ref. 15, reproduced by permission of Gulf Publishing.)

Figure 7.11 from Bardon[15] shows the three-fold increase in permeability resulting of flooding a calcareous core sample by 40 pore volumes of carbonated water. However, reduced injectivity can also occur if liberated fines from the dissolution process plug downstream pore throats. Testing should be considered before any full-scale implementation of the CO_2–water injection process.

7.2 Mechanisms of miscibility

7.2.1 *Phase behavior*

When large amounts (larger than required to saturate the oil) of CO_2 are mixed with oil, intense mass transfers between phases occur. CO_2 condenses into the oil phase, whereas the light and medium components of the oil vaporize into the gas phase.

Menzie and Nielsen[16] observed that oil could be produced successfully by a process of oil vaporization using CO_2. They found that over half of the original stock tank oil could be vaporized during multiple contacts with CO_2.

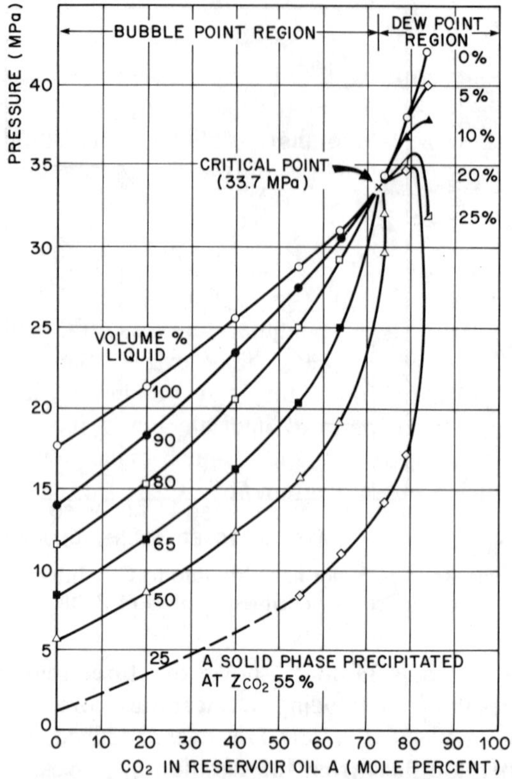

Fig. 7.12 Phase behavior of CO_2–crude oil A mixtures (classical system). (From Ref. 17, reproduced by permission of The Society of Petroleum Engineers.)

The injection–vaporization process leads to a gradation of fluids within the cross-section of the reservoir, ranging from virgin oil to 100% CO_2 vapor. One useful way of displaying this compositional information is through the use of phase diagrams. Figure 7.12 shows a typical phase diagram (P, x), obtained by preparing oil–CO_2 mixtures with different CO_2 contents and by varying the pressure for each mixture.[17] At a given CO_2 concentration lower than the critical point, an increase in pressure reduces the volume of the vapor phase. If the CO_2 concentration is higher than the critical point, a rise in pressure increases the volume of the CO_2 phase because it enhances the transfer of oil components into the CO_2 phase.

Temperature has the opposite effect of pressure but also displaces the critical point toward higher CO_2 concentrations. In the two-phase region, the vapor and liquid phases are separated and analyzed. An original experimental technique has been developed by Orr and Silva[18] to determine the equilibrium phase composition by continuous multiple-contact experiments. In these experiments, densities and compositions are measured simultaneously and rapidly for two phases in equilibrium.

In some cases multiple phases can occur. It was generally considered that such complex systems (up to four phases: 1 vapor, 2 liquid, 1 solid) could exist at low temperatures (<50°C). Reservoirs with temperatures lower than 50°C may extract the crude lower ends forming CO_2-rich liquid mixtures. A number of investigators have observed this phenomenon of multiple liquid phases.[5,6,19] However, similar behavior has been found at much higher temperatures. Figure 7.13 shows such a phase diagram at 85°C.[20]

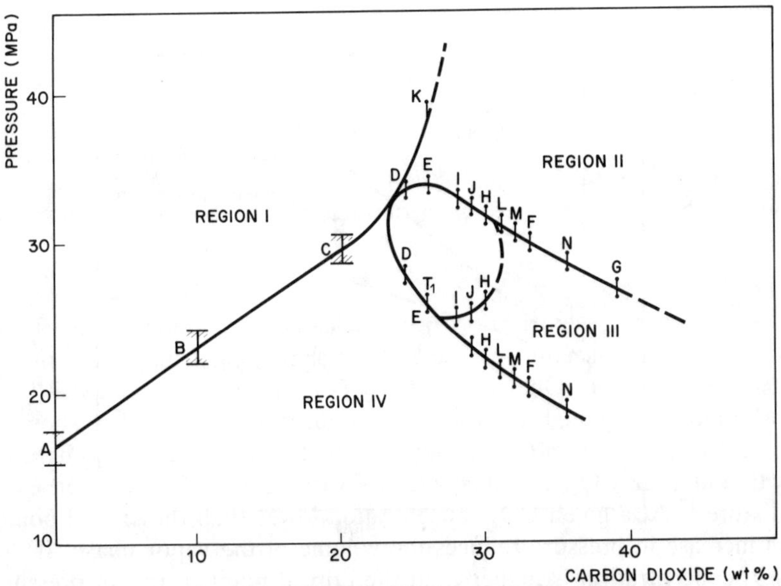

Fig. 7.13 Phase behavior of CO_2–crude oil B mixtures (complex system). (From Ref. 20, reproduced by permission of Agip SpA.)

7.2.2 *Minimum miscibility pressure (MMP)*

Oil displacement by CO_2 injection into a reservoir relies on a number of mechanisms related to the phase behavior of CO_2–crude oil mixtures. This displacement can be miscible or immiscible depending on reservoir pressure, temperature and crude oil composition.

Klins[5] has classified the dominant displacement characteristics for a given CO_2 displacement falling into one of the five regions as shown in Fig. 7.14:

Region I: low pressure applications (generally immiscible)
Region II: intermediate pressure, high temperature applications
Region III: intermediate pressure, low temperature applications
Region IV: high pressure applications (miscible displacement)
Region V: high pressure, low temperature (liquid) applications

Figure 7.14 shows the wide range of possible applications of CO_2 injection. Of course, the most attractive case is when miscibility, by first or multiple contact, can be obtained (see Chapter 6). This is generally true for light and medium oils at pressures much lower than with dry gas.

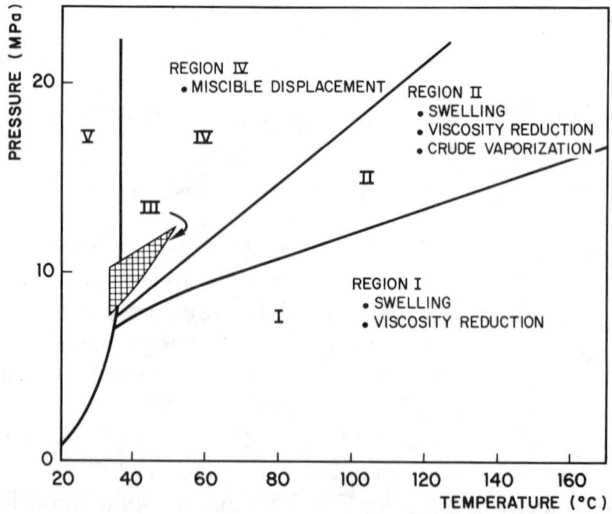

Fig. 7.14 Effect of reservoir temperature and pressure on CO_2 injection displacement mechanisms. (From Ref. 5, reproduced by permission of The International Human Resources Development Corporation.)

(a) *Formation of the miscible front.* The injected CO_2 mixes with the oil by dissolution up to the saturation pressure. Then a mixing zone is developed over a large area. During displacement of CO_2 within the porous medium, a rapid mass transfer between oil and CO_2 takes place by fractionation of oil. The frontal part of the mixing zone thus becomes progressively richer in light hydrocarbon fractions so that part of the gas phase and part of the oil phase will no longer be separated by an interface and will thus become miscible (multiple contact miscibility).

Finally, behind the front the oil becomes progressively heavier, and although saturated in CO_2, it has relatively low mobility due to the loss of all its light and medium components.

If the live oil has a high associated gas content (i.e. a bubble pressure close to reservoir pressure) methane is vaporized ahead of the gas phase and progresses in the reservoir in front of the CO_2–live oil mixing zone.

(b) *Correlations.* As with pure hydrocarbon systems for which Benham (see Chapter 6, Ref. 8) established correlations, many authors have tried to propose correlations for determining the MMP. The general feature is as follows: as the oil specific gravity increases, MMP increases, reflecting the reduced content of vaporizable hydrocarbons in the crude oil. Higher reservoir temperatures result in a higher MMP, with all other parameters being equal.

Studies by Yellig and Metcalfe[21] and Holm and Josendal[22] estimate minimum miscibility pressure as functions of reservoir temperature and oil composition. The Yellig and Metcalfe correlation depends only on the temperature, while Holm and Josendal's criteria determine miscibility as a function of reservoir temperature and molecular weight of the C_{5+} oil fraction. Also, Holm and Josendal point out that heavier gravity oil (as C_{5+} increases) requires notable higher minimum miscibility pressure. Their correlation which is limited to the molecular weight of the C_{5+} oil fraction of around 240 has been extended up to the molecular weight of 340 by Mungan (Figure 7.15).[23] A summary of MMP correlations has been presented by Enick *et al.*[24]

If the gas injected is not pure CO_2 but contains some impurities such as methane or nitrogen, the minimum miscibility pressure is increased all the more as the impurity content is higher. One example by Stalkup[6] shows miscibility pressure increasing from 8·3 to 14·0 MPa

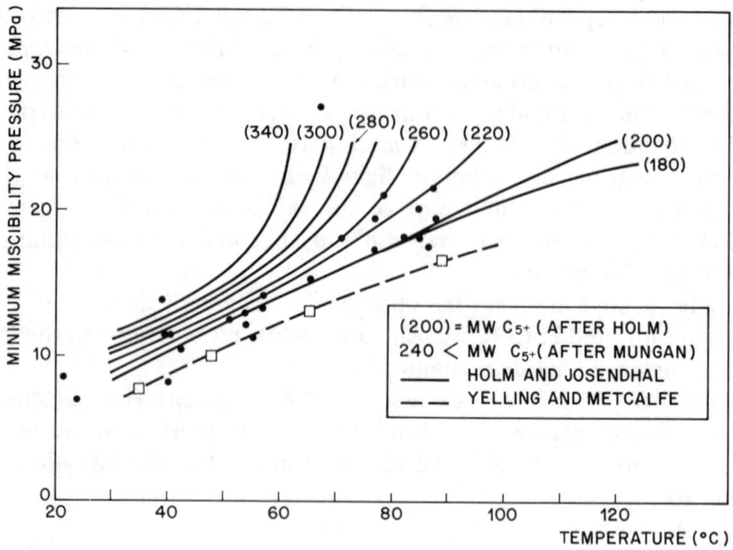

Fig. 7.15 Correlations for minimum miscibility pressure. (From Ref. 23, reproduced by permission of The Canadian Institute of Mining and Metallurgy.)

with a 20 mole per cent contamination of the injected CO_2 gas with methane.

When the crude oil is a paraffinic type, with a high wax content, it has been found that the correlations do not apply at all. For example, in a case where the wax content was as high as 35%, the estimated MMP was supposed to be around 25 MPa, and in fact it was found to be higher than 35 MPa. Hence, the correlations should be used only as a screening tool to decide if further tests are needed to measure the miscibility pressure. The best way for this is by laboratory analysis, in which CO_2 displaces reservoir crudes from sand pack models.

(c) *Determination of the MMP.* Most methods proposed have to do with displacement tests conducted in a long slim-tube packed with a calibrated sand as described in Chapter 6 for hydrocarbon gases. The MMP is defined as the pressure for which the oil recovery is at least 90% after 1·2 PV of solvent is injected. An example of sand pack performance used to estimate MMP is given in Fig. 7.16.

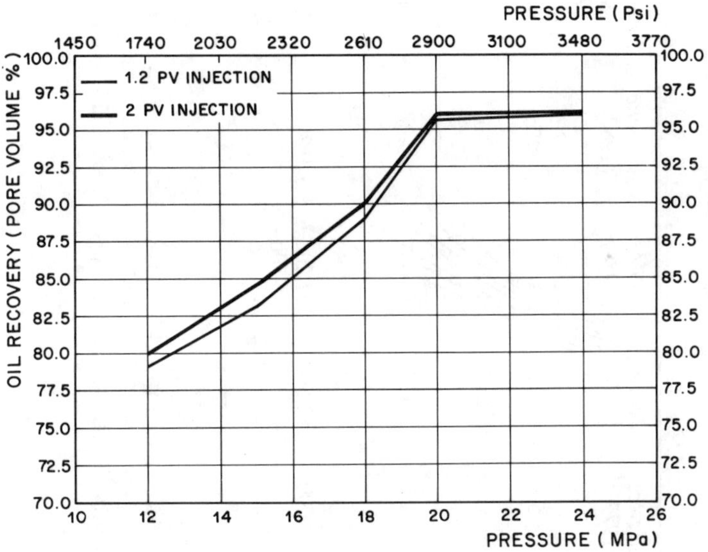

Fig. 7.16 Experimental measurement of minimum miscibility pressure.

7.2.3 *Binary screening guides*

In addition to estimating MMP and deciding if this pressure is obtainable for a particular reservoir, a prospect is screened for CO_2 process compatibility by developing a comparison of key rock, reservoir, and fluid characteristics in technically and economically successful CO_2 floods with those in the candidate reservoir. While this preliminary screening is not a hard and fast rule in predicting the economic viability of an individual CO_2 flood, it may be indicative of a reduced or enhanced chance for success; that is, if a reservoir's properties do not fall in line with any previously successful application of CO_2 injection, forecast of economic implementation may be doubtful. Conversely, a potentially successful flood may not necessarily fulfill all reservoir parameter limitations.

Binary screening guides are inherently outdated from the onset. The number of completed field tests is so statistically small that a significant amount of new information is gathered from each project. Coupled with this dynamic addition of new technology are fluctuating economic factors that affect any decision-making process. However, because of

Table 7.1 Chronology of miscible CO_2 binary screening guides (see Ref. 5, p. 185)

	Study							
Reservoir[a] parameter	*Geffen (1973)*	*Lewin and Associates (1976)*	*NPC (1976)*	*McRee (1977)*[13]	*Iyoho (1978)*	*OTA (1978)*	*Carcoana (1982)*	*Taber and Martin (1983)*
Viscosity (cp at reservoir conditions)	<3	<12	⩽10	<5	<10	⩽12	<2	<15
Gravity (°API)	>30	>30	⩾27	>35	30 to 45	⩾27[b] 27 to 30 ⩾30	>40	>26
Fraction of oil remaining in area to be flooded (before EOR)	>0·25	>0·25	—	>0·25	>0·25	—	>0·30	>0·30
Depth (ft)	—	>3 000	>2 300	>2 000	>2 500	⩾7 200 ⩾5 500 ⩾2 500	<9 800[c]	>2 000
Temperature (°F)	—	NC[d]	<250	—	—	—	<195	NC
Original reservoir pressure (psia)	>1 100	>1 500	—	—	—		>1 200	—
Permeability (md)	—	NC	—	>5	>10	—	>1	NC

[a] SI conversions; 1 cp = 10^{-3} Pa s; °API = $(141.5/\rho) - 131{\cdot}5$ where ρ is the specific gravity (15·56°C/15·56°C); T°F = 1·8T°C + 32;
1 ft = 0·305 m; 1 psia = $6{\cdot}895 \times 10^{-3}$ MPa; 1 md = $9{\cdot}869 \times 10^{-4}$ μm^2.
[b] See depth constraint.
[c] Due to temperature constraint.
[d] NC = Not a critical factor.

their easy application, binary screening guides have become a popular preliminary screening tool (Table 1).

7.3 Partially miscible CO_2 injection

It has been proven that even if miscibility cannot be reached, a high recovery rate can be achieved mainly due to oil swelling, viscosity reduction and, to a lesser extent, to low interfacial tension and diffusion effects.

7.3.1 *'Near miscibility' process with light-to-medium oils*

When the miscibility pressure cannot be reached due to either technical or economic factors, it is recommended to evaluate the efficiency of CO_2 injection under conditions of partial miscibility also called the 'near miscibility' process.

In such a case, the ultimate oil recovery is less than under miscible conditions, but the amount of CO_2 required to produce additional oil is also less important so that the economics of the process may be attractive. As an example, let us consider a medium oil of 29° API, the swelling properties of which are shown in Fig. 7.7. At the field pressure of 18·5 MPa (much lower than MMP estimated at 25 MPa) the swelling factor is 1·8, which means that in swept zones the residual oil saturation (S_{ORG}) is only 1/1·8. So, S_{ORG} varies from 30% to 16%. In addition, this value may be decreased much below 10% due to the exchanges between the two phases (vaporizing and condensing) during the displacement in the reservoir, so that the interfacial tension between the oil phase and the gas phase is strongly reduced. When the interfacial tension is lower than 0·01 mN/m, the efficiency of CO_2 is very close to that of full miscibility.

7.3.2 *Heavy oils*

In the case of heavy oils, primary production is generally very low (less than 10%) due to the high viscosity of the oil. Generally, if the depth of the field is not greater than 1000 m, thermal methods are applied to enhance the recovery (see Chapters 9 and 10). For deeper fields, CO_2

can be used due to the viscosity reduction, especially when the pressure is high enough. Figure 7.8 shows that for an oil of 12·2° API the viscosity of oil decreases from 900 mPa/s to 40 mPa/s as saturation pressure increases to 15 MPa.

7.3.3 *Role of diffusion*

The adverse mobility ratio in the entire process involving gas injection leads to the development of viscous fingering in the reservoir so that much of the efficiency of CO_2 lies in its ability to diffuse in oil and water and thus to contact more oil than that contained in the swept areas. On the other hand, during gas drive in fractured or very heterogeneous reservoirs, the stagnant oil contained in blocks along fractures and drains can be contacted only through a diffusion process. It is important to know if the effect of diffusion will be felt during the project life. In the stimulation process consisting in injecting a limited amount of CO_2 in a well, then producing the well back after a soaking time (a process currently called 'Huff and Puff'), the amount of oil contacted by the small volume injected can be greatly increased by diffusion. Still, molecular diffusion of CO_2 in liquids has been extensively studied for the sake of the chemical industry. The literature provides a few useful correlations between the viscosity of the solvent and the diffusion coefficient of CO_2. However, the experimental conditions differ enormously from usual reservoir conditions, because the experiments were performed at atmospheric pressure. A method has been developed then, to measure the diffusion of CO_2 in oil and water under reservoir conditions.[25] It appears that diffusion of CO_2 in liquids under reservoir conditions is 5 to 10 times faster than what is reported in the literature. The diffusion coefficient of CO_2 dissolving into an oil and water saturated medium is between the diffusion of CO_2 in oil and in water (10^{-8} m^2/s to 5×10^{-8} m^2/s). The diffusion coefficient decreases as pressure falls.

7.4 Sweep efficiency

The main problem in achieving profitable applications of CO_2 flooding has been the very high mobility of the CO_2. The lower density and lower viscosity of CO_2 compared with reservoir oils are responsible for gravity tonguing and viscous fingering, which are more severe than in waterflooding. As a consequence, much of the CO_2 injected flows through the areas already swept. Sometimes the flow is dominated either by viscous fingering or by gravity tonguing.

Several attempts to improve the sweep efficiency have been investigated:

(a) installation of well packers and perforating techniques
(b) shutting in production wells to eliminate low pressure sinks
(c) alternating CO_2–water injections
(d) addition of foaming solutions together with CO_2

Points (c) and (d) aim at reducing the mobility of CO_2.

7.4.1 *Alternate CO_2–water injections*

By combining water injection with gas injection as alternate slugs (WAG process, defined in Section 7.1.3), the volumetric sweep efficiency can be significantly improved by reducing the gas mobility in the reservoir, which becomes close to the mobility of the water. The complete evaluation of the process must take into account the possible effect of hysteresis on relative permeabilities in drainage and imbibition, and the best water/gas ratio must be defined.[26]

7.4.2 *Foams*

Foaming solutions combined with CO_2 injection can generate foams that strongly reduce the mobility of CO_2. The process consists first of selecting the surfactant suited for the particular field application. Surfactants are sensitive to the salinity of the water, the temperature and the oil present in the reservoir. Specific tests must be carried out in the laboratory to define the best conditions of application. This process can be applied for two purposes:

(1) Sweeping the reservoir from injection wells to production wells. This process is still under evaluation because it is difficult to control the stability of the foam during the propagation into the reservoir. This is mainly due to the adsorption of the surfactant on the matrix, especially on clays.

(2) Blocking the CO_2 in the more permeable layers where foams form first. The gas is then pushed toward the more compact layers which had previously been bypassed. A similar application is possible with fissured reservoirs to block the fissures with foam and force the gas to envade the matrix blocks. This is acceptable only if matrix permeability is high enough to allow significant injectivity.

7.5 Field Implementation

7.5.1 *Injection schemes*

Application of CO_2 injection for improved oil recovery has, in some cases, been implemented as a single well stimulation tool similar to steam Huff and Puff (see Chapter 9). However, the vast majority of field tests has been as a driving agent in pattern-type displacements. The CO_2 can be injected as a continuous gas stream or chased with varying amounts of water (WAG process) for mobility control.

In most cases, a total gas volume of 20–40% of the pore volume originally occupied by hydrocarbons (HCPV) is injected in 5–10 equal-volume slugs interspersed with water slugs of similar volume of water. The total volume of CO_2 injected is based on project economics while the water/gas ratio (normally 1/1 to 3/1) is governed by viscous and relative permeability effects.[26] The WAG ratio is then adjusted to optimize recovery.

Several 'hybrid' WAG floods have also been developed. These start injecting at very low WAG ratios and increase the WAG over the life of the flood. This allows the operator to reap the early response benefits of continuous CO_2 injection while maintaining good areal sweep efficiency and a flat gas production profile thus allowing gas processing plant design to be optimized.

7.5.2 *Response*

Field tests with carbon dioxide injection have, in general, recovered 8–12% of the original oil in place, using slug sizes ranging from 20 to 40% of the HCPV. This incremental oil production occurs with or after CO_2 breakthrough rather than being banked ahead of the injected slugs. It appears that early CO_2 breakthrough can be anticipated with significant reinjection of produced gas and an extended period of incremental oil recovery.

Along with the oil response described above, significant volumes of CO_2-produced gas will need to be reinjected. It may or may not be stripped of its heavier hydrocarbon ends depending on their effect on MMP and ultimate project economics. For the moderate CO_2 slug sizes listed above, the project operator will inject 1800 to 2700 sm^3† of CO_2-rich gas to recover 1 m^3 of oil. They will, however, only have to purchase 500–900 sm^3 of CO_2 per m^3 of oil produced with the remaining gas volumes being supplied by reinjecting produced gas.

7.5.3 *Operating problems*

As discussed in previous sections, CO_2–water mixtures form carbonic acid creating a potential corrosion problem. Coated injection tubing and amine-inhibited production wells have usually provided effective prevention. CO_2-hydrate formation is also a consideration for injection line design where triethylene glycol dehydration has been successful.

On the production side, well bore deposits of calcium sulfate, asphaltenes and paraffins have been observed. Phosphonates and solvents have been used to combat these problems. Also, with gas breakthrough, pumping wells will start to flow. Gas locking of pumps, and choke freeze up, are common.

Produced gases can be totally reinjected or separated for sale, as discussed earlier. If CO_2 and/or H_2S are to be separated from the saleable natural gas components, design choices need to be made. These include the use of amine solvents, hot potassium carbonate, membrane separation and distillation techniques. In any case, if CO_2 processing is implemented, the capital cost of installing such a facility will far exceed all other project costs except the purchase of raw CO_2.

7.5.4 *Asphaltene precipitation*

It has been shown that when CO_2 and reservoir oil mix, multiple-phase equilibria may exist including a solid phase (Fig. 7.13). These heavy

† sm^3 means standard cubic meters, i.e. volume at atmospheric pressure (101·325 kPa) and 15°C.

ends of a crude oil become less soluble with the addition of CO_2 and precipitate, causing loss of permeability when the precipitation occurs in the reservoir. More generally, problems of plugging are encountered in production tubings and surface facilities.

It has been observed that it is not only with heavy oils, which contain large amounts of asphaltenes, that problems of precipitation occur, but also with light oils, mostly paraffinic, sometimes containing small amounts of asphaltenes which can precipitate when the pressure and temperature are decreased or when the gas content in crude oil increases. It generally appears that the precipitation occurs when the pressure is close to the bubble pressure. The amount of precipitate and its subsequent effect on oil recovery is difficult to quantify. In addition, CO_2–water mixtures are slightly acidic and it is well known that some acids, when in contact with crude oil, may precipitate asphaltenes.

Depending on the crude oil characteristics, the precipitation of solid material (not only asphaltenes but also waxes and resins) can be a major design problem. This effect should be thoroughly examined in the laboratory before field implementation.

7.5.5 *Concluding remarks*

There is a general agreement with the following observations concerning CO_2 field performance to date:

- Production response from existing floods has been encouraging.
- Early CO_2 breakthrough is in line with predictions.
- Injected water from a previous waterflood or during a WAG cycle does not appear to block CO_2 from contacting residual oil.
- Loss of injectivity due to WAG-relative permeability effects varies but is manageable.
- Simulation capabilities to predict performance are satisfactory.
- Corrosion problems are under control. Problems for maintaining production well hardware integrity are being tackled.
- Total incremental oil recovery estimates using CO_2 injection appear reasonable.

7.6 References

1. Anon., CO_2 and HC injection lead EOR production increase. Annual production report. *Oil and Gas J.*, **88**(17) (23 April 1990) 49–82.
2. Vukalovich, M. P. & Altunin, V. V., Thermophysical properties of carbon dioxide, Collect's Ltd, London, 1968.
3. US Department of Energy, Target reservoirs for CO_2 miscible flooding. DOE/MC/08341, 17 October 1980.
4. Reynolds, W. C., *Thermodynamic Properties in SI,* Stanford Press, Palo Alto, 1979.
5. Klins, M. A., *Carbon Dioxide Flooding,* Int. Human Resources Development Corp., Boston, 1984.
6. Stalkup, F. I., *Miscible Displacements,* Monograph Vol. 8, Henry L., Doherty Memorial Fund of AIME, SPE of AIME, Richardson, Texas, 1983.
7. Reamer, H. H., Olds, R. H., Sage, B. H. & Lacey, W. N., Phase equilibria in hydrocarbon systems. *Industrial and Engineering Chemistry* January (1944).
8. Beeson, D. M. & Orloff, G. D., Laboratory investigation of the water driven carbon dioxide process for oil recovery. *Petroleum Trans. AIME,* **216** (1959) 388–91.
9. Crawford, H. R., Neill, G. H., Bucy, B. J. & Crawford, P. B., Carbon dioxide—A multipurpose additive for effective well stimulation. *J. Petroleum Technol.,* **15**(3) (1963) 237–42.
10. Simon, R. & Graue, D., Generalized correlations for predicting solubility, swelling and viscosity behavior of CO_2–crude oil systems. *Petroleum Trans. AIME,* **234** (1965) 102–06.
11. Simandoux, P., Bardon, C., Denoyelle, L. & Vacelet, O., State-of-the-art of EOR by CO_2 flooding. OAPEC Seminar on EOR, Doha (Nov. 1983) and *Revue Institut Français du Pétrole,* **39**(4) (1984) 403–31.
12. Dodds, W. S., Stutzman, L. F. & Solami, B. J., Carbon dioxide solubility in water. *Industrial and Engineering Chemistry,* **1** (1956) 92–5.
13. McRee, B. C., CO_2, how it works, where it works. *Petroleum Engineer Int.* November (1977).
14. Tumasyan, A. B., Panteleev, V. G. & Meinster, G. P., Influence de l'anhydride carbonique sur les propriétés physiques du pétrole et de l'eau. *Nefteprom. dela,* **2** (1969) 20–3.
15. Latil, M., *Enhanced Oil Recovery,* Gulf Publishing, Houston, 1980.
16. Menzie, D. E. & Nielsen, R. F., A study of the vaporization of crude oil by carbon dioxide repressuring. *J. Petroleum Technol.,* **15**(11) (1963) 1247–52.
17. Simon, R., Rosman, A. & Zana, E., Phase behavior properties of CO_2—reservoir oil systems. *Soc. Petroleum Engineers J.,* **18**(1) (1978) 20–6.
18. Orr, Jr, F. M. & Silva, M. K., Equilibrium phase compositions of CO_2–hydrocarbon mixtures: Measurement by a continuous multiple contact experiment. SPE/DOE Paper No. 10726, Third Symp. on

Enhanced Oil Recovery of Soc. Petroleum Engineers and US Department of Energy, Tulsa, 4–7 April 1982.

19. Orr, Jr, F. M., Silva, M. K. & Lien, Cheng-Li, Equilibrium phase composition of CO_2–crude oil mixtures: Comparison of continuous multiple contact and slim-tube displacement tests. SPE/DOE Paper No. 10725, Third Symp. on Enhanced Oil Recovery of Soc. Petroleum Engineers and US Department of Energy, Tulsa, 4–7 April 1982.
20. Behar, E., Bardon, C. & Vengu, T., Propriétés thermodynamiques d'un pétrole brut en présence de gaz carbonique—Relation avec la pression de miscibilité. 3rd European Meeting on Improved Oil Recovery, Rome (16–18 April 1985), Proc. Agip SpA, pp. 9–18.
21. Yellig, W. F. & Metcalfe, R. S., Determination and prediction of the CO_2 minimum miscibility pressure. *J. Petroleum Technol.*, **32**(1) (1980) 160–8.
22. Holm, L. W. & Josendal, V. A., Effect of oil composition on miscible type displacement by carbon dioxide. *Soc. Petroleum Engineers J.*, **22**(1) (1982) 87–98.
23. Mungan, N., Carbon dioxide flooding—fundamentals. *J. Canada Petroleum Technol.*, **20**(1) (1981) 87–92.
24. Enick, R. M., Reznik, A. A. & Miller, R. A., The statistical and dimensionless time analog to the generalized Dykstra–Parsons method. *SPE Reservoir Engineering*, **3**(1) (1988) 313–19.
25. Denoyelle, L. & Bardon, C. Influence of diffusion on enhanced oil recovery by CO_2 injection. Int. Symp. on CO_2 Enhanced Oil Recovery, IFP-SzKFI, Proc., Budapest, (8–11 March 1983), pp. 203–23.
26. Morel, D. & Latil, M., Basic study of sweep efficiency improvement by water alternate gas injection. 4th European Symp. on Enhanced Oil Recovery, Proc., DGMK, Hamburg, (27–29 October 1987), pp. 723–34.

THERMAL RECOVERY

8 Thermo-chemical alteration of crude oils

Annie Audibert
Institut Français du Pétrole, Rueil-Malmaison, France

and

Shapour Vossoughi
University of Kansas, Lawrence, Kansas, USA

8.1 Introduction 243

8.2 Definition of the involved reactions 245

8.3 Pyrolysis 247
8.3.1 Influence of temperature 248
8.3.2 Effect of the water phase 251
8.3.3 Influence of the mineral matrix 252
8.3.4 Influence of oil composition 253
8.3.5 Kinetics of non-oxidative reactions 255

8.4 Oxidation and combustion 257
8.4.1 Low temperature oxidation 259
8.4.2 Influence of oil composition 259
8.4.3 Kinetics of chemical reactions 261
8.4.4 Kinetic modeling 263

8.5 References 267

8.1 Introduction

During enhanced oil recovery by thermal processes, which are described in Chapters 9 and 10, a crude oil undergoes physical and chemical changes due to the effects of the heat supplied during steam or hot water injection and the heat released during in-situ combustion.[1]

Although any crude oil contains the same basic hydrocarbon compounds, the ratios among the hydrocarbon classes vary widely from one crude oil to another. The hydrocarbon classes include paraffins (*n*-paraffins, branched paraffins), cyclo-paraffins (or naphthenes), and aromatics. The most common nonhydrocarbon compounds in petroleum are derivatives of sulfur, nitrogen and oxygen. The metal compounds include mostly iron, vanadium, aluminum, magnesium and nickel.

The alteration in physical properties (viscosity, specific gravity, interfacial tension, etc.) due to the physical processes associated with thermal recovery techniques (steam distillation, gas stripping, etc.) are described in Chapter 9.

The chemical changes involve different reactions such as:

—Pyrolysis at moderate and high temperatures. Different kinds of reactions may occur, namely cracking, dehydrogenation and condensation, all affected by the presence of water.
—Reactions with oxygen (oxidation and combustion). Depending on the reaction, the hydrocarbon chain may be completely destroyed by combustion or oxygen may be incorporated into the hydrocarbon chain.[2]

All these reactions occur in the presence of a rock matrix, which can have a catalytic effect. Depending on the composition of the reservoir matrix, the mineral phase itself can be affected by heat and the presence of a water phase.[3–5]

The reactions involved in a thermal process (hot water/steam injection and in-situ combustion) will be described according to the relevant range of temperatures. All these reactions have a great influence on the processes, namely:

—Higher recovery factor and easier recovery due to oil lightening.
—Formation of gaseous components such as H_2S or CO_2, which may play a polluting role.[6] However, CO_2 may also play a role in oil recovery. As a matter of fact, (a) CO_2 dissolution decreases oil viscosity (see Chapter 7), and (b) the formation of significant amounts of CO_2 when thermal recovery is applied in carbonate reservoirs may influence reservoir pressure.[7]
—Since insoluble organic matter is formed under some conditions, it may induce formation plugging and cause injection and propagation problems.[8]

Some work has been carried out to characterize the evolution in composition and in properties of the oil produced during thermal recovery, in field operations.[9,10]

Several laboratory techniques have been developed to study the different kinds of reactions.[8,11–17] The variations in oil properties have been characterized according to the heating rate and the plateau of temperature reached, the experimental set-up (closed vessel or flowing medium) and the different phases that may be present, i.e. crude oil, water, mineral phase, gas phases with different compositions. These studies are important not only to determine and correlate oil property evolution with what has been observed in a field operation but also to evaluate the feasibility of a field project or to find some parameters that help in characterizing the oil property evolution in the field. Thus, it is possible to compute the extension of the propagation of the steam or in-situ combustion front into the reservoir.[9,10]

Moreover, some laboratory data can be useful for designing kinetic models that may help in predicting crude oil thermal alteration during thermal recovery in a field and also oil evolution in basin studies.[18–24]

8.2 Definition of the involved reactions

Under conditions prevailing during thermal recovery processes, crude oil undergoes reactions leading to various chemical modifications.

A first group of reactions, called pyrolysis, occurs with heating in the presence of an inert gas and in the presence or absence of a water phase. During these reactions, the molecules can be modified by cracking, dehydrogenation or condensation, depending on the variation in the number of the carbon atoms.

The different reactions are as follows:

—*Cracking*: the cracking involves the rupture of C–C bonds and the formation of molecules with lower molecular weights, thus lightening the oil involved in the process. A typical reaction is as follows:

$$\underset{\text{alkane}}{C_{n+m}H_{2(n+m)+2}} \rightarrow \underset{\text{alkene}}{C_nH_{2n}} + \underset{\text{alkane}}{C_mH_{2m+2}} \qquad (8.1)$$

—*Dehydrogenation*: in dehydrogenation reactions, the number of carbon atoms remains unchanged; only C–H bonds are destroyed, and unsaturated hydrocarbons are formed:

$$\underset{\text{alkane}}{C_nH_{2n+2}} \rightarrow \underset{\text{alkene}}{C_nH_{2n}} + H_2 \qquad (8.2)$$

—*Condensation*: condensation between two hydrocarbons leads to the formation of a molecule with a higher molecular weight. When the reactants are alkanes and alkenes, condensation often leads to the formation of aromatic compounds.

As products are more and more dehydrogenated, they eventually form an insoluble organic deposit having a low hydrogen content. This deposit is generally called pyrobitumen or coke.

The presence of a mineral matrix and a water phase, as described in Section 8.3.3, may affect the rate of the pyrolysis and influence all processes.

A second group of reactions is oxidation, which occurs in the gaseous phase, in the liquid phase between hydrocarbon molecules and dissolved oxygen, or at the surface of the solid. The most important hydrocarbon oxidation reactions are given in Ref. 2. Different kinds of products may be formed:

Complete combustion:

$$C_nH_{2n} + (3n/2)O_2 \rightarrow nCO_2 + nH_2O \qquad (8.4)$$

Incomplete combustion:

$$\begin{array}{c} H \\ | \\ R{-}C{-}R' \\ | \\ H \end{array} + \tfrac{3}{2}O_2 \rightarrow RR' + CO_2 + H_2O \qquad (8.5)$$

$$\begin{array}{c} H \\ | \\ R{-}C{-}R' \\ | \\ H \end{array} + O_2 \rightarrow RR' + CO + H_2O \qquad (8.5\text{bis})$$

Oxidation to carboxylic acid:

$$\begin{array}{c} H \\ | \\ R{-}C{-}H \\ | \\ H \end{array} + \tfrac{3}{2}O_2 \rightarrow R{-}C\begin{array}{l} /\!\!/ O \\ \backslash OH \end{array} + H_2O \qquad (8.6)$$

Oxidation to aldehyde:

$$\begin{array}{c} H \\ | \\ R{-}C{-}H \\ | \\ H \end{array} + O_2 \rightarrow R{-}C\begin{array}{l} / H \\ \backslash\!\!\backslash O \end{array} + H_2O \qquad (8.7)$$

Oxidation to ketone:

$$\begin{array}{c} \mathrm{H} \\ | \\ \mathrm{R{-}C{-}R'} \\ | \\ \mathrm{H} \end{array} + \mathrm{O_2} \rightarrow \begin{array}{c} \\ \\ \mathrm{R{-}C{-}R'} \\ \| \\ \mathrm{O} \end{array} + \mathrm{H_2O} \qquad (8.8)$$

Oxidation to alcohol:

$$\begin{array}{c} \mathrm{H} \\ | \\ \mathrm{R{-}C{-}H} \\ | \\ \mathrm{H} \end{array} + \tfrac{1}{2}\mathrm{O_2} \rightarrow \begin{array}{c} \mathrm{H} \\ | \\ \mathrm{R{-}C{-}OH} \\ | \\ \mathrm{H} \end{array} \qquad (8.9)$$

Oxidation to hydroperoxide:

$$\begin{array}{c} \mathrm{H} \\ | \\ \mathrm{R{-}C{-}H} \\ | \\ \mathrm{H} \end{array} + \mathrm{O_2} \rightarrow \begin{array}{c} \mathrm{H} \\ | \\ \mathrm{R{-}C{-}O{-}O{-}H} \\ | \\ \mathrm{H} \end{array} \qquad (8.10)$$

Combustion reactions (reactions (8.4) and (8.5)) occurring above about 350°C are characterized by a reaction of the CH_2 group with oxygen to form CO_2 or CO and H_2O. Low-temperature oxidation (LTO) (reactions (8.6)–(8.10)) occurs at less than 350°C. LTO occurs in the most reactive fractions of the oil, i.e. resin fraction and then aromatics. The mechanisms of LTO are complex but condensation mechanisms to higher molecular weight molecules are generally thought to occur.

8.3 Pyrolysis

The cracking of the crude oil causes a reduction in oil viscosity, which in turn enhances oil recovery.

Laboratory experiments have been carried out to characterize the reactivity of crude oils and to investigate the effects of thermal alteration in thermal oil recovery processes. The experimental set-up is generally a stainless steel reactor that is pressurized with an inert gas. The catalytic effects of the reactor wall are avoided by working in a glass-lined vessel; or the wall may be considered as pacified by the sufficient sulfur content of some oils.[8,12,15–17] Experiments have been also carried out in sealed gold tubes to limit the catalytic effects of the wall.[13]

Chemical alteration of oil is generally observed through the evolution of its physical properties (density, viscosity) and also through compositional changes in the different phases formed (gas, liquid, solid). The gas phase and the solid phase are respectively analyzed by gas chromatography and elemental analysis. The liquid phase has been characterized by a variety of techniques such as SARA analysis (saturates, aromatics, resins, asphaltenes), simulated distillation gas chromatography, and elemental analysis. The liquid phase generally contains produced light fractions, cracked heavy ones, and unreacted components of the original oil. The oil is generally separated into fractions named by the number of carbon atoms in the hydrocarbon molecules (for example, C_1–C_5, C_6–C_{13}, C_{14}^+).

8.3.1 *Influence of temperature*

Reactions occurring at low temperature (<350°C) are referred to as visbreaking, while high temperature reactions are called cracking. The term 'visbreaking' expresses the reduction in the oil viscosity.

Visbreaking, usually known as a refinery process, is characterized by the formation of products of reaction and their retention in the liquid phase. This implies that production of gas and solid is minimum. It occurs in both thermal recovery techniques, namely in steam injection, in the zone where fluids coexist (liquid water, liquid hydrocarbons, steam), and in in-situ combustion in the cokefaction zone (see Chapter 9, Section 9.5.3 and Chapter 10, Section 10.1.1).

A lot of information exists in the literature relevant to refining processes on the visbreaking of heavy oils. In such a case, the reactions occur at high temperatures for short contact time. However, it must be emphasized that when steam injection and in-situ combustion are applied in oil reservoirs, the oil present in certain regions of the producing formation is subjected to such reactions for long periods of time, up to a few months.

Experimental work has focused on the influence of both temperature and time on the extent of visbreaking;[12] tests implemented with long contact time are reported in Ref. 13.

Cracking is shown to begin at temperatures above 260°C and production of solids and gases is significant above 350°C. High temperature reactions (>350°C) are observed during in-situ combustion in the cokefaction zone. Therefore, chemical changes in oil composition, including deposition of pyrobitumen on the rock matrix, may occur in the case of high pressure steam injection. The deposition of pyrobitumen may have a negative effect on the formation by affecting the permeability of the reservoir.

The distribution of components between different phases (weight per cent in gases, liquids, and solid) may be used to characterize the extent of the pyrolysis reactions. The liquid phase may also be characterized by the carbon number distribution. Both distributions can be combined to yield the total composition of the reacted oil, which is then compared to the initial one.

In the study performed by Fassihi *et al.*[12] the oil was conventionally separated into three fractions: (1) C_1–C_{40}; (2) the liquid phase containing components over C_{40} was plotted as C_{41} and (3) the solid phase was plotted as C_{42}. For each fraction, the composition of the initial oil was substracted from the composition of the reacted oil. The resulting differences (%wt gain or %wt loss) were plotted against the carbon number (Fig. 8.1), where the %wt gain appeared in black and the %wt loss in white.

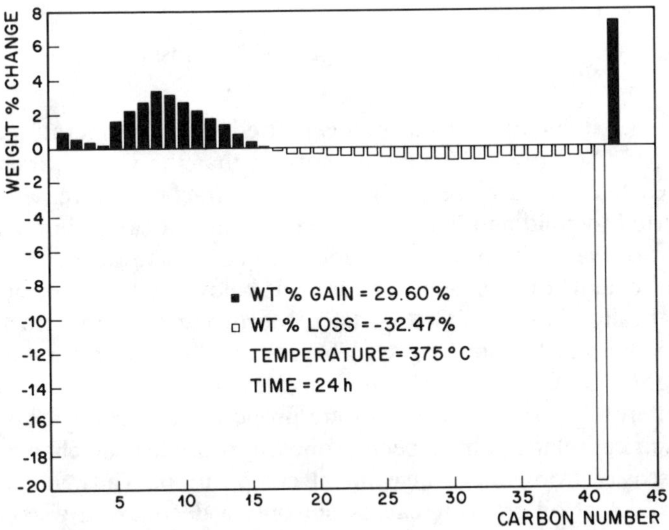

Fig. 8.1 Difference in %wt distribution (oil K). (From Ref. 12, reproduced by permission of The Society of Petroleum Engineers.)

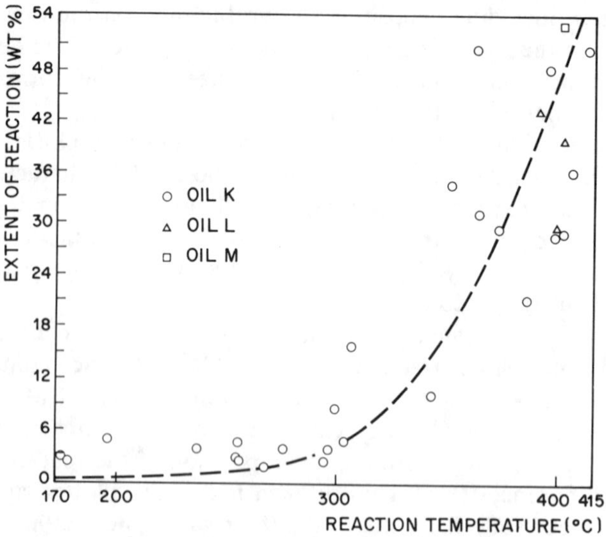

Fig. 8.2 Extent of reaction versus temperature. (From Ref. 12, reproduced by permission of The Society of Petroleum Engineers.)

It is then possible to define the extent of reaction *Er* as:

$$Er = \tfrac{1}{2} \sum (\%\text{wt gain} + \%\text{wt loss}) \tag{8.11}$$

In Fig. 8.2, the extent of reaction is plotted versus temperature.

At 375°C (Fig. 8.1), it may be noticed that the C_5–C_{15} fraction of the reacted oil increases, while the C_{41} fraction decreases, being converted to solid and lighter fractions. In such a case, approximately 10%wt of the oil is converted to solids and 5%wt to gases.

The extent of reaction is minimal below 260°C and increases dramatically above 350°C (Fig. 8.2). More precisely, the cracking of the condensate fraction (C_6–C_{13}) begins at a lower temperature than that of the C_2–C_5 fraction.[13] Kinetic models describing the evolution of the different phases and fractions are proposed in Section 8.3.5.

Empirical relations have been proposed to predict the change in the oil viscosity from the change in oil composition. Thus, it has been observed that oil cracking causes not only a decrease in viscosity but also a reduction in asphaltene content.[17] This is consistent with the observation that, as thermal cracking increases, asphaltenes disappear rapidly, in relation to an increase in coke production.[13]

8.3.2 *Effect of the water phase*

Water in the liquid or vapor phase may participate in the pyrolysis reactions.[8,14–16] Such reactions are called aquathermolysis. For example, the aquathermolysis of organosulfur compounds is the source of H_2S and CO_2, which are effectively produced in certain thermal recovery field operations. This may be written as follows:

$$RCH_2CH_2SCH_3 + 2H_2O \leftrightarrows RCH_3 + CO_2 + H_2 + H_2S + CH_4 \quad (8.12)$$

The organically bound sulfur is removed as H_2S while hydrogen is transferred from the water to the oil phase, leading to an upgrading of the oil.[13–15] This is schematized in Fig. 8.3. From this figure, it may be seen that a large amount of CO_2 is obtained via the water gas shift reaction (WGSR):

$$CO + H_2O \leftrightarrows CO_2 + H_2 \quad (8.13)$$

Moreover, the presence of water may also induce the formation of

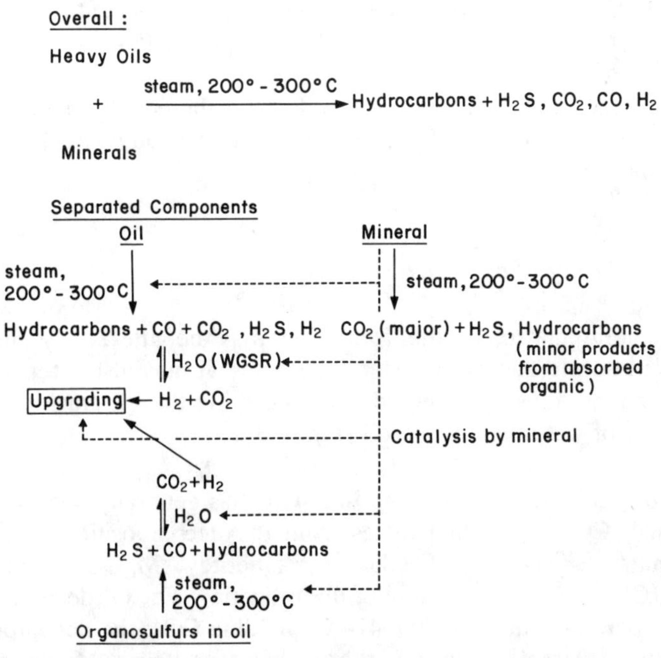

Fig. 8.3 Aquathermolysis. (From Ref. 15, reproduced by permission of The Alberta Oil Sands Technology and Research Authority.)

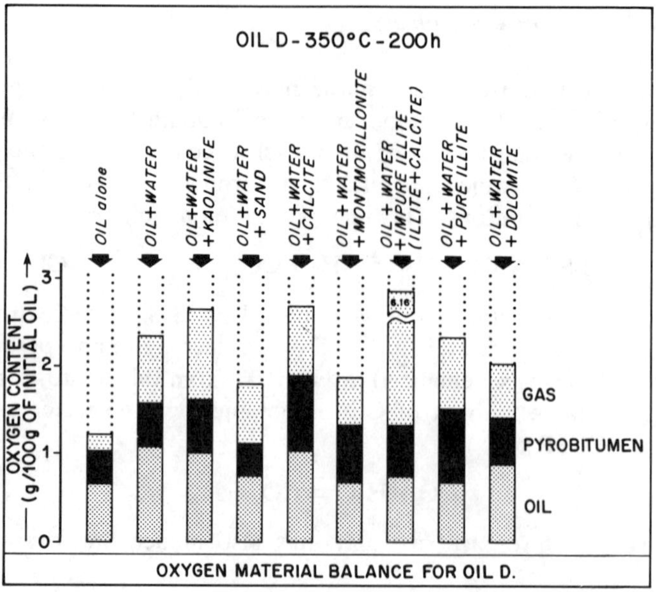

Fig. 8.4 Oxygen material balance. (From Ref. 8, reproduced by permission of The Society of Petroleum Engineers.)

heteroatomic compounds concentrated in the solid phase. In the example shown in Fig. 8.4, no change in the oxygen content of the oil phase has been observed.

8.3.3 *Influence of the mineral matrix*

The mineral matrix of the formation may be altered by the heat provided during thermal recovery processes. It may also interfere with the reactions, thus by catalytic effect affecting the oil cracking or the reactivity of water versus the oil (Fig. 8.3).

(a) *Modifications in rock composition.* It has generally been assumed that high CO_2 production arises from the decomposition of mineral carbonates (siderite $FeCO_3$, magnetite $MgCO_3$, dolomite $CaMg(CO_3)_2$).[5,8,14] For example, upon heating, $FeCO_3$ decomposes in the temperature range of 300–470°C, yielding CO_2; the decomposition begins considerably sooner in the presence of steam. Moreover, any pyrite (Fe S_2) contained in the reservoir may be a source of H_2S;[8,14] if temperature exceeds 300°C, decomposition is appreciable. In such a case, the composition of the effluent gas depends strongly on the temperature of the steam injected.[8] Therefore, the production of H_2S, which may induce corrosion damage in the production facilities, must be controlled.

The thermal alteration of clays, which begins at moderate temperature (~300°C), must also be considered (see Ref. 1, p. 86).

Not only the temperature, but also the injected aqueous fluids may affect the reservoir properties. As a matter of fact, quartz, silico-aluminates, kaolinite and dolomite may be dissolved and analcime, chlorite, montmorillonite, smectite and calcite may be formed under different conditions.[3–5] Such reactions, which are controlled by temperature, pH, and water salinity, affect the porosity and permeability of the reservoir. It has been postulated that an enlargement of fine pores due to mineral dissolution occurs simultaneously with mineral deposition in large pores.[5] Moreover, migration of clay-sized material has also been observed. All these phenomena induce a permeability decrease[4] which may lead to local plugging. All laboratory studies are consistent with the field observations.

(b) *Effects of mineral matrix on cracking.* It has generally been observed that clay minerals have a catalytic effect on cracking reactions, especially on coke formation. Furthermore, clays contribute greatly to the total available surface area. Some transition metals may also have a catalytic effect.

It will be demonstrated in Section 8.3.5 that these catalytic effects have to be taken into account to obtain correct values of activation energies.[18,22] In laboratory studies, the yields of coke are higher when bitumen is cracked in the presence of a sand mixture. This has been studied over a wide range of temperatures[6,22,23] (Fig. 8.5). The differences are not significant in the case of gas yields (Fig. 8.6).

Moreover, the presence of a mineral promotes chemical reactions involving water and reduces the difference in reactivity between crude oils (see Section 8.3.4).

8.3.4 *Influence of oil composition*

Laboratory studies focus on the influence of oil composition on cracking under conditions prevailing during the application of thermal

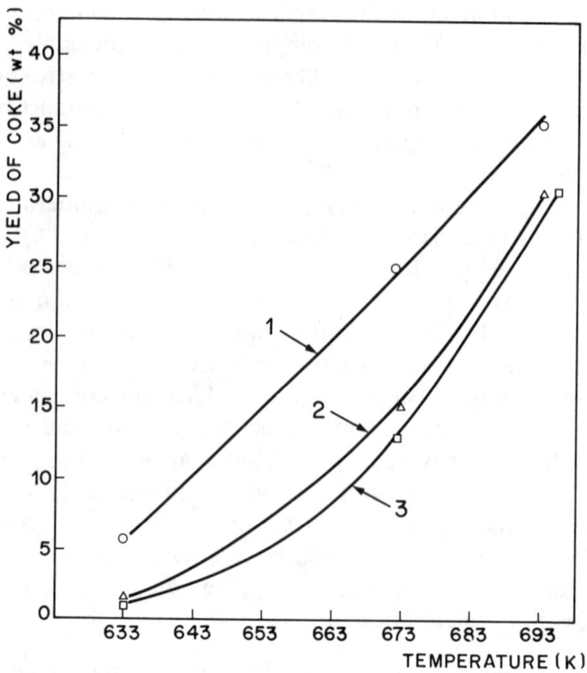

Fig. 8.5 Change of concentration of coke with temperature for 4 h heating. (1) Cracking bitumen–sand mixtures; (2), (3) cracking bitumen only. (From Ref. 23, reproduced by permission of Butterworth-Heinemann Ltd.)

recovery techniques. Results should be useful for predicting how the oil composition will evolve on a field scale.

Some analyses have been carried out on the fluids produced and on the coke residue during in-situ combustion laboratory tests with oils of different compositions.[10] They show that the amount of coke can be related to the oil's thermal reactivity, which is influenced by the oil's geochemical composition. On the contrary, the coke composition, determined by elemental analysis and infrared spectroscopy, depends on the temperature reached in the porous medium and not on the oil properties.[8,11]

It is generally assumed that the coke formation may be written as follows:[8,11,12]

$$\text{Aromatics} \rightarrow \text{Resins} \rightarrow \text{Asphaltenes} \rightarrow \text{Coke} \qquad (8.14)$$

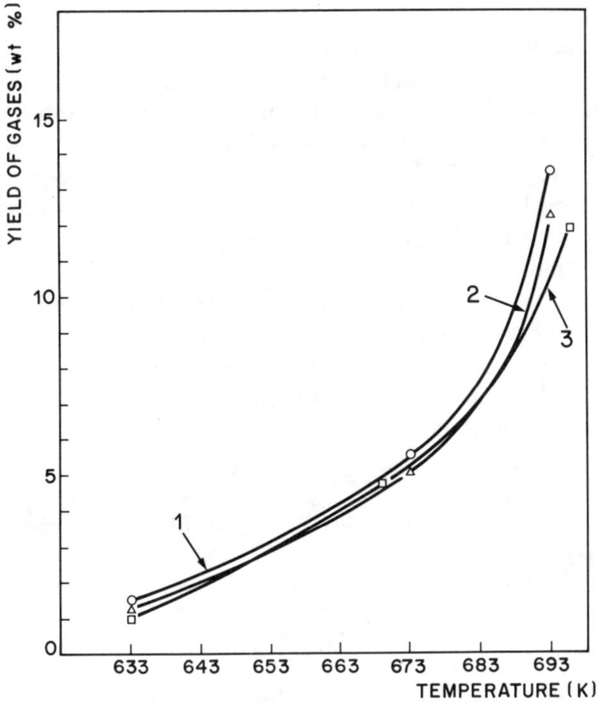

Fig. 8.6 Change of concentration of gases with temperature for 4 h heating. (1) Cracking bitumen–sand mixtures; (2),(3) cracking bitumen only. (From Ref. 23, reproduced by permission of Butterworth-Heinemann Ltd.)

Resins and asphaltenes are the main precursors of coke. Moreover, the small labile species bonded to the main structure of asphaltenes and resins are formed by cracking; thus, the saturated and aromatic fractions are enriched by cracking products. These thermally neoformed hydrocarbons are qualitatively similar to those of the original oil (before biodegradation or other natural actions), but the quantity can be different.[8] More generally, it has been observed that, during a complete high temperature pyrolysis, a crude oil yields from 30 to 40% coke and 70 to 60% CH_4 depending on its initial composition.[13]

8.3.5 *Kinetics of non-oxidative reactions*

Different kinds of reactions have to be considered. These include not only the decomposition of large hydrocarbon molecules into smaller species eventually yielding gases, but also the combining of molecules to yield large amounts of polymeric materials.

Moreover, crude oils are made of a tremendously wide variety of compounds. So, the composition of the oil and gas mixture has to be simplified and described by considering a limited number of pseudo-components as has been done in most models.[18–23] These are generally light, medium and heavy components and solids. Cracking is generally assumed to be a first order reaction with an activation energy of the order of 250 kJ/mole.

Other authors[24] have argued that, because the actual composition of the heavy oil fraction changes during the reaction, the reaction order n has to be considered to be temperature dependent and greater than one. Thus it should decrease with increasing temperature according to the following exponential law:

$$n(T) = a\mathrm{e}^{-bT} \tag{8.15a}$$

where a and b are characteristics of the crude oil system and T the absolute temperature, with

$$-\frac{\mathrm{d}w_\mathrm{h}}{\mathrm{d}t} = k_\mathrm{h}\left(\frac{w_\mathrm{h}}{w_\mathrm{hi}}\right)^{(n(T))} \tag{8.15b}$$

$$k_\mathrm{h} = F_\mathrm{h}\exp\left(-\frac{E_\mathrm{h}}{\mathrm{R}T}\right)$$

where h is the heavy oil fraction, w is the weight fraction, i is the initial condition, F_h is the frequency factor (sometimes called the pre-exponential factor or the Arrhenius constant), E is the activation energy, and R is the universal gas constant. Typical results are given in Table 8.1. Philips *et al.*[23] found that the presence of sand affects the yield of the different pseudo-components, and thus the activation energies for the reactions involved in the cracking of bitumen–sand

Table 8.1 Kinetic data for various crude oils[24]

	Lisbon gas oil	*Athabasca bitumen*	*Crude A*	*Crude B*
a	6·737 3	$1{\cdot}364 \times 10^{3}$	$4{\cdot}482 \times 10^{6}$	$1{\cdot}259 \times 10^{5}$
$b(\mathrm{K}^{-1})$	0·001 808	0·010 10	0·021 93	0·016 7
E_h (kJ/mole)	220·2	223·96	169·3	50·5
F_h (s^{-1})	$2{\cdot}699 \times 10^{12}$	$3{\cdot}899 \times 10^{13}$	$4{\cdot}807 \times 10^{8}$	2·715

mixtures were lower than those involved in the cracking of bitumen alone.

8.4 Oxidation and combustion

Physical and chemical transitions occurring in a given crude oil under an oxidative environment can easily be detected by thermoanalysis techniques.

The differential thermal analysis (DTA) concept is based on the temperature changes that occur in a medium involved in endothermal or exothermal processes. The DTA apparatus is made up of two identical cupels; one containing the reacting sample and the other a reference substance. The cupels may be in a controlled atmosphere, flowing or at rest, e.g. an oxygen-containing gas for the oxidation reactions. The difference in temperature between the two cupels is measured. Differential scanning calorimetry (DSC) is based on the same principle, except that the rate of heat released is recorded instead of the difference in temperature. Thermogravimetric analysis (TGA) entails the continuous recording of the weight changes produced when heating a sample at a constant rate. This technique is complementary to DTA, since the weight variation data are relatable to the DTA peaks. The derivative of the TGA curve (DTG) may also be useful (see Section 8.4.4).

The Effluent gas analysis (EGA) technique, based on the continuous analysis of the effluents from the reactions of a sample subjected to a gas flow and programmed heating, provides a good image of the type of reactions involved.

Figure 8.7 shows different reaction phases occurring during a heating process of crude oil and is a composite plot of TGA, DTG, DSC and CO_2 production curves in an air and nitrogen purge.[25] TGA and DTG curves were generated from analyses in which the crude oil sample was heated at a constant heating rate of 10°C/min with an air or nitrogen purge of 160 cm^3/s. The CO_2 profile was generated from the effluent gas analysis of the TGA run.

- The first peak on the DTG curve in the air purge, which overlaps the distillation peak of the DTG curve in the nitrogen purge, is apparently the distillation of the lighter ends of the crude oil. It starts at room temperature and continues simultaneously with the

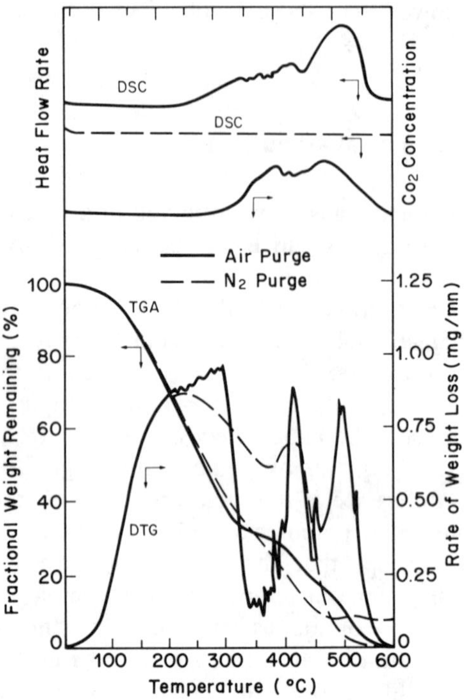

Fig. 8.7 Crude oil reaction phases during a heating process. (From Ref. 25, reproduced by permission of The Society of Petroleum Engineers.)

other reactions throughout the heating period. Slight endothermic behavior is observed on the early part of the DSC curve before the exothermicity of the other reactions is observed.

- The second peak, starting at approximately 350°C, corresponds to low temperature oxidation and to the beginning of combustion; it is exothermic and oxygenated products are formed with some CO and CO_2. The apparent H/C ratio is high. The distinct peak on the DTG curve of the N_2 purge in the same temperature range is an indication of thermal cracking reactions.
- The last peak, which covers a range of 360–600°C, corresponds to exothermic reactions producing essentially CO_2 and some CO. The apparent H/C ratio is low, near to zero. This peak corresponds to the combustion of solid coke.

8.4.1 *Low temperature oxidation*

Low temperature oxidation (LTO) is probably the most thoroughly studied group of reactions involved in in-situ combustion processes. LTO results in partially oxygenated products due to the addition of oxygen to hydrocarbon chains and in CO and CO_2 formation due to the beginning of combustion.

The distinct feature of LTO reactions lies in the fact that oxygen consumption cannot be accounted for by considering CO and CO_2 production. Indeed, elemental analyses made on the oil extracted from rock samples showed higher O/C atomic ratios than the original one.[26] These reactions occur mainly at temperatures lower than 200°C. LTO reactions are believed to induce an alteration of the original crude oil's physical properties and an enhancement of fuel-oil deposition.[27]

The extent of LTO reactions depends on the reactivity of the oil, the temperature, the amount of oxygen available downstream from the front, the characteristics of the mineral matrix and the pressure.

8.4.2 *Influence of oil composition*

Crude oil transformations in an oxidative heating environment are strongly influenced by the composition of the crude oil itself and by the presence of solid particles and their compositions. Burger and Sahuquet[2] used effluent gas analysis to illustrate catalytic effects of some metallic derivatives and to investigate how the properties of both oil and porous media influence crude oil combustion. The effect of the sand's grain size was also investigated (see Ref. 1, p. 260). Two different sizes of sand particles were used in a small reactor. It was noted that CO_2 production was 2·6 times higher for the finer sand, and the corresponding EGA peak was sharper and displayed a higher differential temperature than for coarser sands.

Bousaid and Ramey[28] also found that the amount of fuel oil deposited at a given temperature increased when adding 20%wt of clay to the sand matrix. They performed isothermal cracking of the crude oil/sand mixture in a non-oxidizing gas flow followed by isothermal combustion of the coke.

The effect of the host matrix and solid additives on the combustion of crude oil is best detected by thermal analysis techniques such as

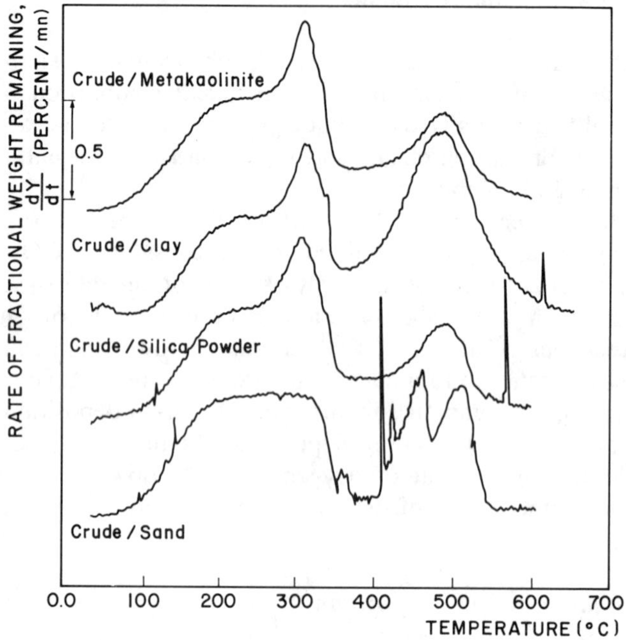

Fig. 8.8 Effect of granular materials on DTG curves for crude oil combustion. (From Ref. 32, reproduced by permission of John Wiley & Sons.)

TGA and DSC. This has clearly been demonstrated by many authors.[25,29–31] Figure 8.8 shows the effect of clay on the DTG curve generated for the Iola crude oil combustion in the presence of air.[32] The changes in the DTG curve are in the form of an increase in the low temperature combustion and in a merging of the two combustion/cracking peaks into one peak.

There is also evidence in the literature[2,27–33] that crude oil combustion can be catalyzed in the presence of clay as is evidenced by the decrease in the activation energy of combustion reactions.

It has been demonstrated[2] that metal traces may affect the reactions (compare Fig. 8.9(a) with Fig. 8.9(b)). Moreover, in addition to surface area, Drici and Vossoughi[25] looked at the effect of heavy metal oxides on crude oil combustion reactions. They used a variety of metal oxides such as titanium, nickel, vanadium, chromium, and cupric and ferric oxides. Although they observed some noticeable catalytic effect

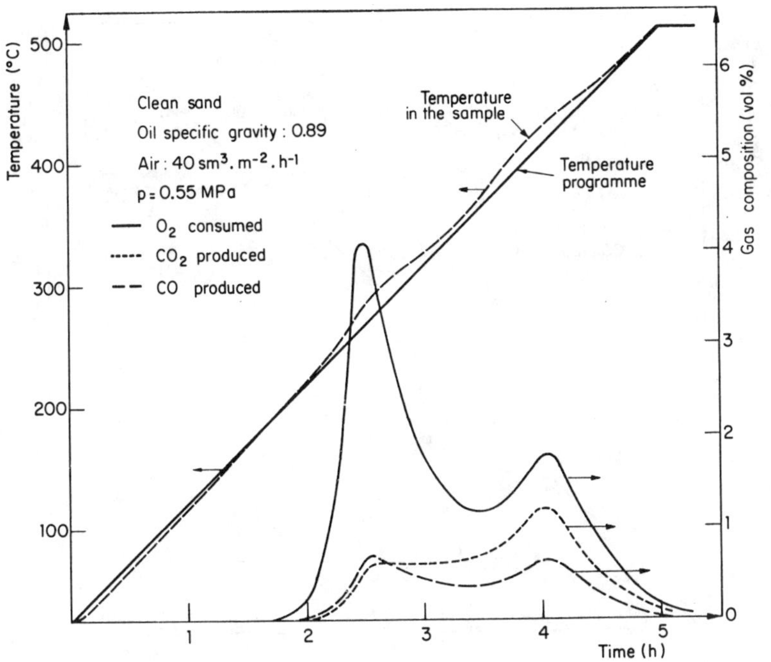

Fig. 8.9a EGA during oxidation of crude oils in porous media—clean sand. (From Ref. 2, reproduced by permission of The Society of Petroleum Engineers.)

specific to each metal oxide, the main conclusion drawn from their numerous TGA/DSC runs was that the surface area effect was predominant. Namely, addition of a small amount of metal oxides to the high-surface-area solid particles such as silica powder or alumina has minimal to no effect on TGA/DTG and DSC curves.

8.4.3 *Kinetics of chemical reactions*

Based on the background established so far, it seems to be reasonable to study the kinetics of the chemical reactions involved in an in-situ combustion process for each group of reactions rather than for each individual reaction. For this reason, most investigators have concentrated on a specific group of reactions. Bousaid and Ramey[28] performed various runs in which isothermal oxidation of precoked,

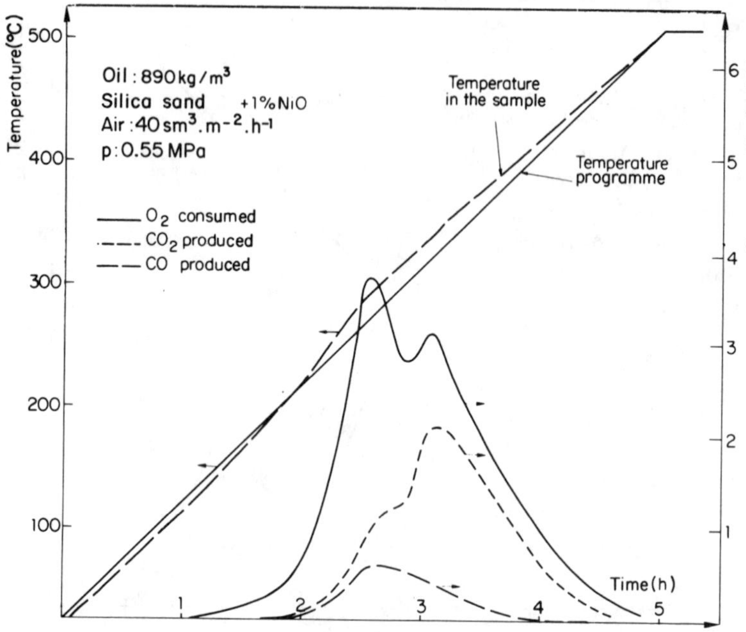

Fig. 8.9b EGA during oxidation of crude oils in porous media—NiO containing sand. (From Ref. 2, reproduced by permission of The Society of Petroleum Engineers.)

unconsolidated sands was carried out in stationary beds. They proposed an expression for the carbon burning rate as a function of carbon concentration, combustion temperature and oxygen partial pressure. They used two types of crude oil and concluded that the burning of carbon was of the first order with respect to carbon concentration and oxygen partial pressure. These results are similar to the observation made by Lewis *et al.*[34] who studied the oxidation of coke in a fluidized bed.

Dabbous and Fulton[35] more specifically investigated the kinetics of low temperature oxidation. Their data produced rate equations for reactions below 260°C for two types of crude oils (density 935 kg/m^3 and 890 kg/m^3) and crushed Berea sandstone as the porous media. The order of reaction with respect to oxygen was found to be between 0·5 and 1·0 depending on the crude oil but independent of the porous media. The activation energy was insensitive to both the crude and the porous media.

Burger and Sahuquet[2] studied the kinetics of the oxidation reactions of both complete and incomplete combustion reactions and concluded that the reaction kinetics formulation is strongly dependent on the physical state of the reactants, i.e. solid, liquid or gas, and on their distribution throughout the porous medium.

Fassihi *et al.*[33] based their kinetic study on a continuous analysis of effluent gas from a combustion cell and proposed a model to differentiate several consecutive reactions which take place concurrently. They distinguished three reactions, namely LTO, middle-temperature reactions and coke combustion. Their kinetic model was based on a power-law type equation with a rate constant dependent on temperature according to the Arrhenius equation. They noticed two differences in kinetic behavior for natural cores compared with clean sand matrices: (1) a reduction in activation energy, and (2) greater fuel deposition. Results of different authors are summarized in Table 8.2.

8.4.4 *Kinetic modeling*

There are basically two approaches to creating a kinetic model describing crude oil combustion: (1) based on effluent gas analysis (EGA), and (2) using thermal analysis techniques. In both cases a power-law type kinetic equation with unknown reaction orders is assumed. Further, the temperature dependency of the rate constant is assumed to follow the Arrhenius equation. A classic example of the first approach is Fassihi's work,[33] while the second approach is illustrated by Vossoughi and El-Shoubary's work.[31]

(a) *Kinetic model from effluent gas analysis.* In this approach, kinetic analysis is based on the analysis of the gas evolving from the reactants while being heated. Assuming power-law type kinetic equation, the combustion rate, R_c, of crude oil in a porous medium is given by,

$$R_c = \frac{dC_m}{dt} = KP_{O_2}^m C_m^n \tag{8.16}$$

where C_m is the instantaneous concentration of fuel, K is the rate constant, P_{O_2} is the partial pressure of oxygen, and m and n are reaction orders. The reaction rate constant is assumed to follow the

Table 8.2 Oxidation and combustion reaction rates

Crude oil density (kg/m^3)	Reaction bed type	Frequency factor ($(Pa\,s)^{-1}$)	Activation energy (kJ/mole)	Pressure (kPa)	Temperature range (°C)	Reference
Metallurgical coke	Fluidized	$7{\cdot}4 \times 10^{-2}$	121·4	1×10^2	300–500	34
Hardwood charcoal	Fluidized	2×10^{-3}	66·5	1×10^2	300–500	
Grafit	Fluidized	$1{\cdot}3 \times 10^3$	205·8	1×10^2	300–500	
—[a]	Sand pack	—[a]	125·5	4×10^3	60–145	36
973·2	Berea sand	$2{\cdot}37 \times 10^{-3}$	61·9	3×10^2	150–540	28
973·2	Berea + clay	$2{\cdot}43 \times 10^{-4}$	48·4	5×10^2	150–540	
921·2	Berea	$9{\cdot}14 \times 10^{-3}$	59·8	2×10^2	150–540	
934·6	Berea sand	$1{\cdot}38 \times 10^{-3}$	58·9	2×10^2	120–250	35
943	Sand pack	6×10^3	135	2×10^2	200–300	33
992	Sand pack	4×10^2	120	2×10^2	200–300	
1 004	Sand pack	2×10^3	133	2×10^2	200–300	
938·3	Mesh sand	—[a]	196·6	1×10^2	100–600	27
938·3	Silica powder	—[a]	159	1×10^2	100–600	
938·3	Sand + clay	—[a]	78·24	1×10^2	100–600	
938·3	Silica sand	$4{\cdot}7 \times 10$	123·5	1×10^2	100–600	31
890	Sand	$1{\cdot}38 \times 10^{-3}$	71·92	$3 \times 10 \rightarrow 10^4$	90–145	1
945	Sand	0·65	84·80	$3 \times 10^3 \rightarrow 10^4$	80-145	
920	Sand	$4{\cdot}74 \times 10^{-3}$	70·92	$3 \times 10^3 \rightarrow 10^4$	50–100	

[a] Not available.

Arrhenius equation for the temperature dependency, i.e.

$$K = F_f \exp(-E/RT) \tag{8.17}$$

where E is the activation energy, and F_f is the frequency factor.

From the change in oxygen concentration, ΔC_{O_2}, related to the change in fuel concentration, it is possible to obtain:

$$\frac{\Delta C_{O_2}}{\left(\int_F^\infty C_{O_2}\,dt'\right)^n} = \beta' \exp(-E/RT) \tag{8.18}$$

where

$$\beta' = \left(\frac{\dot{v}_g}{\nu AL}\right)^{n-1} F_f P_{O_2}^m \tag{8.19}$$

where $\dot{v}_g$ is the gas flow rate, A is the cross-sectional area, L is the length of the pack, and ν is the moles of O_2 reacted with 1 g of fuel. Fassihi *et al.* called the left-hand side of eqn (8.18) the 'relative reaction rate'.[33] A similar equation can be derived for any other oxidation regime in which oxygen is being consumed. Data required to apply eqn (8.18) are ΔC_{O_2} as a function of time which is obtained from effluent gas analysis. The relative reaction rate can be calculated by graphic integration using the ΔC_{O_2} set of data. An Arrhenius type plot, i.e. relative reaction rate versus $1/T$ on a semilog scale, gives E/R as the slope and β' as the intercept (Fig. 8.10). Equation (8.19) indicates that a plot of log β' versus log P_{O_2} can be used to evaluate the reaction order, m, and the frequency factor, F_f.

The advantage of this approach is its ability to produce a kinetic model for LTO reactions. However, the difficulty lies in generating ΔC_{O_2} data, especially for the mid-temperature combustion and LTO regions, which must be obtained by substraction. This makes the data more scattered and less reliable. The best results are obviously obtained if effluent gas analysis (EGA) is combined with a kinetic study using thermal analysis techniques.

(b) *Kinetic model from thermal analysis techniques.* Thermal analysis techniques, such as thermogravimetric analysis and differential scanning calorimetry, measure the energy consumed or produced by the oil as its temperature is increased in an oxidizing gas environment.

As before, a power-law type kinetic model with unknown reaction orders is assumed. The specific rate constant is assumed to be

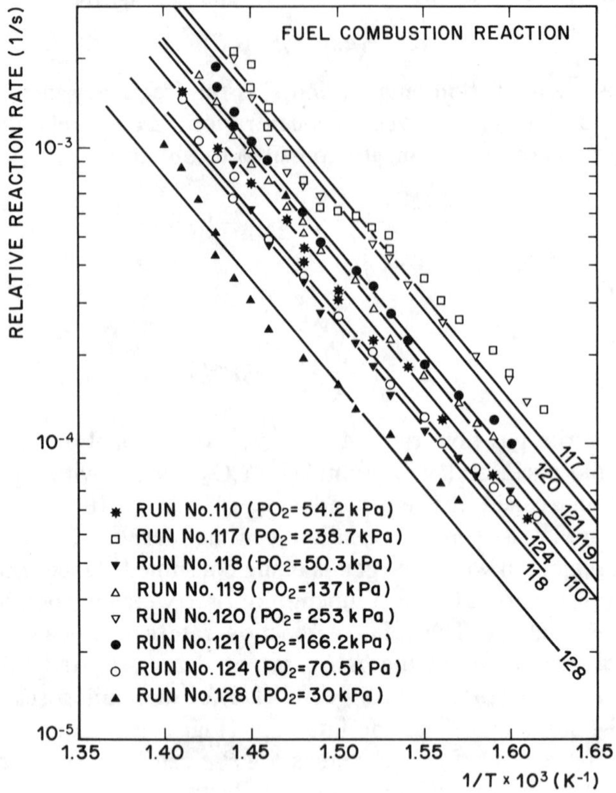

Fig. 8.10 Arrhenius plot for fuel combustion reaction (Huntington Beach Oil). (From Ref. 33, reproduced by permission of The Society of Petroleum Engineers.)

temperature dependent according to the Arrhenius equation.

$$-\frac{d\alpha}{dt} = K\alpha^n P_{O_2}^m A^l \tag{8.20}$$

with

$$\alpha = \frac{W - W_f}{W_i - W_f} \tag{8.21}$$

and

$$K = F_f \exp(-E/RT) \tag{8.22}$$

where A is the specific surface area of the sand grain, n, m and l are power-law constants to be determined, W is the weight of reactant at any time, W_f is the final weight, and W_i is the initial weight of the crude oil/sand mixture sample; α and $d\alpha/dt$ are read directly from the thermogravimetric (TGA) curve and the derivative of the TGA curve (DTG), respectively. For the DSC, the kinetic model is assumed to be:

$$-\frac{dH}{dt} = KH^n P_{O_2}^m A^l \tag{8.23}$$

where dH/dt is the rate of heat released with respect to time, and H is the enthalpy still to be released. It should be noted that H is directly proportional to α, assuming a constant average heat of reaction for each regime of oxidation reactions. H and dH/dt are directly obtained from the DSC curve.

Because of the simplicity of the experimental runs, a large number of experiments can be performed to evaluate the effect of variables with greater accuracy and a greater degree of reliability than for the EGA technique. The activation energies determined by this technique are in agreement with those reported elsewhere.[28,33,35] The reaction order with respect to hydrocarbon content and oxygen partial pressure is also in good agreement with reported values determined by different approaches as reflected in Table 8.2. These values then have to be introduced in numerical models helpful for the study of the different processes (see Chapters 9 and 10).

8.5 References

1. Burger, J., Sourieau, P. & Combarnous, M., *Thermal Methods of Oil Recovery,* Editions Technip, Paris, and Gulf Publishing Company, Houston, 1985.
2. Burger, J. C. & Sahuquet, B. C., Chemical aspects of in-situ combustion: heat of combustion and kinetics. *Soc. Petroleum Engineers J.,* **12** (1972) 410–22.
3. Boon, J. A., Hamilton, T. Holloway, L. & Wiwchar, B., Action between rock matrix and injected fluids in Cold Lake oil sands—potential for formation damage. *J. Canadian Petroleum Technol.,* **22**(3) (1983) 55–66.
4. Kubacki, W., Boon, J., Bird, G. & Wiwchar, B., Effect of mineral transformation in porosity and permeability of dolomite rock during in-situ recovery of bitumen: a preliminary study. *Bull. Canadian Petroleum Geol.,* **32**(3) (1984) 281–8.

5. Bizon, A. E., Boon, J. A. & Kubacki, W., Mineral transformations during in-situ recovery of bitumen from carbonate rock: a statistical experimentation study. *Bull. Canadian Petroleum Geol.*, **32**(1) (1984) 1–10.
6. Prayer, G., Chaziteodorou, G., Muss, H. & Rosskamp, M., Results of a steamdrive pilot project in the Ruehlertwist Field, Federal Republic of Germany. *J. Petroleum Technol.*, **35** (1985) 284–94.
7. Sahuquet, B. & Sitbon, A., Recent results of steamdrive development in Lacq. 4th European Symp. on Enhanced Oil Recovery Hamburg, 27–29 October 1987. DGMK, Hamburg, 1987, pp. 635–44.
8. Monin, J. C. & Audibert, A., Thermal cracking of heavy-oil/mineral matrix systems. *SPE Reservoir Engineering*, **3**(4) (1988) 1243–50.
9. Bojes, J. M. & Wright, G. B., Application of fluid analyses to the operation of an in-situ combustion pilot. 37th Annual Technical Meeting of the Petroleum Soc. of CIM, Paper No. 86-37-61 Calgary, 8–11 June 1986 Petroleum Soc. of CIM, vol. 2, pp. 315–49.
10. Jha, K. N., Lafleur, R., Sawatzky, H., Verkoczy, B., Soueran, D., Reichert, C. & Fuhr, B., The effect of recovery processes on the characteristics of the produced heavy oil. Paper No. 41, Fourth Unitar/UNDP Conf. on Heavy Crude and Tar Sands, Edmonton, 7–12 August 1988.
11. Audibert, A. & Roucaché, J., Modifications in the composition of crude oils during in-situ combustion. *ACS Symp. Series on Oil Field Chemistry No. 396*, ed. J. Borchardt & T. F. Yen, 1989, pp. 408–25.
12. Fassihi, M. R., Meyers, K. O. & Weisbrod, K. R., Thermal alteration of viscous crude oils. SPE Paper No. 14225, 60th Annual Techn. Conf. and Exhibition of Soc. Petroleum Engineers, Las Vegas, 22–25 September 1985.
13. Behar, F., Ungerer, P., Audibert, A. & Villalba, M., Experimental study and kinetic modelling of crude oil pyrolysis in relation to thermal recovery processes. Paper No. 97, Fourth Unitar/UNDP Conf. on Heavy Crude and Tar Sands, Edmonton, 7–12 August 1988.
14. Akstinat, M. H., Gas evolution and change of oil composition during steam flooding of oil reservoirs. *J. Petroleum Geol.*, **5**(4) (1983) 363–88.
15. Clark, P. D. & Hyne, J. B., Steam–oil chemical reactions. Mechanisms for the aquathermolysis of heavy oils. *AOSTRA J. Research*, **1** (1984) 15–20.
16. Clark, P. D., Lesage, K. L. & Hyne, J. B., The behaviour of various oil sand towards high temperature steams. Paper No. 54, Fourth Unitar/UNDP Conf. on Heavy Crude and Tar Sands, Edmonton, 7–12 August 1988.
17. Venkatesan, V. N. & Shu, W. R., Alteration in heavy oil characteristics during thermal recovery. *J. Canadian Petroleum Technol.*, **25**(4) (1986) 66–71.
18. Hayashitani, M., Bennion, D. W., Donnelly, J. K. & Moore, R. G., Thermal cracking models for Athabasca oil sands. SPE Paper No. 7549, 53rd Annual Techn. Conf. and Exhibition of Soc. Petroleum Engineers, Houston, 1–3 October 1978.
19. Abu Khamsin, S. A., Brigham, W. E. & Ramey, Jr, H. J., Reaction

kinetics of fuel formation for in-situ combustion. SPE Paper No. 15736, Fifth Middle East Oil Show of Soc. Petroleum Engineers, Manama, Bahrein, 7–10 March 1987.
20. Ungerer, P., Behar, F., Villalba, M., Heum, D. R. & Audibert, A., Kinetic modelling of oil cracking. *Adv. Org. Geochem.*, **13**(4–6) (1988) 857–68.
21. Shu, W. R. & Hartman, K. J., Thermal visbreaking of heavy oil during steam recovery processes. *SPE Reservoir Engineering*, **1**(5) (1986) 474–89.
22. Millour, J. P., Moore, R. G., Bennion, D. W., Ursenbach, M. G. & Gie, D. H., A simple implicit model for thermal cracking of crude oils. SPE Paper No. 14226, 60th Annual Techn. Conf. and Exhibition of Soc. Petroleum Engineers, Las Vegas, 22–25 September 1985.
23. Phillips, C. R., Haidar, N. I. & Poon, Y. C., Kinetic models for the thermal cracking of Athabasca bitumen. *Fuel*, **64** (1985) 678–91.
24. Lin, C. Y., Chen, W. H. & Culham, W. E., New kinetic models for thermal cracking of crude oils in in-situ combustion processes. SPE Paper No. 13074, 59th Annual Techn. Conf. and Exhibition of Soc. Petroleum Engineers, Houston, 16–19 September 1984.
25. Drici, O. & Vossoughi, S., Catalytic effect of heavy metal oxides on crude oil combustion. *SPE Reservoir Engineering*, **2**(4) (1987) 591–5.
26. Petit, H., In-situ combustion experiments with oxygen-enriched air. *In-situ*, **14**(1) (1990) 49–76.
27. Adegbesan, K. O., Donelly, J. K., Moore, R. G. & Bennion, D. W., Liquid phase oxidation kinetics of oil sands bitumen: Models for in-situ combustion numerical simulators. *AICHE J.*, **32**(8) (1986) 1242–52.
28. Bousaid, I. S. & Ramey, Jr, H. J., Oxidation of crude oil in porous media. *Soc. Petroleum Engineers J.*, **8**(2) (1968) 137–48.
29. Vossoughi, S., TGA/DSC Techniques as research tools for the study of the in-situ combustion process. *Thermochimika Acta*, **106** (1986) 63–9.
30. Vossoughi, S., Bartlett, G. W. & Willhite, G. F., Prediction of in-situ combustion process variables by use of TGA/DSC techniques and the effect of sand-grain specific surface area on the process. *Soc. Petroleum Engineers J.*, **25**(5) (1985) 656–64.
31. Vossoughi, S. & El-Shoubary, Y., Kinetics of crude oil coke-combustion. *SPE Reservoir Engineering*, **4**(2) (1989) 201–06.
32. Vossoughi, S., Willhite, G., El Shoubary, Y. & Bartlett, G., Study of the clay effect on crude oil combustion by thermogravimetry and differential scanning calorimetry. *J. Therm. Anal.*, **27** (1983) 17–36.
33. Fassihi, M. R., Brigham, W. E. & Ramey, Jr, H. J., Reaction kinetics of in-situ combustion, Part 1—Observation; Part 2—Modeling. *Soc. Petroleum Engineers J.*, **24**(4) (1984) 399–416.
34. Lewis, W. K., Gilliland, E. R. & Paxton, R. R., Low-temperature oxidation of carbon. *Ind. & Eng. Chem.*, **46** (1954) 1327–31.
35. Dabbous, M. K. & Fulton, P. F., Low-temperature oxidation reaction kinetics and effect on the in-situ combustion process. *Soc. Petroleum Engineers J.*, **14**(3) (1974) 253–62.
36. Tadema, H. J. & Weijdema, J., Spontaneous ignition of oil sands. *Oil and Gas J.*, **68**(50) (1970) 77–80.

9 Hot fluid injection

Claude Gadelle
Institut Français du Pétrole, Rueil-Malmaison, France

and

Louis M. Castanier
Stanford University, California, USA

9.1 Introduction . . . 272

9.2 Temperature effects on the physical properties of fluids and matrix . . . 272
9.2.1 Viscosity . . . 272
9.2.2 Density . . . 276
9.2.3 Enthalpy and heat capacity . . . 277
9.2.4 Thermal conductivity . . . 278
9.2.5 Interfacial tension–wettability–capillary pressure . . . 280

9.3 Physical mechanisms involved in hot fluid injection . . . 280
9.3.1 Thermal expansion and rock compressibility . . . 280
9.3.2 Effect of the viscosity ratio . . . 281
9.3.3 Evolution of the absolute and relative permeabilities . . . 282
9.3.4 Oil vaporization . . . 283
9.3.5 Relative influence of the various mechanisms . . . 286

9.4 Heat transfer during hot fluid injection . . . 288
9.4.1 Heat loss from surface lines . . . 288
9.4.2 Heat loss in the injection wells . . . 289
9.4.3 Reservoir heating . . . 291

9.5 Implementation of hot fluid injection . . . 295
9.5.1 Hot water injection . . . 295
9.5.2 Cyclic steam injection . . . 298
9.5.3 Steam drive . . . 301
9.5.4 Improvements in steam injection . . . 303

9.6 References . . . 309

9.1 Introduction

Hot fluid injections consist of hot water injection and steam injection. Hot water injection is not widely used and will be discussed only briefly. Steam injection is the most widely used enhanced oil recovery method. In California alone, over 71 000 m^3/day were produced in 1988 as a result of steam injection, representing 72% of the total enhanced oil recovery for the USA. Part of this recovery is due to cyclic steam injection (Huff and Puff), where steam is injected in a well and the same well is subsequently put back into production (Fig. 9.1). Most of the recovery is obtained by steam drive in which the reservoir is swept by steam from well to well (Fig. 9.2). Usually steam drive follows cyclic steam injection.

This chapter on hot fluid injection deals first with the changes in the physical properties of fluids and rocks with temperature, and with the fundamental mechanisms involved in this recovery technique. Then a brief description of heat transfer calculations applied to hot fluid injection is given. The final section describes the implementation of the various methods. Books on thermal methods of oil recovery have been published.[1,2] Answers to specific questions on hot fluid injection can be found in these monographs.

9.2 Temperature effects on the physical properties of fluids and matrix

The temperature of injected fluids can be as high as 350°C. Consequently, the behavior of the physical properties of both reservoir fluids and rock matrix must be known in the temperature range between initial reservoir temperature and the temperature of the injected fluid.

9.2.1 *Viscosity*

The viscosity of real gases tends to increase with pressure. The effect of temperature on dynamic viscosity μ can be expressed by the following empirical law:

$$\mu = CT^{b} \tag{9.1}$$

where C is a coefficient, T is the absolute temperature and, for many gases, b is between 0·7 and 1.[3]

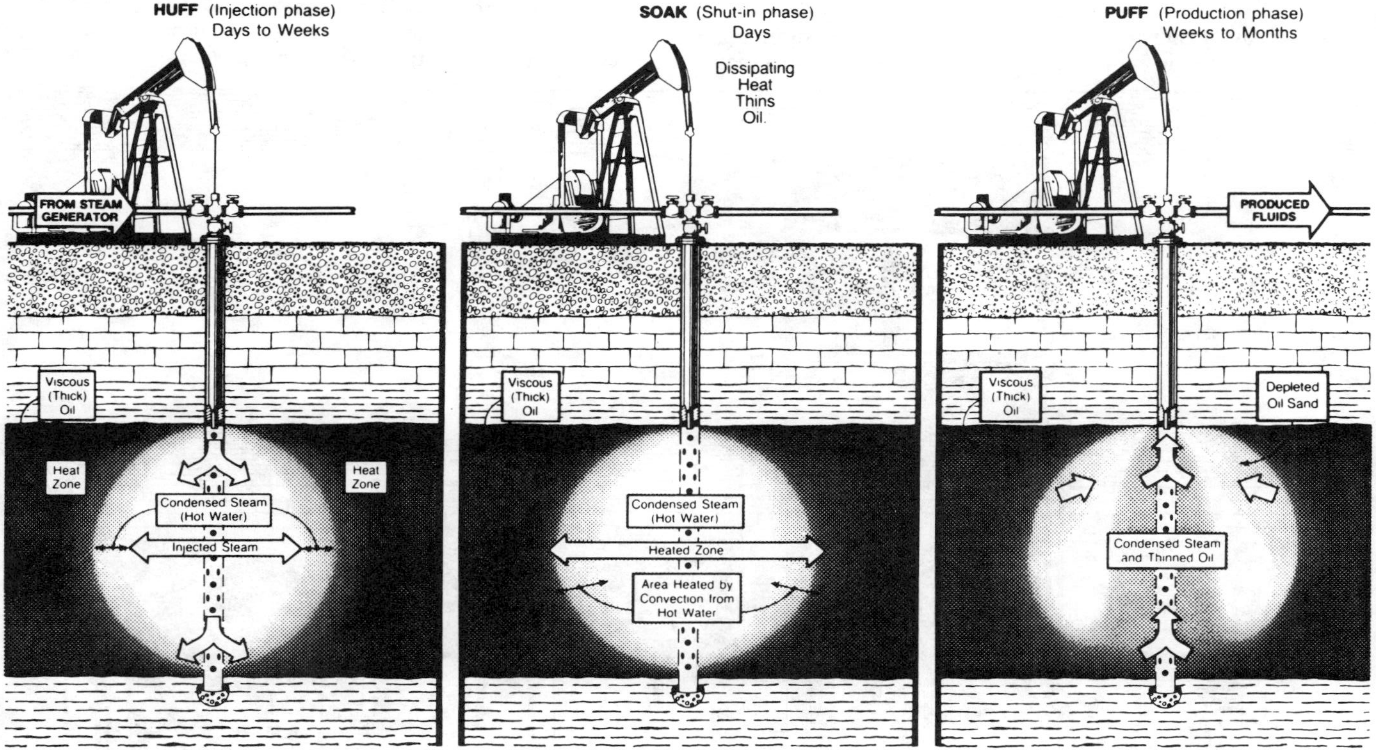

Fig. 9.1 Cyclic steam stimulation. Steam, injected into a well in a heavy-oil reservoir introduces heat that, coupled with alternate 'soak' periods, thins the oil allowing it to be produced through the same well. This process may be repeated until production falls below a profitable level. Schematic portrays one well during the 3 phases of this process. Flow pattern is stylized for clarity. (Reproduced by permission of the US Department of Energy.)

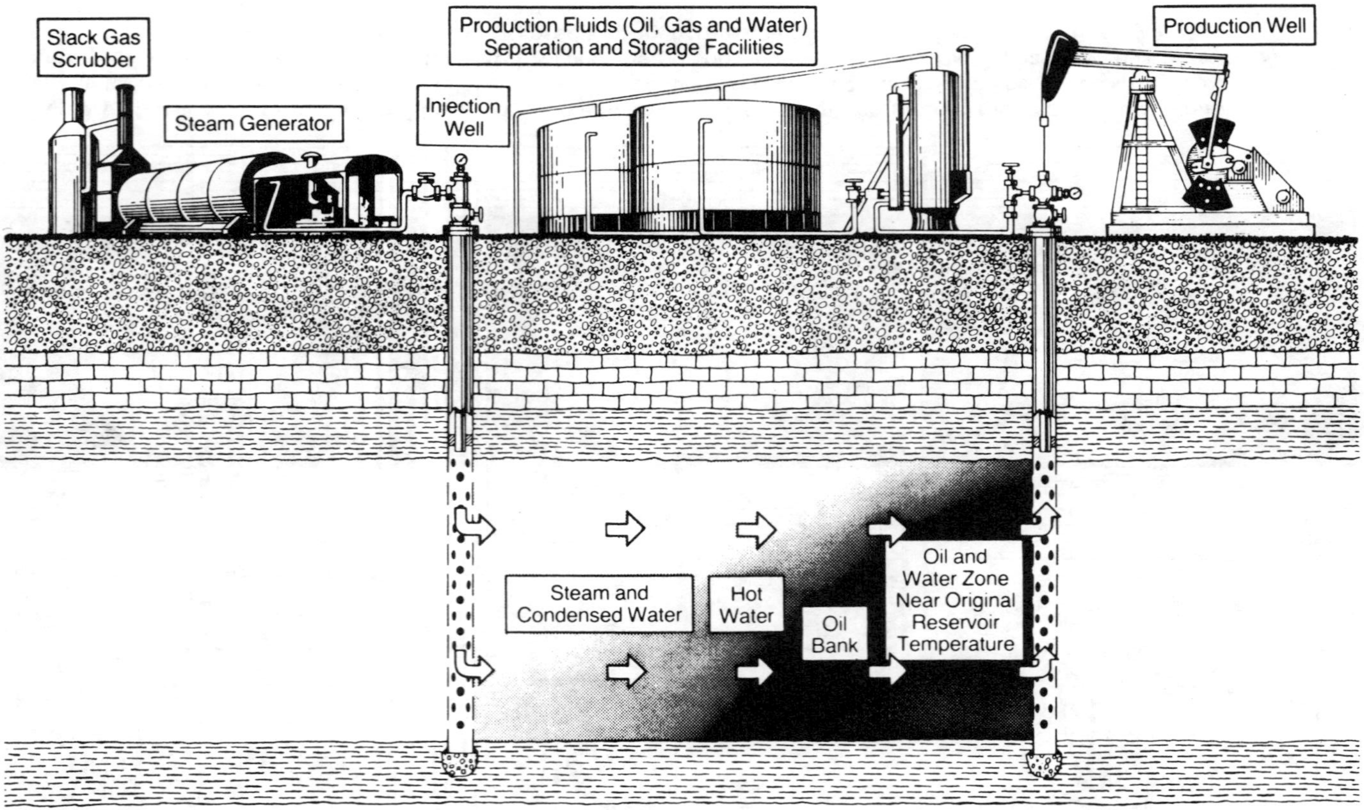

Fig. 9.2 Steam flooding. Heat, from steam injected into a heavy-oil reservoir, thins the oil making it easier for the steam to push the oil through the formation toward production wells. Heat reduces viscosity of oil and increases its mobility. (Reproduced by permission of the US Department of Energy.)

The viscosity of a liquid decreases very rapidly as the temperature increases. As a general rule, the more viscous a liquid is, the more significant its viscosity decrease is for a given increase in temperature. One equation for predicting the variation of dynamic viscosity μ with temperature is that of Andrade:

$$\mu = b \exp(C/T) \tag{9.2}$$

where b and C are coefficients, and T is the absolute temperature.

For the viscosity of liquid hydrocarbons, a better representation, used to establish the ASTM viscosity–temperature diagram (Fig. 9.3), is as follows:

$$\log(\log(\nu + f)) = c - b \log T \tag{9.3}$$

in which ν is the kinematic viscosity, T the absolute temperature and b, c and f are coefficients.

At high temperatures, the viscosity must be measured in the

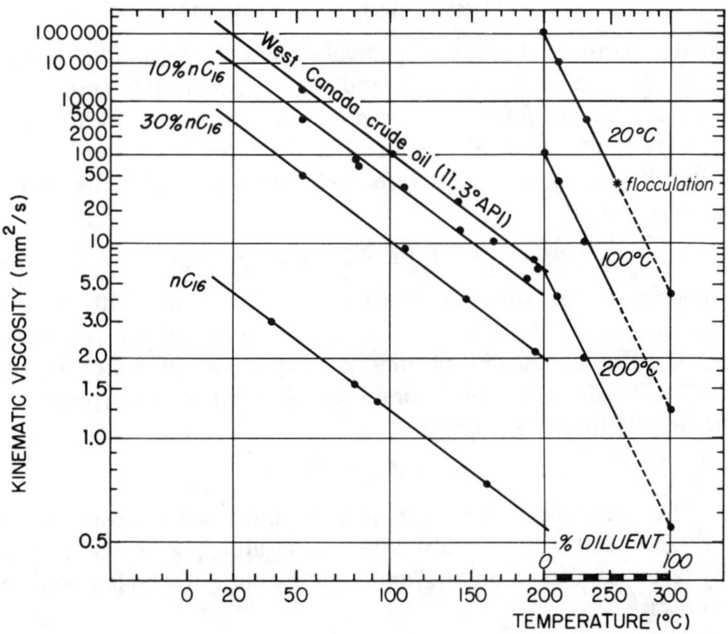

Fig. 9.3 Combined effects of temperature and dilution on the viscosity of a crude oil. (From Ref. 4, reproduced by permission of The World Petroleum Congresses.)

presence of an inert atmosphere under pressure so as to prevent oil oxidation and to avoid any noticeable change of composition of the liquid phase (by distillation or by gas dissolution). At a given temperature, the viscosity change of a blend of two miscible hydrocarbon mixtures can also be evaluated using a linear correlation[4] in the ASTM diagram (Fig. 9.3).

9.2.2 *Density*

The density ρ and its converse, the specific volume v, depend on pressure and temperature. Their influence is defined by two coefficients:
the thermal coefficient

$$\beta = (1/v)(\mathrm{d}v/\mathrm{d}T)_P = -(1/\rho)(\mathrm{d}\rho/\mathrm{d}T)_P$$

the compressibility coefficient

$$\kappa = -(1/v)(\mathrm{d}v/\mathrm{d}P)_T = (1/\rho)(\mathrm{d}\rho/\mathrm{d}P)_T$$

Thermal expansion of rocks is relatively small. According to measurements performed on quartz and sandstone between 25°C and 1000°C, the average value of β for such rocks should be of about $5 \times 10^{-5}\ \mathrm{K}^{-1}$.[5]

The density of liquids at pressure P and temperature T is calculated from density ρ_o, β_o and κ_o at reference conditions P_o, T_o using a first order approximation:

$$\rho/\rho_o = 1 - \beta_o(T - T_o) + \kappa_o(P - P_o) \tag{9.4}$$

The coefficient of thermal expansion β is around $10^{-3}\ \mathrm{K}^{-1}$ for hydrocarbons at 20°C. For crude oils, β increases when ρ decreases (Fig. 9.4). The coefficient of thermal expansion of water is about $2 \times 10^{-4}\ \mathrm{K}^{-1}$, which is about 5 times less than for hydrocarbons.

The density of gases is given by:

$$\rho = (MP)/(ZRT) \tag{9.5}$$

where M is the molecular weight, R is the universal gas constant, P is the pressure and T is the absolute temperature; Z is the compressibility factor as a function of P and T (the values of which are given in the literature).

Based on properties ρ_o and Z_o at reference pressure P_o and absolute temperature T_o, the gas density ρ at P and T is:

$$\rho = (Z_o\, P\, T_o\, \rho_o)/(Z\, P_o\, T) \tag{9.6}$$

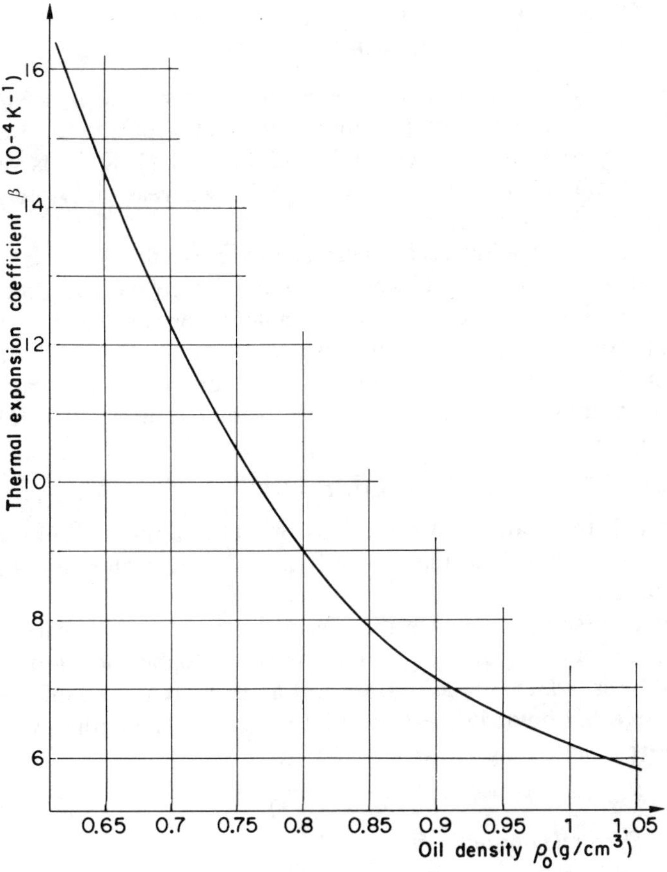

Fig. 9.4 Thermal expansion coefficient of crudes.

9.2.3 *Enthalpy and heat capacity*

When there is no chemical reaction and no change in state, the variation in enthalpy H with temperature at constant pressure can be expressed by:

$$(\mathrm{d}H) = c_p\,\mathrm{d}T \tag{9.7}$$

As a general rule the heat capacity per unit mass of solids, liquids and gases increases with temperature and frequently varies with pressure.

The effect of temperature can be expressed as follows:

$$c_p = b + cT + fT^2 \tag{9.8}$$

where c_p is the specific heat at constant pressure and b, c and f are coefficients. At room temperature, the heat capacity for sand and natural dry rocks lies between 0·75 and 0·88 kJ kg^{-1} K^{-1}. Its value for a crude oil is around 1·8 kJ kg^{-1} K^{-1} whereas it is equal to 4·18 kJ kg^{-1} K^{-1} for water.

Enthalpies are additive for perfect mixtures. The same is true for heat capacities. The heat capacity of a mixture can be determined experimentally, but it can also be calculated considering that the heat capacity of a complex mixture is equal to the sum of the heat capacities of its constituent elements. The volumetric heat capacity of a fluid-saturated porous medium is evaluated using the following equation:

$$(\rho c)^* = (1-\Phi)(\rho_\sigma c_\sigma) + \Phi S_o(\rho_o c_o) + \Phi S_w(\rho_w c_w) + \Phi S_g(\rho_g c_g) \tag{9.9}$$

where Φ is the porosity, S represents the saturations, the subscripts σ, o, w and g represent the matrix, the oil, the water and the gas, respectively.

Water possesses desirable thermal properties. Its heat capacity and its latent heat of vaporization are amongst the highest for liquids. On a weight basis, water is able to transport a much greater quantity of heat than can other fluids in the same physical state. The enthalpy of water in different states is given in Fig. 9.5.

9.2.4 *Thermal conductivity*

Heat transfer in a porous medium often occurs by thermal conduction. As long as the flow rate in the porous medium is not too great, the fluids and the reservoir rock are at the same temperature. Under this condition, it is possible to consider the real medium as a continuum medium having an equivalent thermal conductivity. The effective thermal conductivity depends on the thermal conductivity of the matrix, on the conductivity of the saturating phases and on their relative distribution in the medium. The conductivities are not additive, and the theoretical models developed for the evaluation of the equivalent thermal conductivity of actual porous media are only approximations. It is preferable to use experimentally determined values.[6]

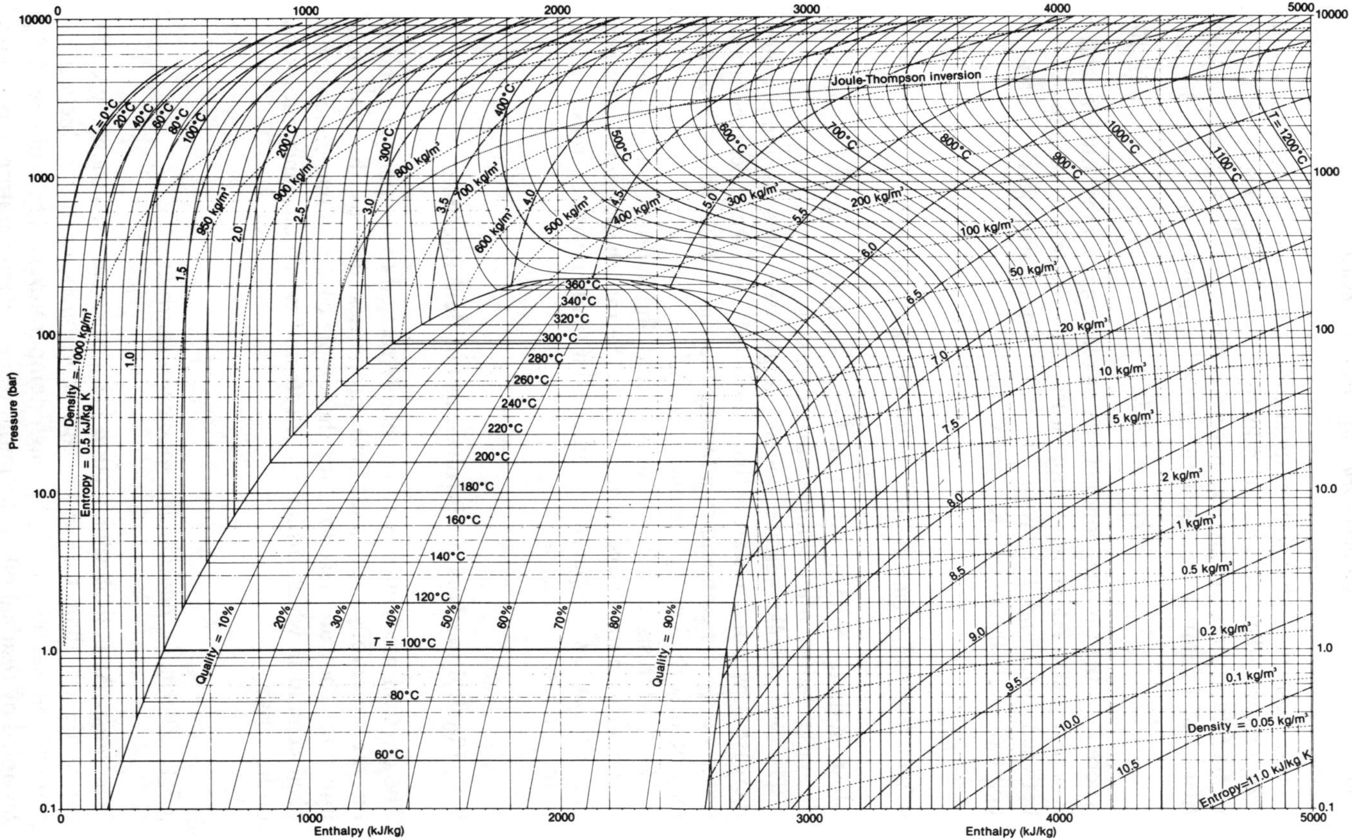

Fig. 9.5 Pressure–enthalpy chart for water and steam.

The thermal conductivity of most natural rocks changes slightly with temperature. The thermal conductivity of gases varies little with temperature, pressure, or the nature of the gas. The thermal conductivity λ of a liquid decreases with increasing temperature, according to:

$$\lambda/\lambda_o = 1 - b(T - T_o) \tag{9.10}$$

where λ_o is the thermal conductivity at reference temperature T_o and b is a constant. At room temperature, the thermal conductivity lies between 0·13 and 0·15 W m^{-1} K^{-1} for liquid hydrocarbons and is approximately 0·6 W m^{-1} K^{-1} for water.

9.2.5 *Interfacial tension–wettability–capillary pressure*

Most authors believe that interfacial tension is constant or may decrease slightly when temperature increases.

The water–solid surface contact angle decreases as temperature increases, which indicates that the rock surface becomes more water-wet (polar compounds affecting the way a rock surface is wetted may be desorbed at high temperatures).[7]

The hysteresis between drainage and imbibition capillary curves decreases with increasing temperature. The ratio between the capillary pressures at low temperature and at high temperature decreases when water saturation increases.[7]

9.3 Physical mechanisms involved in hot fluid injection

The physical mechanisms are intimately related to temperature effects on the reservoir rock and fluid properties due to the hot fluid injection. They are described for the simple case of a one-dimensional displacement in which instability effects and sweep efficiency problems do not arise.

9.3.1 *Thermal expansion and rock compressibility*

(a) *Fluids.* As indicated in Section 9.2.2, thermal expansion is greater for crude oil than for water and contributes to the reduction of residual oil saturation S_{or} at high temperatures. The volume ΦS_{or} occupied by residual oil at the initial reservoir temperature T_i becomes

$\Phi S_{or}(1 + \beta(T - T_i))$ at temperature T (Φ is considered to be independent of temperature). If residual oil saturation is assumed to be constant at any temperature, oil recovery due to thermal expansion would increase by an amount equal to $\Phi S_{or}\beta(T - T_i)$.

The effect of thermal expansion is very important for fractured reservoirs. A significant amount of oil can be produced even in the absence of imbibition.[8]

(b) *Rocks*. Rocks are characterized by very low values of the thermal expansion coefficient and of the compressibility coefficient. The variations in their petrophysical properties, porosity and permeability versus temperature and mechanical stress, are of great importance to the petroleum engineer. All authors agree on the fact that, for a given temperature, permeability tends to decrease when the effective stress increases. If permeability depends only on the geometry of the porous medium, its evolution with temperature will be a function of the thermal expansion coefficient which has a very low value, and so will be insignificant. Results concerning the influence of temperature at a given stress are contradictory. Depending on test conditions, it can be observed that absolute permeability increases, is constant or decreases when temperature increases. It should be noted that the various studies have been carried out using quite different materials and fluids. Therefore each case is a specific one. The phenomena observed may be explained as follows: (a) the thermal expansion of solids can cause a change in the size of pore throats in low permeability media; (b) interactions between the injected fluids and the matrix can be temperature sensitive; (c) chemical changes in the constituents of heated rocks can modify the porosity and permeability of the medium (see Chapter 8).

(c) *Compaction*. The compaction of some formations when pore fluid pressure declines due to fluid extraction may be a production mechanism. This is the case, for example, in the Lake Maracaibo area in Venezuela.[9]

9.3.2 *Effect of the viscosity ratio*

The effect of viscosity reduction is a key factor in displacement mechanisms in thermal recovery processes. Decreasing the oil viscosity would affect the mobility ratio M defined as

$$M = (k_{rw}\mu_o)/(k_{ro}\mu_w) \tag{9.11}$$

Assuming that the extreme saturations and the relative permeabilities k_{rw} and k_{ro} are independent of temperature, the mobility ratio depends on the viscosity ratio μ_o/μ_w. Generally, oil viscosity μ_o is reduced faster than water viscosity μ_w as temperature increases and the greatest viscosity reduction occurs with the most viscous oils. The viscosity ratio reduction is effective in decreasing the fractional flow of hot water, consequently increasing hot water displacement efficiency. The situation is different for light oils because in some cases the viscosity ratio μ_o/μ_w increases with temperature.

9.3.3 *Evolution of absolute and relative permeabilities*

Temperature appears to have a definite effect on permeabilities, and extensive experimental efforts have been devoted to this problem.

As indicated in Section 9.3.1, the experimental results showing the evolution of absolute permeability versus temperature are quite contradictory. Doubt must be expressed about the general validity of results. Explanations may be found in the validity of Darcy's law in relation to experimental conditions, in changes in interactions between phases and even in experimental artifact.

Extensive research has been carried out to study the effect of temperature on relative permeabilities (Table 9.1).[10] As for the absolute permeability, the results are not in good agreement. Most authors agree that at high temperatures, the relative permeability to oil increases or does not change, while relative permeability to water is less affected. It seems that the viscosity ratio μ_o/μ_w is an important parameter. If this ratio is not temperature dependent, the relative permeability curves are not affected. Conversely, when this ratio decreases with increasing temperature, relative permeability to oil increases with temperature.

It is generally agreed that residual oil saturation S_{or} decreases and irreducible water saturation S_{wi} increases with increasing temperature (Fig. 9.6).[11] Besides the viscosity ratio, the factors that influence the evolution of saturations are wettability and interfacial tension, which can be temperature dependent.

When the rock is water-wet, the interfacial properties affected by temperature changes are water–oil interfacial tension and the thickness of the wetting film. When the rock is preferentially oil-wet, the adsorption equilibria of polar compounds change with increasing temperature towards desorption. Because of the complexity of the mechanisms involved, special care must be exercised in selecting appropriate values for relative permeabilities and saturations when trying to simulate or predict oil recovery from a thermal project by modeling.

9.3.4 *Oil vaporization*

An important phenomenon during steam injection is the steam distillation of the light components of the crude oil. If the pressure is lower than the sum of the partial pressures of the water and the oil, the liquid mixture will boil and give off a vapor phase composed of steam and organic compounds. Like steam, some hydrocarbons will condense ahead of the steam front and then become mixed with the crude oil.

The consequences of the presence of these light hydrocarbons are as follows:

- The hydrocarbon phase downstream from the steam front is enriched in light components. The viscosity of the crude encountered by the advancing steam front is reduced (this effect is much more pronounced with heavy oils); the result will be a decrease in residual oil saturation and an increase in oil recovery.
- The residual oil in the steam zone has an increasing content of heavy fractions that are less and less volatile, and the oil saturation decreases with time, causing the very low values of S_{or} which have been reported for some field operations.
- Sometimes, due to the increase in light constituents, heavy components such as asphaltenes may precipitate inside the porous medium. Changes in permeabilities and wettability can occur.
- In other cases, the lighter oil produced makes possible the elimination of residues deposited near the wellbore during cold production, again affecting permeabilities and wettability.

Table 9.1. Temperature effects on two-phase relative permeability measurements by various researchers

	Porous media	*Effect on* $S_{w_{irr}}$	*Effect on* S_{o_r}	*Effect on* S_{g_r}	*Effect on* k_{r_w}	*Effect on* k_{r_o}	*Effect on* k_{r_g}	k_{r_w}/k_{r_o}	k_{r_g}/k_{r_o}	*Suggested influence*
Miller and Ramey	Berea sandstone	Unchanged	Unchanged		Increase at $S_{w_{irr}}$			Unchanged		Fundamental flow properties, temperature independent
Montgomery	Sand pack	Unchanged	Unchanged		None			Unchanged		
	Sand pack		Decreased							
Okandan, *et al.*	Sandstone	Increased								
	Limestone	Decreased								
Poston, *et al.*	Sand pack	Increased	Decreased		Increased	Increased		Increased		Wettability change, reduction in capillary forces
Samaroo and Guerrero	Sandstone	Decreased for N_2-brine, increased for oil-brine								Viscous forces, capillary forces
Sanyal, *et al.*	Sandstone	Increased	Decreased			Increased				Pore structure alteration, wettability change
Sinnokrot, *et al.*	Sandstone	Increased	Decreased		Decreased	Increased		Decreased		Wettability change
Shilobod	Limestone	Unchanged	Unchanged		None	None		Unchanged		
	Berea sandstone				None	None				
Torabzadeh and Handy	Berea sandstone	High if T unchanged Low if T increased	Decreased		High if T increased Low if T decreased	Increased		Decreased		Interfacial tension reduction, wettability change
Weinbrandt	Boise sandstone	Increased	Decreased		Increased at flood out, decreased below room temperature S_{o_r}	Increased		Decreased		Wettability change

Wilson	Sand pack				None	None				
Combarnous and Pavan	Sand pack		Decreased							Viscosity ratio
Davidson	Sand and gravel pack		Decreased				Decreased near S_{o_r} Increased at low S_g	Increased at low S_w	Increased at low S_g	H_2O–oil case: interfacial tension Gas-oil case: Klinkenberg effect
Edmondson	Berea sandstone	No effect	Decreased		Affected	Affected		Affected		Interfacial tension, viscosity reduction
Goffen, *et al.*	Nellie Bly sandstone	Unchanged		Unchanged						
Gomaa, *et al.*	Monarch sandstone				Increased	Decreased	Increased			
Habowski	Berea sandstone	Increased						Affected		
Hjmeland and Larrondo	$CaCO_3$ crystals									Decrease in adhesion tension Wettability change, interfacial tension, film formation
Lo and Mungan	Teflon (oil-wet), Berea sandstone (H_2O–wet)	Increased	Decreased		Increased	Increased				Viscosity ratio, interfacial tension
Maini and Batycky	Sandstone	Increased	Decreased		Increased	Increased				Thermal expansion of oil, wettability alteration
McNeil and Moss			Decreased							

(From Ref. 10, reproduced by permission of World Oil, Gulf Publishing Co.)

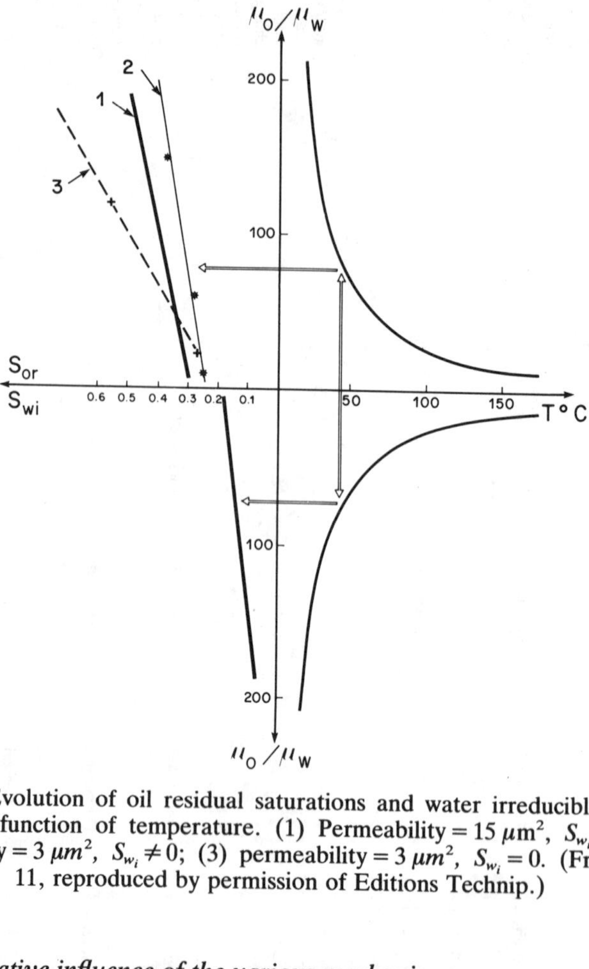

Fig. 9.6 Evolution of oil residual saturations and water irreducible saturations as a function of temperature. (1) Permeability = 15 μm^2, $S_{w_i} \neq 0$; (2) permeability = 3 μm^2, $S_{w_i} \neq 0$; (3) permeability = 3 μm^2, $S_{w_i} = 0$. (From Ref. 11, reproduced by permission of Editions Technip.)

9.3.5 *Relative influence of the various mechanisms*

Figure 9.7[2] represents, in a very schematic form, the relative influence of the different mechanisms involved in hot fluid injection.

When the temperature increases for light oils, thermal expansion and vaporization of the light components are the most important factors for oil recovery, whereas reduction of the viscosity ratio becomes much less important.

For heavy crudes, the reduction in the viscosity ratio with temperature plays the most important role in oil recovery. Changes in

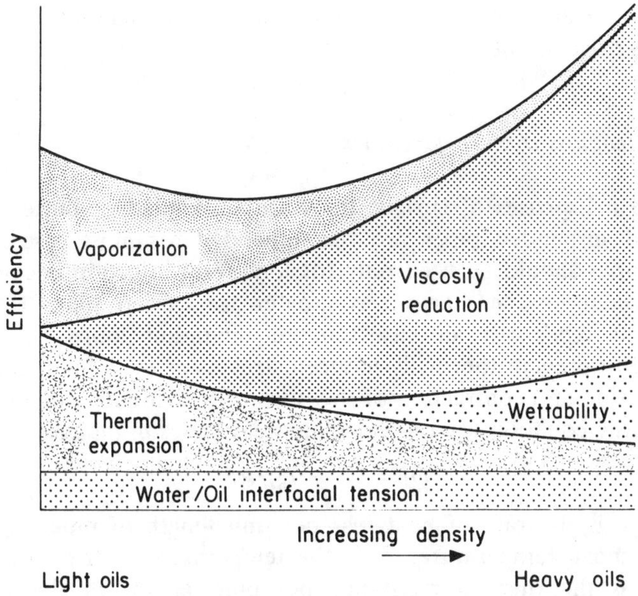

Fig. 9.7 Contributions of the different mechanisms to the improvement of oil displacement by a hot fluid instead of unheated water: schematic presentation. (From Ref. 2, reproduced by permission of Editions Technip.)

wettability can also have a favorable influence. Vaporization and thermal expansion are of less importance.

9.4 Heat transfer applied to hot fluid injection

Since hot fluid injection consists of injecting a hot fluid into an oil-bearing strata, heat losses have to be taken into account in order to calculate the amount of heat that effectively reaches the formation. The engineer usually knows the characteristic parameters of the fluids at the generator (flow rate, temperature, pressure, enthalpy). In order to know the useful fraction of enthalpy reaching the reservoir, heat losses must be calculated in surface lines, in the wells and towards the overburden and underburden formations. Because most of the relevant thermal parameters are not usually known with excellent accuracy, for practical purposes approximate calculations can be used to obtain an estimate of the order of magnitude of the heat losses.

Prats[1] and Burger *et al.*[2] give summaries of the state-of-the-art for heat loss calculations.

9.4.1 *Heat loss from surface lines*

The two main cases of surface lines are suspended or buried pipes. Insulated suspended pipes are recommended for most cases, while buried pipes should be used only for long-term injection because the rate of heat loss at start-up is greater for buried pipes than for suspended pipes. In all cases ample dilatation loops should be designed to prevent thermally-induced stresses or even failures caused by expansion and contraction of the lines with changing temperatures.

The basic equation used to calculate heat losses in pipes is:

$$Q_{ls} = (T_b - T_a)/R_h \tag{9.12}$$

where Q_{ls} is the rate of heat loss per unit length of pipe, T_a is the ambient mean temperature, T_b is the temperature of the bulk steam and R_h is the thermal resistance per unit length. In the case of steady-state heat transfer, the inverse of R_h may be expressed as the product of the area exchanging heat with the surroundings and of a heat transfer coefficient. Therefore eqn (9.12) can be expressed as follows for steady-state heat losses per unit length along a pipe with radius r:

$$Q_{ls} = 2\pi r\mathbf{h}(T_b - T_a) \tag{9.13}$$

where $\mathbf{h}$ is the overall heat transfer coefficient between the pipe and the outside.

Determination of thermal resistance and heat transfer coefficients is given in Refs 1 and 2. Most of the calculations assume steady-state heat losses. For short injection times, McAdams[12] details transient calculations in certain configurations of lines.

9.4.2 *Heat loss in the injection wells*

For designing a hot fluid injection process, not only heat losses but also temperatures need to be estimated for the various components of the well. The calculation of the casing temperature is of great significance, since it provides guidelines to select the appropriate casing material and cement composition, in consideration of thermal stresses.

Contrary to heat losses in surface lines which can be considered to be in the steady state, heat losses in wells never reach the steady state. In fact, after a few days of injection, heat transfer in a well can be calculated by the following hypothesis:

—heat transfer within the well is in a pseudosteady-state mode and
—heat losses in the formation are transient.

Ramey[13] and Willhite[14] discuss this point in detail and give an equation for the time-dependent function describing the rate of heat loss to the formation. This equation is a measurement of how fast the heat is conducted away from the well into the earth. The basic eqn (9.12) for heat losses in surface pipes is still valid with the provisos that now:

—T_a represents the geothermal temperature and hence varies with depth, and
—R_h is time dependent.

To illustrate the typical elements entering into heat loss calculations from wells, let us take the example of an insulated tubing centered in a cemented casing. Figure 9.8 is a schematic representation of the various resistances to heat transfer. The thermal resistance per unit length can be expressed as:

$$R_h = \frac{1}{2\pi}\left[\frac{1}{r_1 \mathbf{h}_1} + \frac{\ln[r_1'/r_1]}{\lambda_1} + \frac{\ln[r_{ins}/r_1']}{\lambda_{ins}} + \frac{1}{r_{ins}(\mathbf{h}_C + \mathbf{h}_R)} + \frac{\ln[r_2'/r_2]}{\lambda_2} + \frac{\ln[r_3/r_2']}{\lambda_{cem}} + \frac{f(t)}{\lambda_F}\right] \quad (9.14)$$

In this formulation, the successive terms are the heat transfer between the fluid and the tubing wall expressed through a film coefficient, the transfer through the tubing wall, the conduction through the insulation, the resistance to convection and radiation in the annulus space, the transfer through the casing wall, then through the cement. It should be mentioned that the terms relative to heat transfer between the fluid and the tubing wall and through tubing and casing walls are negligible compared with the other contributions to thermal resistance. The symbols are defined in Fig. 9.8.

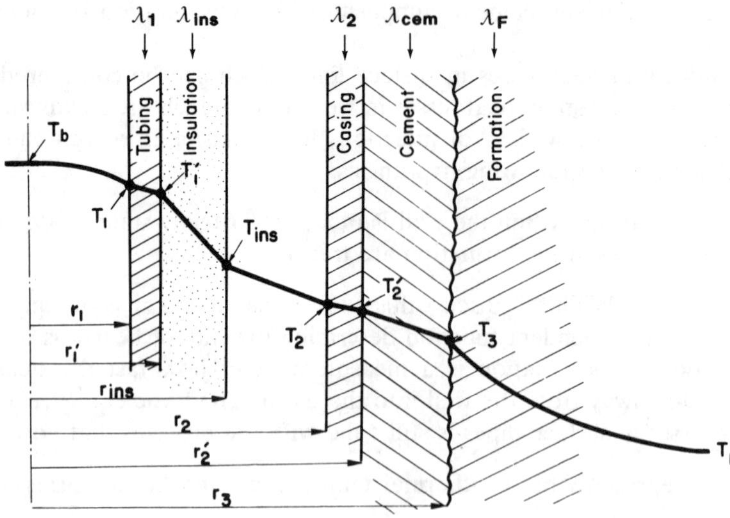

(1) Quasi steady-state in the drill-hole ($r \leqslant r_3$):

$Q_l = 2\pi r_1 \mathbf{h}_1 (T_b - T_1)$	transfer between the fluid vein and the inside surface of the tubing;
$= 2\pi \dfrac{\lambda_1}{\ln[r'_1/r_1]}(T_1 - T'_1)$	conduction through tubing wall;
$= 2\pi \dfrac{\lambda_{ins}}{\ln[r_{ins}/r'_1]}(T'_1 - T_{ins})$	conduction through insulation (conductivity λ_{ins});
$= 2\pi r_{ins}(\mathbf{h}_C + \mathbf{h}_R)(T_{ins} - T_2)$	convection–radiation in the annular space;
$= 2\pi \dfrac{\lambda_2}{\ln[r'_2/r_2]}(T_2 - T'_2)$	conduction through the casing wall (conductivity λ_2);
$= 2\pi \dfrac{\lambda_{cem}}{\ln[r_3/r'_2]}(T'_2 - T_3)$	conduction through the cement (conductivity λ_{cem}).

(2) Transient state in the ground:

$$Q_l = 2\pi \frac{\lambda_F}{f(t)}(T_3 - T_F)$$

Fig. 9.8 Heat losses in a well. Fluid injection into insulated tubing. (From Ref. 2, reproduced by permission of Editions Technip.)

In eqn (9.14), the last term is the time-dependent function reflecting the thermal resistance of the earth. If the time t is greater than one day, the transient function $f(t)$ can be calculated from the following equation:

$$f(t) = 0{\cdot}5 \ln t + \ln(2(\alpha_F)^{0{\cdot}5}/r_e) - 0{\cdot}29 \tag{9.15}$$

where α_F is the thermal diffusivity of the ground, which is considered to be constant, and r_e is the radius at the cement–formation interface.

The previous equations make it possible to calculate the effect of the geometry and of the materials used in the well on heat losses to the formation. In particular, tubing and casing diameter increases will cause increased heat losses. Insulating the tubing reduces the heat losses, but the cost and operational problems associated with insulating the tubing may make this solution less attractive. To reduce heat transfer through the annulus, the presence of water must be avoided. This can be achieved by filling the annulus with an inert gas under pressure or with a thixotropic organic liquid, or by using a downhole packer. Typical results of heat loss calculations in wells are reported in Refs 1, 2, and 14. Figure 9.9 shows a typical example of these results. Of particular importance is the determination of the minimum rate of injection at which the process becomes inefficient. When the hot fluid injected is condensing steam, the percentage of heat loss is inversely proportional to the injection flow rate.

9.4.3 *Reservoir heating*

Assuming that calculations of heat losses in the surface lines and down the injection well have been performed, we know the basic information about the fluid injected at the sandface (temperature, pressure and enthalpy). Some of the heat injected into the reservoir is lost through the adjacent formations and some is lost through the produced fluids. In order to estimate the heated volume of the reservoir, which in turn allows calculation of the enhanced production, the amount of heat remaining in the oil-bearing strata has to be calculated.

To raise the temperature of a unit volume of reservoir from initial temperature T_i to steam temperature T_s an amount of heat Q has to be provided equal to

$$Q = (\rho c)^*(T_s - T_i) = (\rho c)^*\Delta T \tag{9.16}$$

where $(\rho c)^*$ is the isobaric heat capacity of the unit volume of reservoir. The value of this parameter should be selected carefully

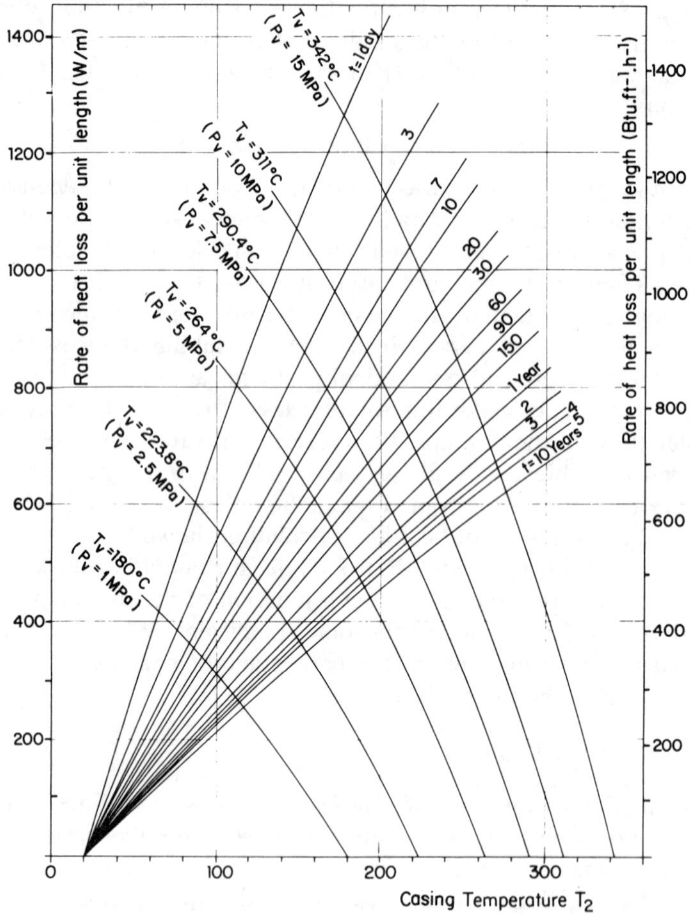

Fig. 9.9 Rate of heat loss per unit length and the casing temperature during hot fluid injection into the well. (From Ref. 2, reproduced by permission of Editions Technip.)

The hot fluid is injected into the tubing; a packer is used to insulate the annular space. Tubing 2 7/8″; casing 7″, 26 lb/ft. The emissivities of tubing and casing surfaces are 0·9. The annular space is filled with air at atmospheric pressure. The conductivity of the cement is assumed to be equal to that of the formation: $\lambda_{cem} = \lambda_F = 2{\cdot}44\ W\,m^{-1}\,K^{-1}$ (this results in a slight overestimation of the heat loss). The diffusivity α_F is equal to $1{\cdot}028 \times 10^{-3}\ m^2/s$ and we take $T_F = 20 + 1{\cdot}5\ P_v$ (in °C); P_v (in MPa) is the water vaporization pressure at the temperature T_v of the fluid. The calculations were carried out as described in this chapter.

depending on the problem to be addressed. In most simple steam injection models, assuming that all the heat injected remains in the steam zone, the value of $(\rho c)^*$ to be used in the calculations should be calculated using the saturations existing in the steam zone rather than an average for the whole reservoir.

A number of authors have calculated the growth of the steam-heated volume during heat injection. Marx and Langenheim's model,[15] although one of the simplest, is the best known and gives insight into understanding some important features of steam injection. They consider the temperature distribution in the reservoir as a step profile with the steam volume at injection temperature T_s and the rest of the reservoir at initial temperature T_i (Fig. 9.10). From a bulk heat balance for a horizontal reservoir with uniform and constant properties, we obtain:

$$Q_i = (\rho c)^* h(T_s - T_i)\, dA/dt + Q_l \qquad (9.17)$$

where Q_i is a constant heat injection rate, h is the reservoir thickness, T_s and T_i are the steam and initial temperatures, $(\rho c)^*$ is the volumetric heat capacity, Q_l is the rate of heat loss to the adjacent formations and dA/dt is the rate of change of the heated area A with time. Ramey[16] has shown that this method also applies to nonconstant injection rates. Assuming heat losses only by vertical conduction to the

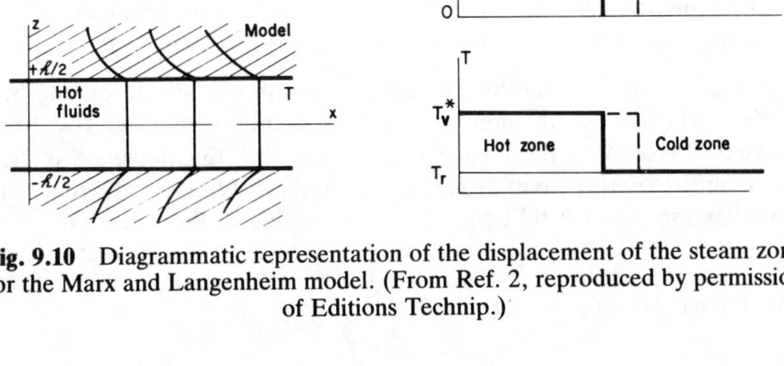

Fig. 9.10 Diagrammatic representation of the displacement of the steam zone for the Marx and Langenheim model. (From Ref. 2, reproduced by permission of Editions Technip.)

adjacent formations, an equation for Q_l can be obtained such as:

$$Q_l = Q_i(1 - e^{t_D} erfc(t_D)^{0.5}) \quad (9.18)$$

where *erfc* is the complementary error function† and t_D a dimensionless time given by:

$$t_D = 4((\rho c)_s/(\rho c)^*)^2 \alpha_s t/(h)^2 \quad (9.19)$$

α is the thermal diffusivity and s refers to the adjacent formations; it is assumed that the overburden and underburden have the same thermal properties.

The growth of the heated zone dA/dt can then be calculated as:

$$dA/dt = (Q_i/(\Delta T(\rho c)^* h))e^{t_D} erfc(t_D)^{0.5} \quad (9.20)$$

Let us define G as:

$$G = 2(t_D/\pi)^{0.5} - 1 + e^{t_D} erfc(t_D)^{0.5} \quad (9.21)$$

The areal extent of the heated zone is given by:

$$A = (Q_i(\rho c)^* h/(4\Delta T \alpha_s(\rho c)_s^2))G \quad (9.22)$$

The heat remaining in the reservoir is:

$$Q_R = (Q_i(\rho c)^{*2} h^2/(4\alpha_s(\rho c)_s^2))G \quad (9.23)$$

Values of the above-mentioned functions can be found in Ref. 2.

The following conclusions may be drawn from the results of Marx and Langenheim's model:

- For a given amount of heat, the hot volume decreases when the reservoir thickness decreases and when the steam temperature and pressure increase.
- The growth of the heated zone becomes slower with time.
- The fraction of the heat injected at the sandface that is lost to the surrounding formations increases with time and becomes excessively high when the reservoir thickness is less than 5–10 m.

Let us note that this method assumes a piston-like displacement (no gravity override or channeling). It also assumes that all the heat injected is stored in the steam zone, neglecting the presence of hot water ahead of the steam zone. Later models demonstrate that this simplification is not valid beyond a critical value of the injection time.

† The complementary error function *erfc* x is defined as: *erfc* $x = 1 -$ *erf* $x = 1 - (2/(\pi)^{0.5})\int_0^x \exp(-\alpha^2)\, d\alpha$.

Farouq Ali[17] proposed relations applicable when the thermal properties of the overburden and underburden formations are different. Prats[18] has discussed the validity of the idealized assumptions by Marx and Langenheim and those of Lauwerier.[19] He developed expressions for the heat efficiency that can be applied to any thermal recovery process. Willman *et al.*[20] describe laboratory research confirming the validity of the analytical models discussed above and discuss the effect of the hot water zone ahead of the condensation front. Computed values of the steam zone area beyond the critical time (i.e. when one must consider the hot water zone ahead of the steam in the calculations) have been derived by Hearn.[21] An approximate expression based on Hearn's model has been proposed in Ref. 2 to calculate the steam zone area. Myhill and Stegemeier[22] proposed an estimate of the amount of heat stored in the steam zone itself. More recently, Yortsos and Gavalas[23] have derived analytical solutions for the growth of the steam zone in multidimensional reservoirs and have improved the critical time definition. The work by Lauwerier[19] and Malofeev[24] can also be used to estimate bottomhole temperatures in producing wells. The heat lost through produced fluids has to be estimated under bottomhole conditions to complete the heat balance over the whole area of interest. In fact, only numerical or physical simulations suffice for the exact determination of the temperature variations of produced fluids in time.

Laboratory data and/or field experience can be used to determine residual oil saturation in the hot zone of the reservoir. The oil production schedule can be calculated by using the previously described models, as it will be seen in Section 9.5.3.

9.5 Implementation of hot fluid injection

Hot fluid injection has been used in a variety of processes. By far the most common is steam injection either in a cyclic mode or as a displacement process. In some fields, hot water injection has been attempted.

9.5.1 *Hot water injection*

Figure 9.11(a) shows the water saturation and temperature profiles during one-dimensional displacement of oil by hot water. The ex-

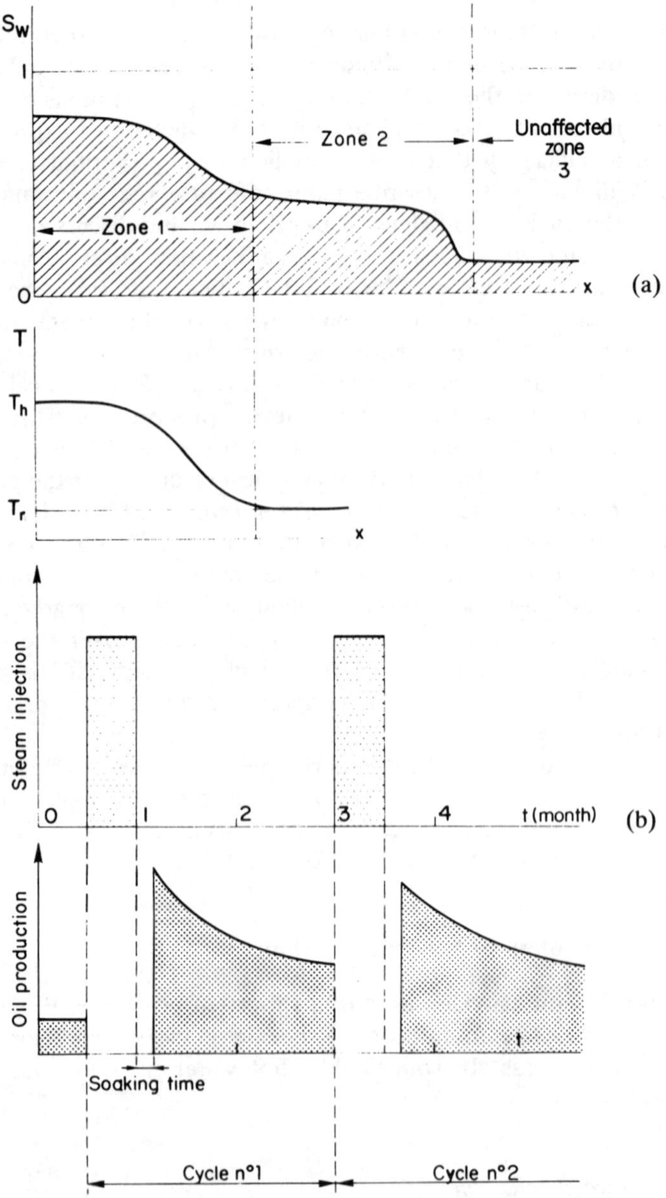

Fig. 9.11 (a) Water saturation and temperature profiles during oil displacement by hot water. (b) Schematic representation of two successive cycles of steam stimulation. (c) Temperature, steam and water saturation profiles during oil displacement by steam.

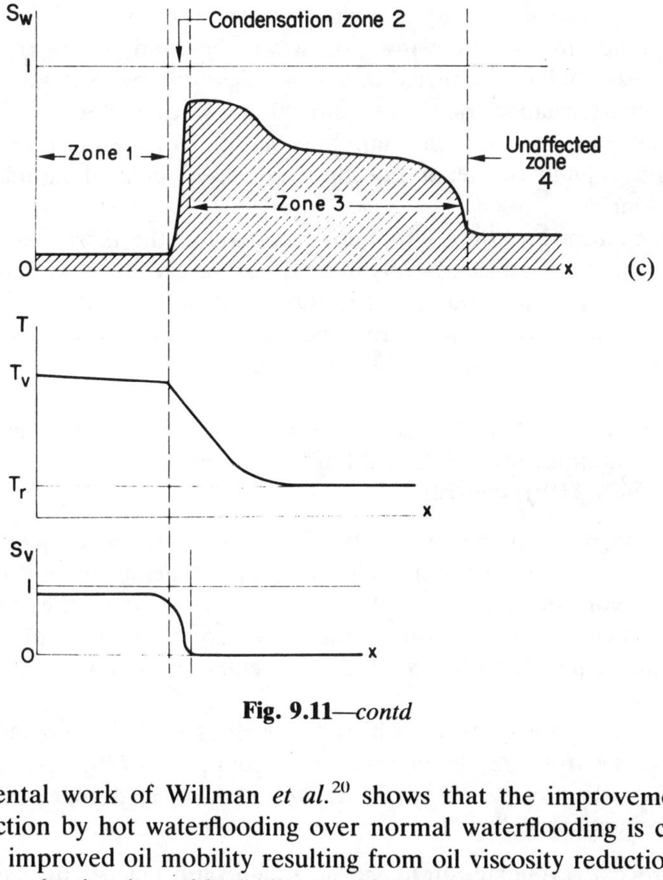

Fig. 9.11—*contd*

perimental work of Willman *et al.*[20] shows that the improvement in production by hot waterflooding over normal waterflooding is caused by the improved oil mobility resulting from oil viscosity reduction and by the reduction in residual oil saturation at high temperature. Prats[1] and Burger *et al.*[2] give calculations and design considerations for hot waterflooding. However, field applications of hot waterflooding have been plagued by severe channeling and high water/oil ratios. It appears that poor sweep efficiency makes waterflooding of very viscous crudes uneconomic. The only field test showing economic success is the Schoonebeek flood where oil viscosity at reservoir conditions is moderate.

9.5.2 *Cyclic steam injection*

Prior to discussing steam injection, it is important to note that 'steam' is in fact a two-phase mixture of liquid water and steam produced from a generator. Steam quality is defined as the ratio by weight of gas to gas + liquid. In the following, steam will be used to designate the steam–water mixture injected in a well unless otherwise stated.

The performance of steam injection is characterized by the oil/steam ratio, which is the ratio between the volume of oil produced and the volume of steam injected, expressed in cold liquid water equivalent.

Cyclic steam injection is the most common of the thermal recovery processes. It is used extensively in heavy oil reservoirs as well as in tar sands and bitumen resources. It is also almost always used to improve injectivity prior to steam drive and in some cases prior to in-situ combustion projects, at least in the producers.

(a) *Description of the process and mechanisms.* Another name of cyclic steam injection, 'Huff and Puff', describes the process. A typical cycle (Fig. 9.11(b)) consists of:

- An injection period during which a slug of steam is injected into an oil bearing stratum. The slug size depends mainly on the reservoir thickness but also on the location and the operator practice, and it may vary from a few thousands cubic meters (of water equivalent) to 15 000 cubic meters or sometimes more per cycle.
- A soak period during which the well is shut in to allow heat transfer from the steam zone to the cold parts of the reservoir.
- A production period when the well is made to produce until the production declines to an uneconomic level.

The process is repeated until the oil/steam ratio reaches its economic limit.

The mechanisms involved in oil production by steam stimulation are complex. It is obvious that the reduction in oil viscosity caused by the increase in temperature makes for easier oil flow during the production period. Field observations have shown that water is preferentially retained in the reservoir, probably because of changes in surface forces and wettability. This causes a change in relative permeability to water and needs to be further studied because its origin and magnitude are still unknown. Thermal expansion and distillation of part of the oil may also improve the process. Another contributing factor can be near-wellbore clean-up by removal of cold oil and/or heavy residues by the steam slug.

For a cyclic steam project to be effective, a driving force must be present in the reservoir to help the oil flow during the production period. Gravity drainage, solution gas drive and compaction are often important driving mechanisms, completing the effect of the steam slug on well stimulation. The changes in gas solubility caused by the increased temperature often increase the magnitude of the solution gas drive effect.

(b) *Performance prediction.* The best prediction technique for cyclic steam injection remains the history matching of existing projects run in the field in question. Because the above-mentioned changes in water relative permeability are usually not well defined, performance prediction in an area where no steam injection has been performed is difficult. However, cyclic steam injection is relatively easy to plan and implement in the field. It is then possible to use production data to obtain the relevant reservoir parameters by history matching and to use that information to predict the behavior of subsequent wells.

Simplified models can be useful in getting production estimates. Because of the fact that these models consider only the most easily obtained parameters and make approximations as to the nature of the heat and fluid flows, caution should be exercised in the analysis of the results. These models, however, can be very useful in screening prospective projects or for performing sensitivity studies.

A simplified analytical method is described by Boberg and Lantz.[25] This model calculates the size of the heated region using Marx and Langenheim's technique. Then the temperature history of the heated region after termination of the steam injection is computed using an approximate energy balance, taking into account conduction towards the cold part of the reservoir and heat removed by the produced fluids. Fluid flow is treated by steady-state radial equations with the effect of stimulation taken into account through the viscosity ratio between cold oil and oil at the average temperature of the hot zone. The average temperature T_{av} within the hot zone, which does not change in size, is given by:

$$T_{av} = T_i + (T_s - T_i)(f_{HD} f_{VD}(1 - f_{pD}) - f_{pD}) \tag{9.24}$$

where T_i is the initial reservoir temperature, T_s is the injected steam temperature, f_{pD} is a time-dependent dimensionless quantity accounting for the removal of heat by the produced fluids, and f_{HD} and f_{VD}

time-dependent dimensionless quantities accounting for heat loss by conduction in the horizontal and vertical directions respectively.

The viscosity of the oil in the hot zone can be calculated once the average temperature of this zone is known, and the ratio of productivity indices after and before stimulation can be derived as:

$$I/I_o = (\ln(r_d/r_h)/\ln(r_d/r^*) + (\mu_{o,Tav}/\mu_{o,Ti})(\ln(r_h/r^*)/\ln(r_d/r^*)))^{-1} \quad (9.25)$$

where r_d is the drainage radius of the well, r_h is the radius of the hot zone, r^* is the equivalent well radius taking skin effects into account and $\mu_{o,Tav}$ and $\mu_{o,Ti}$ are the oil viscosity at T_{av} and T_i, respectively. It must be noted that the only mechanism taken into account is the reduction in viscosity caused by the steam. In the equivalent radius taking skin effect into account other beneficial effects can be included such as wellbore cleaning.

Gontijo and Aziz[26] describe a model based on a realistic distribution of the steam in the reservoir, which is more practical for gravity drainage reservoirs or very viscous oils.

(c) *Field observations.* Cyclic steam injection has been used extensively in California, Venezuela and Canada. The large number of projects worldwide leads to some observations:

- The amount of steam injected is usually between 40 and 100 t/m of reservoir height with individual cases varying from 15 to 300 t/m of reservoir height. Measuring the steam injected by unit height of the reservoir is appropriate because, of course, more steam tends to be injected into a thicker reservoir than into a thin one. However, steam tends to enter only the top part of thick sands (gravity override). In some cases barriers between two producing sands cause entry of the steam only into the top sand. Improving sweep efficiency is a major component of today's research on thermal recovery. This will be discussed in Section 9.5.4.
- The soaking time varies from two days to one month depending on the operator. Its effect is still controversial, and in some cases it has very little effect on the cycle performance.
- Thick, gravity drainage dominated reservoirs can sustain numerous cycles, while solution gas drive reservoirs can usually sustain only three or four cycles before reaching their economic limits.

- Most of the few failures were caused by either poor well location (gas cap or water zones), low initial oil saturation (in which case no thermal recovery technique will be acceptable), or high shale and clay contents. Unacceptably high heat losses either in the injection wells (too deep or poor insulation) or in the reservoir (pay zone too thin) have also accounted for poor results.

The literature on field projects involving cyclic steam injection is very broad.[1,2] Rather than citing all the relevant papers, Refs 27–30 are compilations of field tests in various regions. Figure 9.12 shows an example of oil production by cyclic steam injection.[29]

9.5.3 *Steam drive*

Steam drive or steam flooding is a continuous process similar to waterflooding. The ultimate oil recovery achieved by steam drive is higher than the recovery possible by cyclic steam injection but the oil/steam ratio is usually lower (in the range of 0·1 to 0·4 m^3/m^3). As most cyclic injection projects mature, steam drive is rapidly becoming the major thermal recovery method, especially in California. A 1990 status report on steamflooding in the US is given in Ref. 31.

(a) *Description of the process and mechanisms.* Steam drive involves injection of steam from well to well. It is usually designed as continuous flooding of the reservoir by steam until the oil/steam ratio

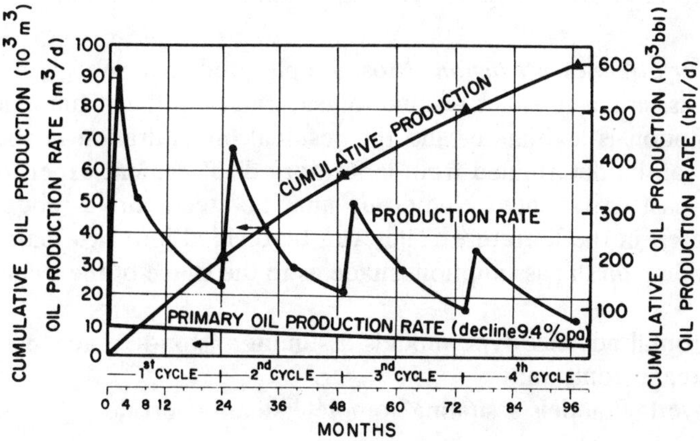

Fig. 9.12 Oil production by cyclic steam injection. (From Ref. 29, reproduced by permission of The Canadian Institute of Mining and Metallurgy.)

decreases below the economic limit. Pattern spacing depends on the reservoir characteristics.

Figure 9.11(c) gives the temperature, steam and liquid water saturation profiles during one-dimensional displacement of oil by steam.

Because of the presence of a condensible gas phase, the behavior of steam drive is very different from the behavior of water injection projects. The presence of the gas phase causes distillation of the light components of the oil and their movement forward towards the cold part of the reservoir. When the steam condenses, these light components do likewise, thus generating a 'solvent' bank at the condensation front. Of course the viscosity of the oil is reduced by the increase in temperature, and its mobility is improved. Very low values of residual oil saturations ($<5\%$) have been reported in the field. It is clear that the displacement efficiency in steam drives can be very high providing that a good sweep is obtained.

Because of the condensible nature of the gas phase of the steam, steam drive is a stable displacement process. Indeed if small steam fingers are generated, they will soon loose heat to the cold volume surrounding them and be destroyed by condensation. In a steam drive, oil is contacted by cold water, then by hot water and by the condensed light ends, and finally by steam. The displacement of very viscous oil by cold water ahead of the heat front is difficult and may cause blocking in the reservoir. For this reason it is recommended that the producers be steam cycled before a steam drive operation.

(b) *Performance prediction.* Most simple predictive models for steam drive assume a constant residual oil saturation in the steam zone. This assumption is legitimate, and the residual oil saturation in the steam zone can be determined from laboratory displacement experiments or from past experience. Numerous analytic steam drive models have appeared in the literature. They can be divided into two main groups depending on the assumptions made as to the shape of the steam zone:

—frontal advance type models assuming piston-like advance of the steam front, and
—overlay models assuming complete steam override.

The volume of the steam zone can be found using one of many analytical models such as those of Marx and Langenheim,[15] Myhill and

Stegemeier,[22] Yortsos and Gavalas,[23] or Van Lookeren.[32] Other references can be found in Burger *et al.*[2] and in Prats.[1] To determine oil production, the first step is to calculate the volume of the steam zone and the amount of oil displaced from it. This can be written as:

$$N_p = (S_{oi} - S_{or})\Phi(h_n/h)V_s E_c \tag{9.26}$$

where N_p is the total oil produced, S_{oi} is the initial oil saturation, S_{or} is the residual oil saturation, Φ is the porosity, h_n/h is the ratio of net over gross reservoir thickness, V_s is the volume of the steam zone, and E_c is a capture factor representing the ratio of oil produced to oil displaced. All the production parameters can be calculated on the basis of the developments in Section 9.4.3. Economic indices such as the oil/steam ratio can be evaluated at this stage.

(c) *Field observations*. Steam drive has been extensively practiced in California (Kern River, San Ardo, Midway Sunset, Huntington Beach, etc.) and in Venezuela (Tia Juana) as well as at other locations. An example of the production performance of steam drive[33] is given in Fig. 9.13. Recoveries have ranged from 5% of the original oil in place to 73% with oil/steam ratios varying from 0·05 to 0·61 m^3/m^3. Chieh Chu[34] and Farouq Ali and Meldau[35] surveyed the steam drive projects. Some general comments can be made at this stage:

- Most steam drive projects involve reservoirs of high porosity (25% or more), high permeability and high residual oil saturation after primary production.
- Steam drive is usually applied to viscous oils but can also be effective in displacing light oils.
- Spacing is usually limited to between 0·25 and 10 hectares per injector. Well location needs to take into account the specific characteristics of each reservoir.
- Sweep efficiency is critical to the success of a steam drive project.

9.5.4 *Improvement of steam injection*

Steam is lighter and more mobile than oil. As a result, one of the major problems facing steam injection is the poor sweep efficiency caused by gravity override and/or channeling of the steam through the most permeable parts of the reservoir. As an example, Blevins and

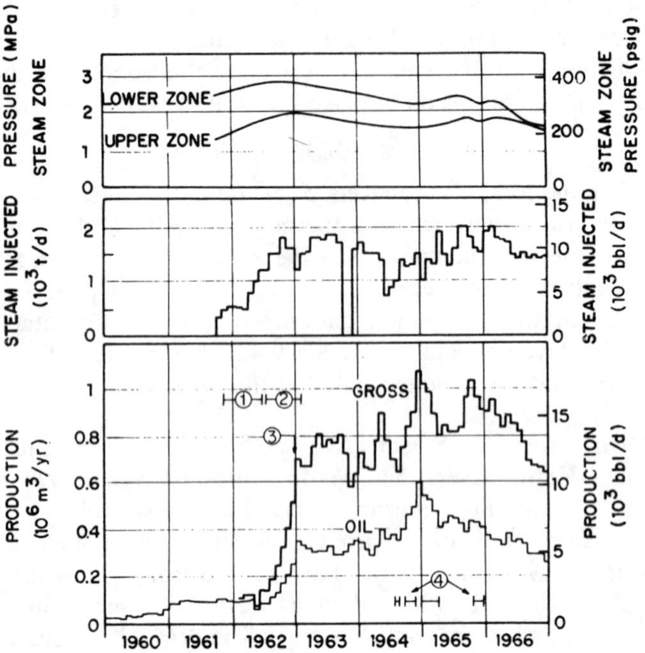

Fig. 9.13 Performance of the Tia Juana C3/C4 project. (1) Increased water cut in most wells. (2) Increased temperature in most test wells. (3) First water cut increase in the first row of surrounding wells. (4) Steam stimulation of some of the production wells. (From Ref. 33, reproduced by permission of The Society of Petroleum Engineers.)

Billingsley[36] give some data on the Kern River Field showing both override and channeling of the steam. In the same field Brigham and Strom[37] performed an analysis of a five-spot pattern and found a sweep efficiency factor of 38%. It is obvious that the economic success of steam injection methods depends on the sweep efficiency. Some reservoirs also have problems of poor injectivity or bad interwell communication.

Improvements can be broadly divided in two categories:

- operational changes such as fracturing, use of horizontal wells, pressure cycling, selective completions or injection in underlying aquifers, and
- use of additives injected with the steam.

(a) *Improved operations.* Selective completions for the injectors have been tried in California and Venezuela. The goal was to be able to inject steam into the lower part of the interval of interest in order to reduce gravity override. Although some improvements in oil/steam ratios have been reported in the literature, the results on the whole have not been excellent. This is probably caused by the fact that selective completions have mostly affected near wellbore flowpaths and have been unable to modify the rest of the reservoir.

Injection of steam into an underlying aquifer has been attempted in a few fields. Oil/steam ratios decrease because of the heat lost in heating the water in the aquifer. This process is limited to reservoirs with very low injectivity in the oil-bearing areas underlain by thin aquifers.

Fracturing, either before steam injection or caused by the injection of high pressure steam, has been shown to be effective in bitumen fields as well as in heavy oil reservoirs.

Horizontal wells have been used in several projects and seem to be a promising technique for improving sweep efficiency in thermal recovery operations as shown by numerical simulations.[38]

(b) *Use of additives.* Various combinations of gases have been suggested for use as an additive to steam. It seems that no field test has shown technical or economic success. Some tests in which air was injected with steam seem to have improved the oil/steam ratio, probably by in-situ combustion or at least exothermic oxidation of part of the oil. Theoretically, a combination of CO_2 and steam could be beneficial to projects with highly viscous oils by reducing the oil viscosity via CO_2 dissolution.

Water-soluble additives, mostly surfactants, have been used extensively to improve the sweep efficiency of steam injection operations or to try to modify the interfacial properties of the oil/water system and reduce residual oil saturation. Of course extensive screening has to be performed to find chemicals that are stable at steam temperature and can be propagated efficiently during steam injection. Particular attention has to be given to the adsorption and the partitioning of the products in the reservoir concerned.

(c) *Use of foams.* In addition to the fact that surfactants may improve near-wellbore oil flow by modification of the oil/water relative permeability curves and/or possible removal of solids, a reduction of steam mobility can be seen by the formation of foams.

Foams are dispersions of gas bubbles separated by liquid films. These films are generated in porous media by snap-off, leave behind and division mechanisms. A detailed description of the foam formation processes in porous media is outside the scope of this book, but the reader can be referred to the work of Owete *et al.*[39] or Hirasaki and Lawson.[40] In porous media, foam structure is strongly dependent on the porous medium geometry. In fact, after a few inches of porous medium, one cannot find any difference between the flow behavior of an externally generated foam and a foam obtained by coinjecting gas and surfactant solution in the porous medium.

Liquid films are normally unstable and will break quickly. If surfactants are added to the liquid, stability is improved and foams can persist almost indefinitely. Surfactants suitable for foam generation are usually used in aqueous solutions. As such they should have a smaller molecular weight than the surfactants used in micellar polymer operations. Of course, problems such as adsorption, sensitivity to oil and brine and temperature stability should be considered.

Before looking at the mobility control characteristics of foam, it is necessary to briefly review some basic ideas.

FOAM CHARACTERIZATION. Foams can be characterized by three main parameters:

(1) The first is gas volume expressed as a fraction of the total volume. This is the foam quality, and it varies with both temperature and pressure. Foam quality can range from less than 50% to 97% and more. Note that quality is expressed by a volume ratio and is different from steam quality expressed as a mass ratio.

(2) Average bubble size or texture is a very important parameter for flow through a porous medium. If the average bubble size is much smaller than the pore size, the flow of foam should appear as bubbles dispersed in the channels. If the average bubble size is larger than the pores, the liquid will be present as lamellae separating gas bubbles. These films will progress, breaking and reforming at the pore necks. In practice, for petroleum applications the second case is by far the more likely.[40]

(3) The range of bubble size is the third parameter to be considered. As diffusion of the gas from small bubbles to large ones contributes to foam collapse it is obvious that foams with a large bubble distribution range will be more unstable.

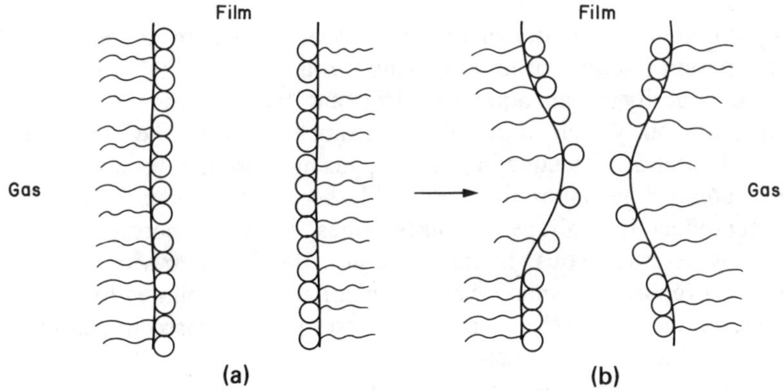

Fig. 9.14 Gibbs-Marangoni effect.

FOAM STABILITY. Let us examine a liquid film separating two gas bubbles. With no surfactant the film will drain quickly under the influence of gravity or other external forces. When a surfactant is present in solution in the film (Fig. 9.14(a)), the surfactant molecules will preferentially orient themselves at the gas/liquid interface, with the polar group toward the interior of the film and the nonpolar tail toward the gas phase. The gas–liquid surface tension is a decreasing function of the difference between surface and bulk concentration. Suppose that an external force causes the film to thin (Fig. 9.14(b)). The surface area of the film will increase locally, the surface concentration will decrease and the surface tension will consequently increase. The difference in surface tension along the film boundaries will cause the film to go back to its original shape. This is known as the Gibbs-Marangoni effect. The process is described in detail in Overbeek.[41] In the absence of external forces the film will be in equilibrium due to the balance between the van der Waals forces between the molecules of the film (attractive) and the forces of the electrical layer inside the film boundary (repulsive). If the film becomes smaller than the equilibrium thickness, it will collapse. Film drainage and gas diffusion from small to large bubbles will contribute to film thinning and ultimately to foam collapse. Besides these mechanisms, external effects such as local heating, the presence of contaminating molecules in the films (oil in particular), local pressure gradient across the films or unfavorable wettability conditions contribute to foam collapse in porous media.

MOBILITY CONTROL. Early research[42] showed that foam can drastically reduce the mobility of the gas phase. In fact, measurements of the apparent viscosity of foam flowing through sandpacks[43] correspond to the viscosity of an aqueous surfactant foam in the 50–80 mPa s range, obviously well above the viscosity of any of the constituent fluids. Foams are effective in reducing gas mobility in a broad range of permeabilities but laboratory and field observations seem to suggest a greater effect in high permeability zones when the porous medium tested is heterogeneous. In the literature, the flow behavior of foams has been reported in many ways. Fractional permeabilities to gas and water, ratios of effective permeability to apparent viscosity, mobility reduction factors, and relative permeability ratios can all be used to describe the flow behavior of viscous foam in porous media. However, the reduction in mobility caused by foam can be viewed as either a single phase flow of increased viscosity, or as a decrease in gas relative permeability. The low mobility of gas in foam flow can be caused by an increase in trapped gas saturation and the blocking of pore throats due to liquid lamellae. These two mechanisms probably coexist during foam flow in porous media.

MODELING FOAM FLOW. Two main approaches for modeling foam flow can be distinguished:

(1) *Population balance models* describe the reduction of mobility of the gas phase as a function of the number and texture of bubbles created minus the number of bubbles destroyed in a given porous medium.
(2) *Empirical* simulators usually consider the reduction in gas mobility to be caused by the increase in viscosity of the gas phase when foam is present in the porous medium. A non-Newtonian relationship for the gas viscosity (power law) can be added for a better match of the laboratory data.

Despite research spanning over thirty years, the modeling of foam flow through porous media is still not adequate. To optimize its use, knowledge of the rheological properties of foam flowing through porous media is needed.

Hirasaki[44] has reviewed the state-of-the-art in steam foam technology. Although most of the exact mechanisms of the process are not well known, it is now obvious that the use of steam foam can improve the oil/steam ratio by increasing the sweep efficiency[45] (Fig. 9.15).

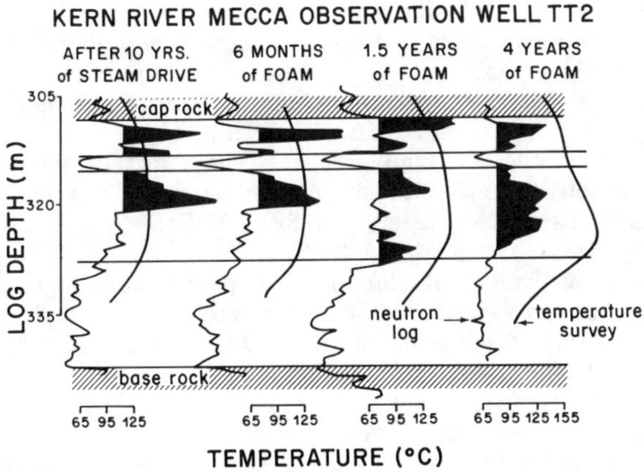

Fig. 9.15 Improved vertical sweep efficiency by steam foam, 30 m from injector. (From Ref. 44, reproduced by permission of The Society of Petroleum Engineers.)

Among the important parameters affecting steam–foam efficiency are the type and concentration of the surfactant, the porous medium structure, flow rates, steam quality and oil composition. Addition of a non-condensible gas phase has been shown to increase the pressure gradient by helping to stabilize the foam. Castanier[46] reviewed 18 field tests of steam with additives performed in California and Venezuela. Generally positive field results have been obtained that are successful both technically and economically.

This technology needs to be optimized but will, without doubt, be used routinely in the future.

9.6 References

1. Prats, M., *Thermal Recovery,* Soc. Petroleum Engineers AIME, Richardson, Texas, Monograph vol. 7, 1982.
2. Burger, J., Sourieau, P. & Combarnous, M., *Thermal Methods of Oil Recovery,* Editions Technip, Paris, and Gulf Publishing Company, Houston, 1985.
3. Licht, Jr, W. & Stechert, D. G., The variation of the viscosity of gases and vapors with temperature. *J. Phys. Chem.*, **48** (1944) 23–47.
4. Burger, J. & Robin, M., Combinaisons de l'injection de solvant et des méthodes thermiques pour la production des huiles très lourdes. *Proc. Eleventh World Petroleum Congr.*, John Wiley and Sons, Chichester, Vol. 3, 1984, pp. 251–60.

5. Somerton, W. H. & Selim, M. A., Additional thermal data of porous rocks. Thermal expansion and heat of reaction. *Soc. Petroleum Engineers J.*, **1**(4) (1961) 249–53.
6. Somerton, W. H., Some thermal characteristics of porous rocks. *Trans. Soc. Petroleum Engineers AIME,* **213** (1958) 375–8.
7. Torabzadeh, S. J. & Handy, L. L., The effect of temperature and interfacial tension on water/oil relative permeabilities of consolidated sands. SPE Paper No. 12689, 4th SPE/DOE Symp. on Enhanced Oil Recovery, Tulsa, 15–18 April 1984.
8. Quettier, L. & Eymard, R., Mephisto: un modèle numérique pour simuler les injections de vapeur dans les réservoirs très hétérogènes. Proc. of Third European Meeting on Improved Oil Recovery, Rome, April 1985. Agip SpA, Rome, 1985, pp. 317–28.
9. Merle, H. A., Kentie, C. J. P., Van Opstal, G. H. C. & Schneider, G. M. G., The Bachaquero study. A composite analysis of the behavior of a compaction drive/solution gas drive reservoir. *J. Petroleum Technol.*, **28** (1976) 1107–15.
10. Honarpour, M., Degroat, C. & Manjnath, A., How temperature affects relative permeability measurement. *World Oil,* **202**(5) (1986) 116–26.
11. Koci, Xh., Robin, M., Gadelle, C. & Quintard, M., Effet de la température sur les propriétés de déplacement polyphasique en milieu poreux. *Revue Inst. Français du Pétrole,* **44**(6) (1989) 763–83.
12. McAdams, W. H., *Heat Transmission,* McGraw Hill, New York, 1954.
13. Ramey, Jr, H. J., Wellbore heat transmission. *J. Petroleum Technol.*, **14** (1962) 427–35.
14. Willhite, G. P., Over-all heat transfer coefficients in steam and hot water injection wells. *J. Petroleum Technol.*, **19** (1967) 607–15.
15. Marx, J. W. & Langenheim, R. H., Reservoir heating by hot fluid injection. *Trans. Soc. Petroleum Engineers AIME,* **216** (1959) 312–15.
16. Ramey, Jr, H. J., Discussion of reservoir heating by hot fluid injection. *Trans. Soc. Petroleum Engineers AIME,* **216** (1959) 364–5.
17. Farouq Ali, S. M., *Oil Recovery by Steam Injection,* Producers Publ. Co., Bradford, 1970.
18. Prats, M., The heat efficiency of thermal recovery processes. *J. Petroleum Technol.*, **21** (1969) 323–32.
19. Lauwerier, H. A., The transport of heat in an oil layer caused by injection of hot fluid. *Applied Sci. Res.*, **5** (section A) (1955) 145–50.
20. Willman, B. T., Valleroy, V. V., Runberg, G. W., Cornelius, A. J. & Powers, L. W., Laboratory studies of oil recovery by steam injection. *J. Petroleum Technol.*, **13** (1961) 681–90.
21. Hearn, C. L., Effect of latent heat content of injected steam in a steam drive. *J. Petroleum Technol.*, **21** (1969) 374–5.
22. Myhill, N. A. & Stegemeier, G. L., Steam drive correlation and prediction. *J. Petroleum Technol.*, **30** (1978) 173–82.
23. Yortsos, Y. C. & Gavalas, G. R., Analytical modeling of oil recovery by steam injection—Part 1: Upper bounds; Part 2: Asymptotic and approximate solutions. *Soc. Petroleum Engineers J.*, **21**(2), (1981) 162–90.
24. Malofeev, G. E., Calculation of the temperature distribution in a

formation when pumping hot fluid into a well. *Neft. Gaz,* **3** (7) (1960) 59–64.
25. Boberg, T. C. & Lantz, R. B., Calculations of the production rate of a thermally stimulated well. *J. Petroleum Technol.,* **18** (1966) 1613–23.
26. Gontijo, J. E. & Aziz, K., A simple analytical model for simulating heavy oil recovery by cyclic steam in pressure-depleted reservoirs. SPE Paper No. 13037, 59th Annual Techn. Conf. and Exhibition of Soc. Petroleum Engineers, Houston, 16–19 September 1984.
27. Burns, J., A review of steam soak operations in California. *J. Petroleum Technol.,* **21** (1969) 25–34.
28. Carrigy, M. A., Thermal recovery from tar sands. *J. Petroleum Technol.,* **35** (1983) 2149–57.
29. Borregales, C. J., Steam-soak on thè Bolivar coast. *The Oil Sands of Canada–Venezuela,* CIM Special Volume No. 17, ed. D. A. Redford & A. G. Winestock. The Canadian Institute of Mining and Metallurgy, Montreal, 1977, pp. 561–83.
30. Farouq Ali, S. M., Current status of steam injection as a heavy oil recovery method. *J. Canadian Petroleum Technol.,* **13**(1) (January-March 1974) 54–68.
31. Blevins, T. R., Steamflooding in the US: A status report. *J. Petroleum Technol.,* **42**(5) (1990) 548–54.
32. Van Lookeren, J., Calculation methods for linear and radial steam flow in oil reservoirs. *Soc. Petroleum Engineers J.,* **23**(3) (1983) 427–39.
33. De Haan, H. J. & Schenk, L., Performance analysis of a major steam drive project in the Tia Juana Field, Western Venezuela. *J. Petroleum Technol.,* **21** (1969) 111–19.
34. Chu, C., State-of-the-art review of steamflood projects. *J. Petroleum Technol.,* **37** (1985) 1887–902.
35. Farouq Ali, S. M. & Meldau, R. F., Current steamflood technology. *J. Petroleum Technol.,* **31** (1979) 1332–42.
36. Blevins, T. R. & Billingsley, R. H., The ten pattern steam flood, Kern River Field, California. *J. Petroleum Technol.,* **27** (1975) 1505–14.
37. Brigham, W. E. & Strom, J. L., An engineering and economic analysis of a steamflood plus surfactant field project. Doc. US Department of Energy, DE-ACO3-80SF11445, May 1985.
38. Renard, G., Petit, H. & Valentin, E., Comparative technical and economic evaluation of steam injection for horizontal-well and vertical-well patterns. Fifth European Symp. on Improved Oil Recovery, Proc., Hungarian Hydrocarbon Inst., Budapest, 25–27 April 1989, pp. 585–94.
39. Owete, O. S., Castanier, L. M. & Brigham, W. E., Flow behavior of foam in porous media. Proceedings of AIChE Annual Meeting, Las Vegas, Nevada, September 1982.
40. Hirasaki, G. J. & Lawson, J. B., Mechanisms of foam flow through porous media. SPE Paper No. 12129, 58th Annual Techn. Conf. and Exhibition of Soc. Petroleum Engineers, San Francisco, 5–8 October 1983.
41. Overbeek, J. Th. G., Colloids and surface chemistry, a self-study subject, study guide. Center For Advanced Engineering Study and Department of Chemical Engineering, MIT, Cambridge, MA, 1972.

42. Fried, A. N., The foam drive process for increasing the recovery of oil. US Department of the Interior, Report of Investigation No. 5866, 1960.
43. Treinen, R. J., Brigham, W. E. & Castanier, L. M., Apparent viscosity measurements of surfactant foam in porous media. SUPRI TR 48, US Department of Energy, Report DOE/SF/11564-13 (DE 86000260), 1985.
44. Hirasaki, G. J., The steam–foam process. *J. Petroleum Technol.*, **41**(5) (1989) 449–56.
45. Patzek, T. W. & Koinis, M. T., Kern River steam–foam pilots. *J. Petroleum Technol.*, **42** (1990) 496–503.
46. Castanier, L. M., Steam with additives: Field projects of the eighties. *J. Petroleum Sci. Engineering*, **2** (1989) 193–206.

10 In-situ combustion

Jacques Burger
Comité d'Etudes Pétrolières et Marines, Paris, France

10.1	**Introduction**	313
10.1.1	Principles of dry forward combustion	314
10.1.2	Forward combustion combined with water injection or water-fireflooding	316
10.2	**Heat balance considerations**	318
10.2.1	Oxidizing-gas requirement and fuel content	318
10.2.2	Minimum fuel content for dry combustion	320
10.2.3	Water/oxidizer ratio in wet combustion	321
10.3	**Main parameters involved in in-situ combustion processes**	324
10.3.1	Fuel-content determination under simulated heating	324
10.3.2	Combustion-tube tests	326
10.4	**Process design and performance prediction**	333
10.5	**In-situ combustion operations**	335
10.6	**References**	336

10.1 Introduction

In-situ combustion is a recovery process in which an oxygen-containing gas is injected into an oil-bearing formation and where it reacts with organic components from the oil. The thermal and hydrodynamic effects help recover the unburned fractions of the crude.

Air is most frequently used for sustaining the process, but enriched air and even oxygen can be considered. Combustion is ignited in the injection well, so that the front propagates in the same direction as the fluid flow. This process is often known as dry forward combustion or fireflooding. Variations of the process have been developed, consisting of combining forward combustion with water injection.

10.1.1 *Principles of dry forward combustion*

In this procedure, the combustion front pushes the unburned fractions of the crude oil along ahead of it. The heavier fractions are transformed into a carbonaceous deposit with a low hydrogen content, which is often improperly called 'coke'. This deposit is burned with the oxygen from the injected gas.

It is convenient to describe the process by considering four main zones, which the author has numbered in the direction upstream to downstream (see Fig. 10.1).

(a) *Zone 1.* Combustion has already taken place in this part of the layer, which no longer contains any organic material. The injected gas heats up upon contact with the matrix and recovers a small amount of the heat released by the combustion. This zone constitutes a sort of heat exchanger with the temperature decreasing towards the inlet end.

(b) *Zone 2.* The combustion zone: oxygen is used up by the combustion reactions of the hydrocarbons and of the 'coke' retained on the rock. The temperature reached in this zone depends essentially on the nature and quantity of the various organic components remaining in the porous medium; the peak temperature is generally in the 400–600°C range.

(c) *Zone 3.* The cokefaction zone: heavy fractions that have been neither displaced nor vaporized are pyrolyzed in this zone. This pyrolysis is combined with oxidation if all the oxygen has not been used up in the combustion zone.

(d) *Zone 4.* When the temperature is sufficiently low, no further chemical changes are observed provided that the flowing gas does not contain unconsumed oxygen. The following phenomena are observed:

- (i) In the region nearest to the reaction zones, successive vaporization and recondensation of connate water and the light fractions of oil take place. Water from the combustion also condenses. This phenomenon of successive vaporization–condensation accelerates the downstream heat transfer.
- (ii) In the region where the temperature is lower than the condensation temperature of water, a water bank is observed in which

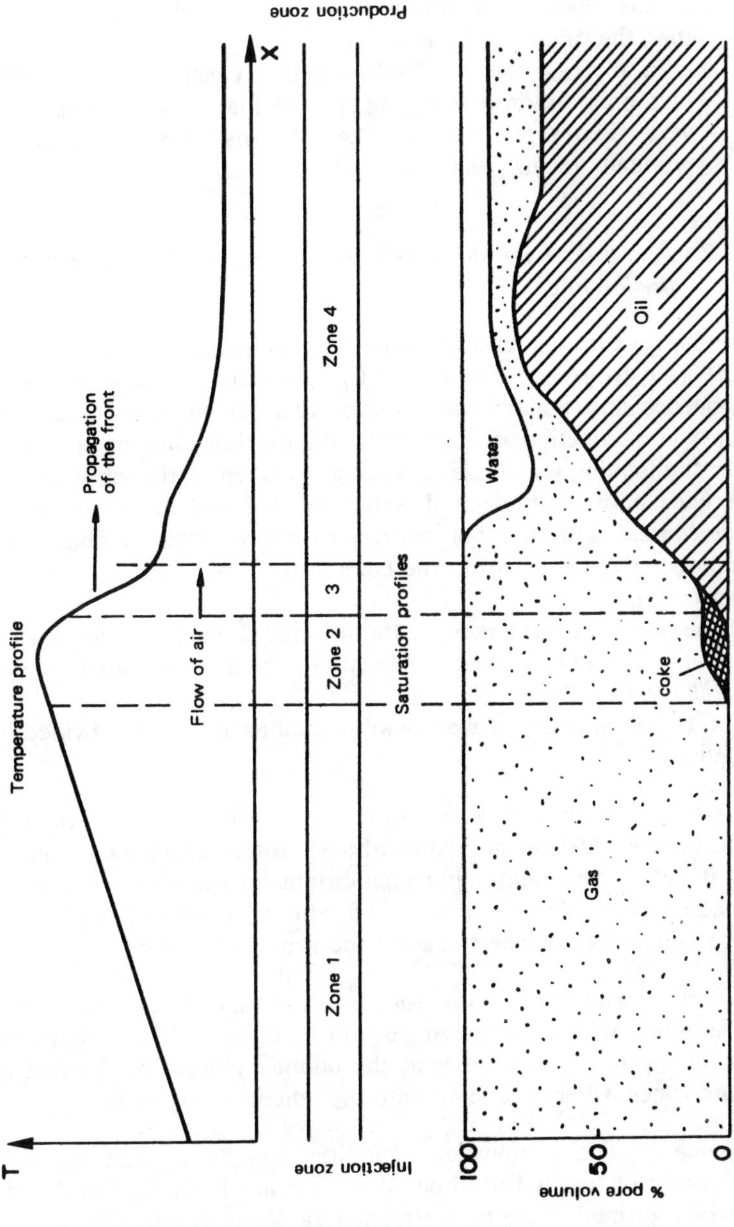

Fig. 10.1 Dry forward combustion: temperature and saturation profiles.

water saturation is greater than the original saturation. Ahead of this zone, an oil bank appears, with an oil saturation greater than the original oil saturation.

In any case the two banks together constitute a zone with a high pressure drop. Plugging up of the layer may even occur for very viscous oils. Beyond the oil bank, the porous medium gradually regains its initial characteristics.

10.1.2 *Forward combustion combined with water injection or water-fireflooding*

With dry forward combustion, any heat generated in the matrix, which is not used to preheat the injected gas, is lost by conduction to the overburden and underburden. Better thermal efficiency is achieved when water injection is combined with the injection of air. In this water-fireflooding technique, advantage is taken of the great enthalpy difference between the liquid water injected and the steam formed upon contact with the hot matrix, to recover the thermal energy accumulated upstream from the combustion front and to transport it downstream.

If the water/oxidizer ratio of the injected fluids is not too high we have what is known as wet combustion, which is described diagrammatically in Fig. 10.2.

The region affected by wet forward combustion can be divided into five zones.

(a) *Zone 1.* Zone 1, having already been swept by the combustion front, contains little or no hydrocarbons. However, since the temperature is below the liquid/vapor equilibrium temperature for water we find a two-phase flow. Since there is a saturation of liquid water, a part of the injected water never reaches the vaporization front.

(b) *Zone 2.* The water in this zone is in the vapor phase, so the pores are saturated with the injected gas and water vapor. The vaporization front of the injected water is at the boundary between the first and second zones, where the temperature gradient is very steep.

(c) *Zone 3.* The combustion zone: the oxygen is used up in the combustion of the hydrocarbons that have not been displaced and of the 'coke' formed in the downstream area of this zone.

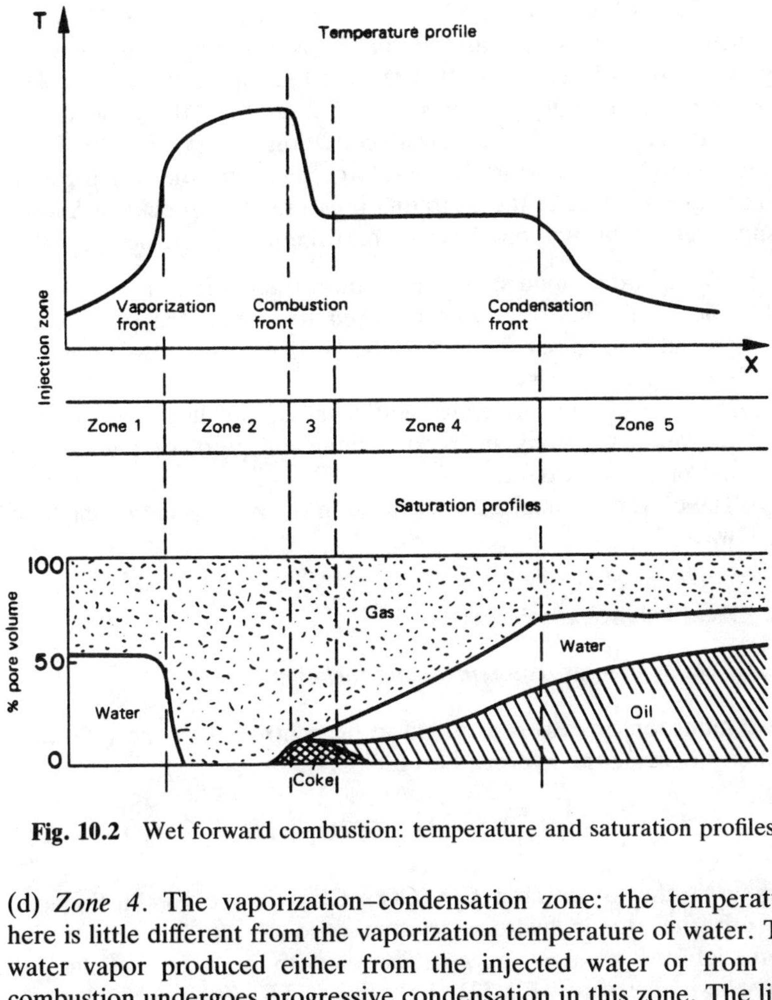

Fig. 10.2 Wet forward combustion: temperature and saturation profiles.

(d) *Zone 4*. The vaporization–condensation zone: the temperature here is little different from the vaporization temperature of water. The water vapor produced either from the injected water or from the combustion undergoes progressive condensation in this zone. The light and intermediate oil fractions are vaporized and carried downstream. Some chemical reactions may also take place if the extent of this zone and its temperature are sufficient.

(e) *Zone 5*. Downstream from the vaporization–condensation plateau, a sharp pressure drop is observed, which is due to the existence of the water bank and then the oil bank. Farther on, the porous medium gradually regains its initial characteristics.

As the water/oxidizer ratio increases, the energy given off by the combustion and built up in the matrix is no longer sufficient to vaporize all the injected water that reaches the vaporization front. The high temperature region (Zones 2 and 3 in Fig. 10.2) eventually disappears completely. Under certain conditions it is possible to obtain a hot zone, at a uniform temperature, close to the vaporization temperature of water at the operating pressure. This process is known as superwet combustion and has the following characteristics:

(a) Water in the liquid state is present in the entire hot zone.
(b) The exothermic reactions required to sustain the process are spread throughout the part of the hot zone where hydrocarbons and oxygen coexist.
(c) Because of the moderate heat level of the hot zone, partial oxidation reactions are predominant and pyrolysis phenomena are of limited extent.
(d) The oil is not completely displaced in the zone swept by the heat wave.

10.2 Heat balance considerations

10.2.1 *Oxidizing-gas requirement and fuel content*

The combustion process is assumed to be controlled by the following reaction involving the hydrocarbon fuel CH_x:

$$CH_x + \left(1 - \frac{\Gamma}{2} + \frac{x}{4}\right)O_2 \rightarrow (1-\Gamma)CO_2 + \Gamma CO + \frac{x}{2}H_2O \quad (10.1)$$

where Γ is the molar ratio $CO/(CO + CO_2)$ in the combustion gas and x is the atomic hydrogen/carbon ratio of the fuel. Since the volume of one kmol of oxygen, assumed to be a perfect gas, at atmospheric pressure (101·325 kPa) and 273 K (0°C) is 22·414 m^3, the consumption of 1 kg of CH_x requires the following volume of oxygen determined at temperature T:

$$v_{O_2,T} = 11{\cdot}207 \frac{2 - \Gamma + \dfrac{x}{2}}{(12 + x)} \frac{T}{273} \quad (10.2)$$

Let y_{O_2} be the mole fraction of oxygen in the oxidizing gas. Based on eqn (10.2), shows that the standard volume of gas, v_g, at atmospheric pressure and at temperature $T = 288$ K (i.e. 15°C or about 60°F), required for the stoichiometric combustion of a mass m_c of fuel, is calculated as follows:

$$v_g = \frac{b}{y_{O_2}} \frac{2 - \Gamma + \frac{x}{2}}{12 + x} m_c = \frac{C}{y_{O_2}} m_c \qquad (10.3)$$

where $b = 11{\cdot}82\ \mathrm{sm^3/kg} = 189\ \mathrm{sft^3/lb}$ and $C = ((2 - \Gamma + \frac{x}{2})/(12 + x))b$. For air ($y_{O_2} = 0{\cdot}21$), $(b/y_{O_2}) = 56{\cdot}3\ \mathrm{sm^3/kg}$ or $901\ \mathrm{sft^3/lb}$.†

Since the atomic hydrogen/carbon ratio of the fuel available for in-situ combustion is usually between 1 and 2, and the molar ratio $CO/(CO + CO_2)$ ranges from 0·05 to 0·2, between 10 and 12 $\mathrm{sm^3}$ of air is necessary to burn 1 kg of fuel.

Let us consider the one-dimensional propagation of a combustion process in a porous medium, with an oxygen-containing gas being injected at constant flow rate. The combustion front, assumed to be perpendicular to the gas flow, moves at velocity u_b. Let V_{gf} be the gas flux feeding the combustion front, i.e. the gas flow rate per unit surface area of the combustion front, determined under standard conditions.

The ratio V_{gf}/u_b is directly related to ratio v_g/v representing the volume of gas required for the combustion front to sweep the unit bulk volume of the porous medium:

$$\frac{V_{gf}}{u_b} = \frac{1}{X_u} \frac{v_g}{v} \qquad (10.4)$$

X_u is the oxygen utilization, i.e. the fraction of the oxygen content of the gas that is actually used up in the combustion reaction. V_{gf}/u_b is the gas requirement at the combustion front.

Hence, combining eqns (10.3) and (10.4), we get:

$$\frac{V_{gf}}{u_b} = \frac{b}{y_{O_2}} \frac{2 - \Gamma + \frac{x}{2}}{X_u(12 + x)} \frac{m_c}{v} = \frac{C}{y_{O_2} X_u} \frac{m_c}{v} \qquad (10.5)$$

where $b = 11{\cdot}82\ \mathrm{sm^3/kg}$ and $C = \dfrac{2 - \Gamma + \frac{x}{2}}{12 + x} b$

† $\mathrm{sm^3}$ means standard cubic meters, i.e. volume at atmospheric pressure (101.325 kPa) and 15°C.

m_c/v is the fuel content, i.e. the amount of fuel per bulk unit volume of porous medium. The fuel available for the process consists of the coke-like residue and undisplaced heavy ends of the crude (Section 10.1). It should be mentioned that a fraction of the injected gas builds up behind the moving combustion front. Hence the gas flux V_g at injection is higher than the gas flux actually feeding the combustion front. It may be shown that the gas requirement at injection, V_g/u_b, does not differ significantly from the gas requirement at the front for air-fireflooding, whereas the discrepancy becomes significant for high-pressure oxygen-fireflooding due to gas storage in the swept zone.[1]

10.2.2 *Minimum fuel content for dry combustion*

The overall heat balance can be used to predict the temperature of the combustion front for a given value of the fuel content. The minimum values of the fuel and oxidizing-gas requirements for viable combustion can thus be evaluated.

The following assumptions will be considered: (a) the propagation of combustion is strictly one-dimensional, without any effect of gravity or transverse heat transfer; (b) a steady hydrodynamic regime prevails at the inlet of the control volume; (c) a steady-state thermal regime is achieved; (d) the pressure is assumed to be constant within the control volume; and (e) the porous medium is homogeneous.

The temperature profile exhibits a horizontal tangent at the peak temperature T_B and at the initial reservoir temperature T_r. Assuming that the whole profile propagates at velocity u_b, a simplified heat balance can thus be written for the control volume indicated in Fig. 10.3, characterized by zero heat loss by conduction across the inlet face 1 and the outlet face 2:

$$\rho_g V_{gf} \bar{c}_g (T_B - T_r) + Q u_b \frac{m_c}{v} = (\overline{\rho c})^* (T_B - T_r) u_b \tag{10.6}$$

where ρ_g is the density of the injected gas under standard conditions (101·325 kPa, 288 K), V_{gf} is the gas flux (standard conditions) across the combustion front, $\bar{c}_g$ is the average heat capacity between T_r and T_B of the unit mass of gas, $(\overline{\rho c})^*$ is the equivalent heat capacity per unit bulk volume of the porous medium, and Q is the heating value

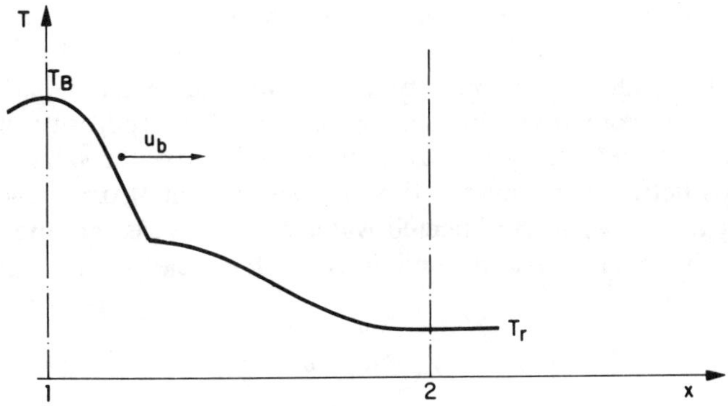

Fig. 10.3 Dry combustion: schematic temperature profile downstream from the temperature peak.

of the fuel. Hence, using eqn (10.5) to eliminate the fuel content m_c/v, we have:

$$\frac{V_{gf}}{u_b} = \frac{\overline{(\rho c)}^*(T_B - T_r)}{\frac{y_{O_2}X_u}{C}Q + \rho_g\bar{c}_g(T_B - T_r)} \quad (10.7)$$

The heat of combustion Q (with H_2O condensed) per unit mass of fuel burned, may be calculated for reaction (10.1) as a function of the atomic hydrogen/carbon ratio x and of the ratio $\Gamma = CO/(CO + CO_2)$:[1]

$$Q = \frac{393\,200 - 283\,800\Gamma + 130\,450x}{12 + x} \quad \text{kJ/kg} \quad (10.8)$$

Since the formation of the coke-like fuel occurs above 350°C, a minimum value of the temperature difference $(T_B - T_r)$ is likely to be necessary to enable a viable combustion process to be obtained in the porous media. Equation (10.7) can be used to calculate the corresponding value of the minimum gas requirement at the combustion front.

Assuming $X_u = 1$, $x = 1{\cdot}5$, $\Gamma = 0{\cdot}1$, $\rho_g\bar{c}_g = 1{\cdot}23$ kJ sm^{-3} K^{-1}, $\overline{(\rho c)}^* = 1800$ kJ m^{-3} K^{-1} and a minimum value of $T_B - T_r$ equal to 325°C, the minimum air requirement is calculated to be about 140 sm^3/m^3, and the corresponding fuel requirement to be 12·7 kg/m^3. We will see in Section 10.3.2 that this result is in good agreement with observations in adiabatic combustion cells.

10.2.3 *Water/oxidizer ratio in wet combustion*

When an oxygen-containing gas and water are injected under one-dimensional conditions into a porous medium (porosity Φ) that initially contains only gas, a zone with uniform water saturation S_{w1} develops between the inlet end and a water front that advances at a velocity u_{w1}. If V_w is the injected water flux and V_g is the injected gas flux, both under standard conditions, in the steady-state regime we have:

$$V_w = \Phi S_{w_1} u_{w_1} \tag{10.9}$$

and

$$\frac{V_w}{V_g} = \Phi S_{w_1} \frac{u_{w_1}}{V_g} \tag{10.10}$$

Let us consider a hot zone moving in the same direction as the fluids, initially situated downstream from the injection area and bounded by a given temperature profile.

Moreover, we assume that on the upstream part of this profile, the point where the temperature is equal to water vaporization temperature advances at a constant velocity u_v (Fig. 10.4(a)). Some of the injected water will be vaporized if the water front reaches the thermal front (Fig. 10.4(b)).

If the thermal front is that of dry combustion, we take the velocity u_v as equal to ξu_b, where u_b is the velocity of the combustion front and ξ is a constant that is less than 1. Part of the injected water is vaporized and carried along to the high temperature zone if:

$$\frac{V_w}{V_g} > \xi \Phi S_{w_1} \frac{u_b}{V_g} \tag{10.11}$$

By applying eqn (10.11) to various experimental conditions, it can be shown that, for wet combustion with air, the minimum value of the water/air ratio for the injected water to reach the combustion front is between $0{\cdot}2 \times 10^{-3}$ and $0{\cdot}8 \times 10^{-3}$ m^3/sm^3 depending on the case.[1] When the injected water/air ratio (WAR) is less than this value, water injection has no effect on forward combustion.

However, since the vaporization of the water reaching the combustion front relies upon the enthalpy stored behind the front, the water

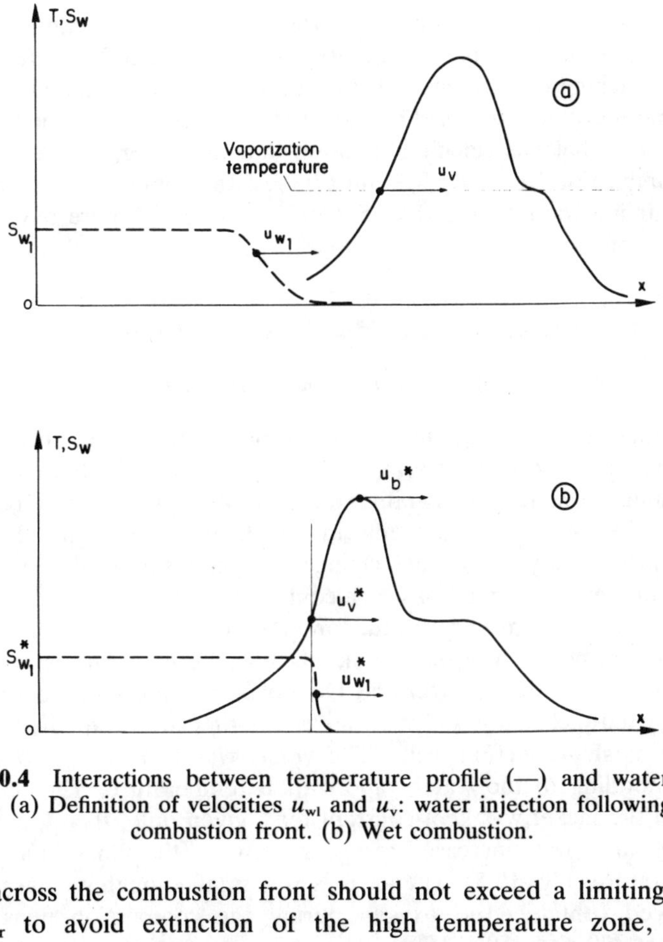

Fig. 10.4 Interactions between temperature profile (—) and water front (---). (a) Definition of velocities u_{wl} and u_v: water injection following a dry combustion front. (b) Wet combustion.

flux across the combustion front should not exceed a limiting value $(V_{wf})_{cr}$ to avoid extinction of the high temperature zone, which corresponds to $\xi^* > 1$. Hence, for stable wet combustion to be possible, the injected water flux must satisfy the following inequality (with $\xi^* = 1$):

$$V_w < \Phi S^*_{w_1} u^*_b + (V_{wf})_{cr} \tag{10.12}$$

where u^*_b is the propagation velocity of the wet combustion front and S^*_{w1} is the water saturation behind the vaporization front. Hence:

$$\frac{V_w}{V_g} < \Phi S^*_{w_1} \frac{u^*_b}{V_g} + \frac{(V_{wf})_{cr}}{V_g} \tag{10.13}$$

During the steady-state adiabatic propagation of wet combustion, the upper limit for the enthalpy available per unit time to heat the gas and water reaching the combustion front up to the peak temperature T_B^* is the amount of heat released by the combustion reaction per unit time.[1]

The heat-balance calculation shows that the water/oxidizer ratio at the combustion front $(V_w/V_g)_f$ must be less than about 1×10^{-3} m^3/sm^3 with air injection or less than 5×10^{-3} m^3/sm^3 with pure oxygen[1] for wet combustion to be viable.

10.3 Main parameters involved in in-situ combustion processes

10.3.1 *Fuel-content determination under simulated heating*

Experiments simulating the approach of a combustion front in an impregnated sand or rock sample can be carried out in the laboratory to evaluate the fuel availability for forward combustion. Given the complexity of the phenomena involved, we can hardly hope to reproduce exactly the various changes in composition, phase distribution and temperature; all of which contribute to fixing the fuel content. Simulation runs are very useful for getting a rough idea of the fuel availability, but such results should always be used warily.

In the Alexander, Martin and Dew experiments, a cylindrical core in a heated jacket was swept by air circulating from the outer surface to the axial production well.[2] The vessel was first heated to a level corresponding to the water vaporization temperature. The temperature of the jacket was kept constant for a given time, then the control temperature was increased progressively to the final combustion temperature (Fig. 10.5). The air flow was small enough so that the gas produced contained no oxygen during the combustion phase, i.e. between 200 and 450°C (Fig. 10.5). A concentration gradient existed within the sample, similar to that observed in Zones 2 and 3 in Fig. 10.1. The amount of carbon burned is calculated from the area under the CO and the CO_2 concentration curves and from the air flow rate.

The value of the mass of carbon burned per unit mass of rock w_C, have been correlated to the following characteristics of crude oil: the density, viscosity, hydrogen/carbon ratio and Conradson residue. The fuel availability increases with density, viscosity and Conradson residue; it decreases as the hydrogen/carbon ratio of the oil increases.[2] However, it can be seen from the spread of the results in

(a)

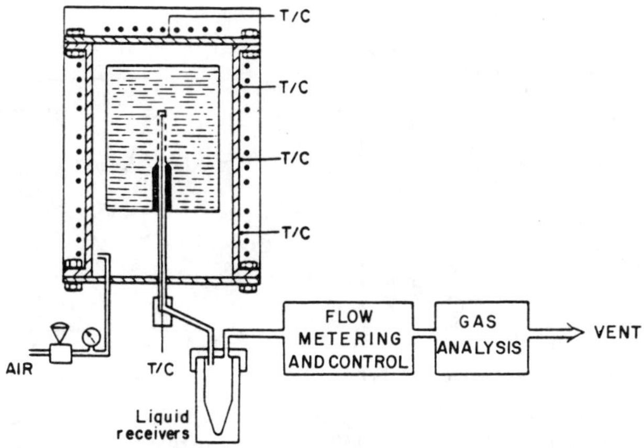

(b)

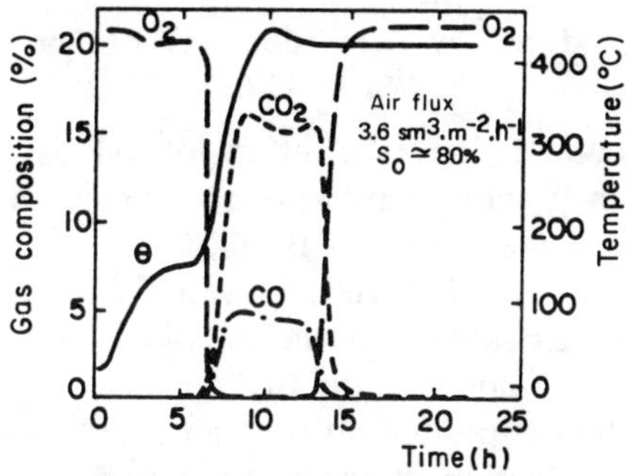

Fig. 10.5 Experimental determination of fuel requirement for forward combustion. (a) Experimental device; (b) temperature and effluent gas composition versus time for a test. (From Ref. 2, reproduced by permission of The Society of Petroleum Engineers.)

Fig. 10.6 that there is not a very strong correlation between w_C and any one particular physical property of the oil, even in tests carried out on the same rock.

Correlation with the Conradson residue gives the smallest spread, because it is a measure of the coke-forming capacity of the oil. More or less complex empirical formulae have been proposed to determine w_C from various physical properties of the oil.[2] In fact, thermogravimetric and differential thermal analyses, which are discussed in Chapter 8, have shown just how difficult it is to characterize the reactivity of crude oils. Moreover, the findings of Alexander *et al.*[2]

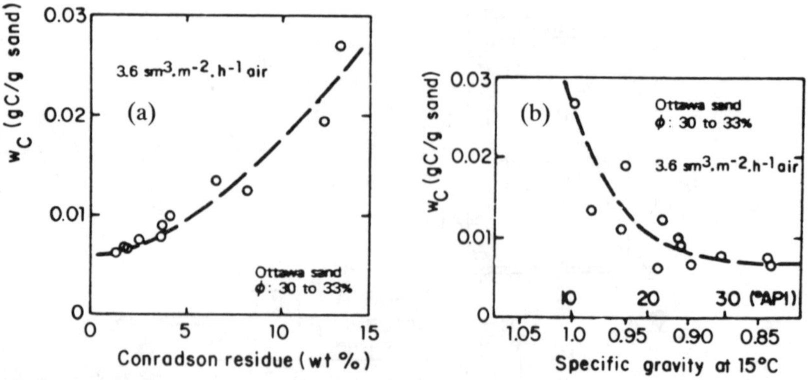

Fig. 10.6 Influence of oil properties on the amount of coke deposit. Maximum heating temperature: 430°C. (a) Correlation with Conradson residue. (b) Correlation with the specific gravity (or API gravity) of the oil. (From Ref. 2, reproduced by permission of The Society of Petroleum Engineers.)

would tend to indicate that the effect of oil saturation on fuel availability is much greater than what is actually found in forward combustion runs in a one-dimensional cell.

On the other hand, the influence of various characteristics of the matrix, e.g. its surface area and the catalysts it may contain, can significantly affect the amount of the fuel deposited.[3,4]

The mass of carbon deposited on the unit mass of rock, w_C, can be used for a rough evaluation of the amount of fuel available for the in-situ combustion process. The mass of fuel CH_x available per unit bulk volume of the porous medium is evaluated as follows:

$$\frac{m_c}{v} = \rho_s(1-\Phi)\left(1+\frac{x}{12}\right)w_C = \rho_s(1-\Phi)w_{CH_x} \qquad (10.14)$$

where x is the atomic hydrogen/carbon ratio of the fuel, ρ_s is the density of the solid rock; w_C and w_{CH_x} are the masses of carbon and of fuel deposited per unit mass of rock. The density ρ_s should be expressed in the same units as m_c/v (e.g. in kg/m^3).

10.3.2 *Combustion-tube tests*

(a) *Equipment.* The most common practice for determining the fuel content and air requirement is to conduct one-dimensional combustion tests in the laboratory. Such tests are carried out in combustion tubes equipped with thermocouples.[5–9]

The tubes are generally constructed of thin-walled stainless steel and are mounted within a pressure jacket, with the annulus between both tubes being filled with an insulating material and pressurized at a pressure level adjusted to the pressure in the combustion tube. Some investigators use heating collars along the combustion tube, with the power input being adjusted manually[6] or being computer-controlled,[8,9] so as to approach one-dimensional propagation in adiabatic conditions. Typical equipment is shown in Fig. 10.7.

The combustion-tube tests are generally performed in sand packs, either with a sand–oil–water mixture being tamped into the tube, or with oil and water being injected with an appropriate sequence into the sand pack. Laboratory investigations in naturally-consolidated cores have also been described.[10]

Combustion tests provide the values of the main parameters of the propagation of the combustion process, from which the oxidizing-gas requirement and the fuel consumption can be derived. The propagation velocities of the fronts are calculated from the temperature measurements versus time. The combustion reaction is characterized by the composition of the produced gas. The efficiency of the process

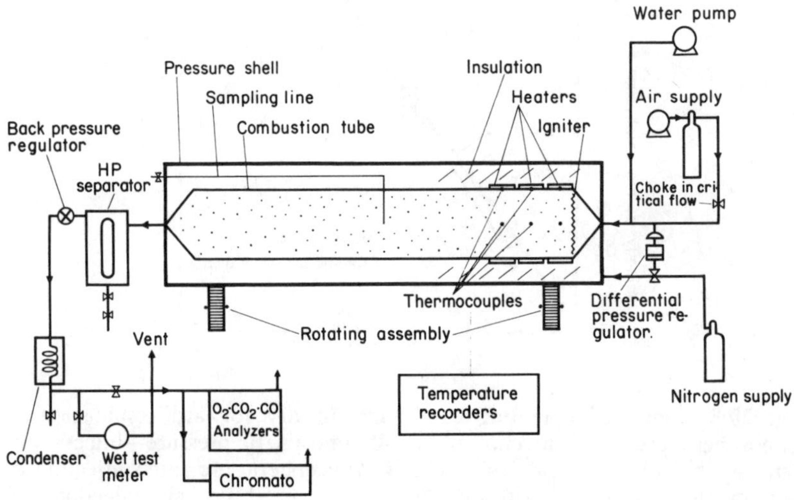

Fig. 10.7 Laboratory equipment for in-situ combustion. (From Ref. 6, reproduced by permission of The Society of Petroleum Engineers.)

is determined by measuring the amounts and properties of the oil and water produced.

(b) *Results in dry combustion.* Figure 10.8 shows the influence of pressure and air flux (injected air flow rate in standard conditions per unit surface area of the combustion front), expressed in $sm^3\,m^{-2}\,h^{-1}$, on the velocity of the forward combustion front and the maximum

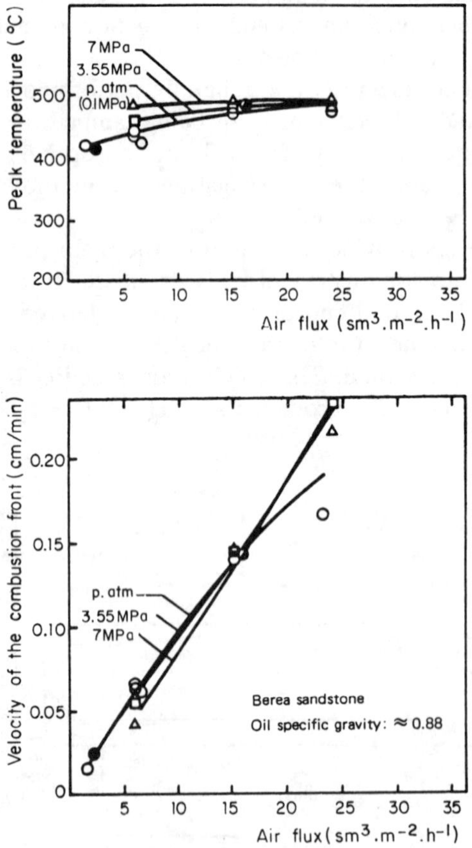

Fig. 10.8 Effect of operating conditions on dry forward combustion ○ Atmospheric pressure—injection of air; ● atmospheric pressure—injection of $6\,sm^3\,m^{-2}\,h^{-1}$ of a gas with 7·4% O_2. ◐ Atmospheric pressure—injection of $6\,sm^3\,m^{-2}\,h^{-1}$ of a gas with 54·5% O_2. □ 3·55 MPa—air injection. △ 7·0 MPa—air injection. (From Ref. 5, reproduced by permission of The Society of Petroleum Engineers.)

temperature achieved. The results in Fig. 10.8 have been reported by Wilson *et al.*[5] for a mixture of Berea sand and a lubricating base stock with specific gravity 0·883. At constant pressure, it can be seen that basically the velocity of the front increases linearly with air flux, whereas the maximum temperature is relatively unaffected. Furthermore, the effect of pressure is very slight. Subject to various restrictions which shall be discussed later on, for a given oil–matrix mixture we can write approximately:

$$\frac{\text{air flux}}{\text{velocity of the combustion front}} \approx \text{constant},$$

$$\text{for any pressure and air flux} \quad (10.15)$$

The ratio of the injected air flux to the velocity of the combustion front is the amount of air actually injected in order for the combustion front to sweep a unit volume of the porous medium (air requirement).

Due to the low sensitivity of the air requirement to operating conditions, it is possible to scale up from the results obtained in the laboratory to determine the air requirement of a future field application, provided that the differences in pressure and air flux from laboratory to field are not too great.

In fact, eqn (10.15) is not valid at high air flux, especially at low pressure, because the residence time of the oxygen in the high-temperature zone is not sufficient for complete consumption to be obtained. In some oil–matrix mixtures, the air requirement tends to increase with pressure at low gas flux.

Specific tests have been performed to determine the effect of enrichment by oxygen on the dry combustion process.[8,9,11] The oxygen requirement and fuel consumption may not be significantly affected by the oxygen content (see an example in Fig. 10.8). However, in some cases the oxygen requirement tends to increase with the oxygen partial pressure. With a medium-gravity crude oil, it was observed that combustion could not be sustained with air or 30% oxygen, whereas combustion was maintained with an oxygen content of 40% or higher.[11] The influence of the enrichment by oxygen on the performances of *in-situ* combustion may be explained by: (a) the effect of oil reactivity and operating conditions on the quality of the residual organic matter, (b) the influence of the oxygen partial pressure on the kinetics of the reactions involved, and (c) oxygen storage in the swept zone (see Section 10.2.1).

As we have already pointed out in Section 10.3.1, the amount of carbon burned w_C under simulated heating depends on the characteristics of the oil and of the matrix. The same is true for the fuel content and air requirement in fireflooding. The values of both parameters are generally higher for heavy oils than for light oils. Their range of variation is shown in Fig. 10.9 for a specific series of runs performed in fine sand. Stable propagation of the combustion front was found to be impossible when the specific gravity of the oil was too low, due to insufficient fuel deposition. However, a combustion process propagated in a 40° API crude (specific gravity: 0·825) with a fuel availability of only 12·3 kg/m^3 corresponding to an air requirement of 140 sm^3/m^3 and a peak temperature of 345°C. These minimum values for viable combustion are consistent with those resulting from the heat balance calculations in Section 10.2.2.

However, there is no single correlation of the air requirement and fuel availability with the oil density alone[1] since the extent of the reactions involved depends on the characteristics of the oil–matrix system (oil oxidability, surface area of the matrix, possible catalytic effects). The influence of the matrix is particularly important for light oils, which are generally only slightly cokefiable. Their fuel content may be increased enough in clay-containing rocks so as to make propagation of in-situ combustion possible where normally, in a clean medium with low specific area, it would be completely impossible, e.g. in pure silica.[1,3,4] Therefore any laboratory test designed to evaluate

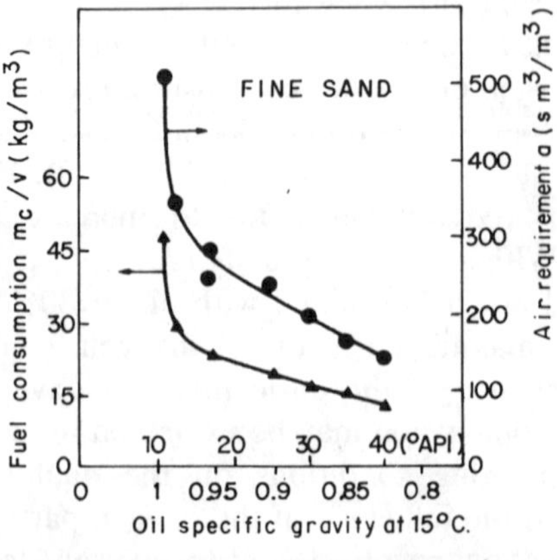

Fig. 10.9 Influence of oil gravity on fuel consumption and air requirement (dry forward combustion). (From Ref. 12, reproduced by permission of The Society of Petroleum Engineers.)

the feasibility of a combustion project in the field should be carried out on oil and matrix samples from the field or by restoring their original characteristics, so that the oil–matrix interactions may be accounted for.

For air-combustion tests, the air requirement generally ranges between 150 and 400 sm^3/m^3 depending on the operating conditions, with a fuel content ranging respectively from some 15 to 40 kg/m^3. If combustion takes place efficiently, the produced gas contains no or little oxygen, 12–15% CO_2 and 0·5–3% CO. For oxygen-combustion tests, the CO_2 content in the produced gas may range between 60 and 80%.

(c) *Results of wet combustion tests.* As mentioned in Sections 10.1.2 and 10.2.3, the wet combustion regime is obtained from a moderate water/oxidizer ratio. The effect of water injection on the combustion process has been studied experimentally with air injection[6,7] or for enriched-air combustion.[8] The lengthening of the steam plateau and the reduction of the extent of the high temperature zone are shown in Fig. 10.10.

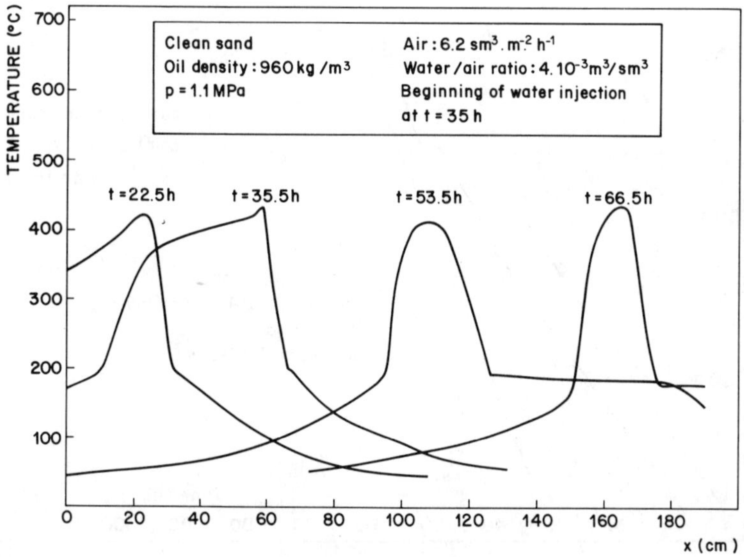

Fig. 10.10 Temperature profiles for wet combustion after an initial dry-combustion phase.(From Ref. 6, reproduced by permission of The Society of Petroleum Engineers.)

The large volume of condensing steam that sweeps the pore volume ahead of the combustion front enhances the displacement efficiency. This may lead to lower oil saturation ahead of the combustion zone and to an increase in the velocity u_b^* of the wet combustion front with respect to the velocity u_b obtained in dry combustion.[6] This effect is illustrated in Fig. 10.11, where we also see that the steady-state velocity u_b^* is slightly greater than the velocity of the vaporization front u_v^*. The high temperature zone thus retains a fairly constant thickness, and the residence time of oxygen in the high temperature zone is long enough for the combustion reaction not to be partially quenched by the approach of the cold water bank. Effectively, the oxygen utilization is complete and analyses of samples taken after the test from the zone swept by the combustion front reveal no residual organic material, just as for dry combustion.[6]

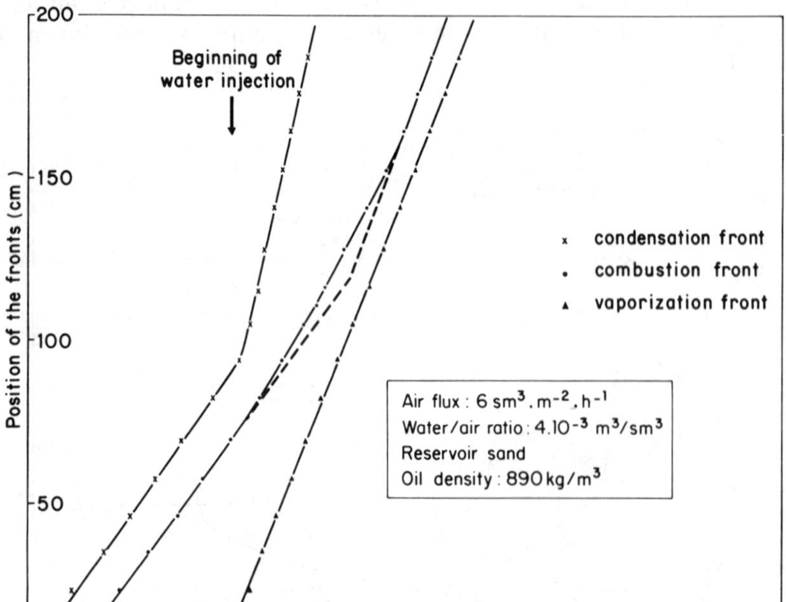

Fig. 10.11 Progression of the fronts in wet combustion: operating pressure = 1·1 MPa. (From Ref. 6, reproduced by permission of The Society of Petroleum Engineers.)

The acceleration of the combustion front, which is generally observed when water is injected along with the oxidizing gas in the wet combustion regime, results in a significant decrease in the oxidizing-gas requirement and in fuel consumption.[6–8]

Oil production is accelerated, and the recovery achieved by wet combustion may be even higher than that obtained with dry combustion.

If the water/oxidizer ratio is higher than a critical value, the vaporization front tends to overrun the combustion front and extinction of the high temperature zone occurs.[6] This point is discussed in Section 10.2.3 for one-dimensional steady-state water-fireflooding. Under such conditions it may be shown that, when air is the oxygen carrying gas, the maximum water/air ratio *at injection* for the wet combustion regime is in the range of 0·002–0·004 m^3/sm^3. As mentioned in Section 10.1.2, the superwet combustion regime may be sustained at higher water/air ratios if operating pressure and oil oxidability are sufficient.[7,13,14]

10.4 Process design and performance prediction

The design of a combustion field test is based: (a) on the estimation of fuel availability and oxidizing-gas requirement determined in the laboratory or by correlations, and (b) on an appropriate injection program taking into account the characteristics of the candidate reservoir.

Three-dimensional numerical models are available for predicting the propagation of the combustion process as a function of time.[15,16] However, simplified evaluation methods have often been used for designing a combustion project.

Since the combustion front is sustained by a high-temperature reaction between the flowing oxidizer and the fuel available in the rock material, several theoretical studies on the propagation of a vertical heat wave have focused on calculating the oxidizing-gas flow rate leading to extinction.[17,18]

It may be concluded from these studies that, for dry in-situ combustion with air, the combustion front should be fed by an air flux higher than a minimum value, $V_{g\,min}$, which is between 0·2 and 0·6 $sm^3\,m^{-2}\,h^{-1}$, depending on fuel availability and reservoir thickness. This evaluation has been substantiated by the analysis of field test results.[19,20]

It must be recognized that the optimization of the design of an actual field test depends largely on the volumetric distribution of the oxidizing gas in the reservoir. In particular, gravity forces play an important role, so that there is always a tendency for the injected gas to override the oil column. In the most frequent case of air-fireflooding characterized by rather high gas injection rates, gravity override is predominant when the reservoir thickness is greater than about 5 m. As a matter of fact, the thickness of the burned zone determined by temperature measurements or post-burn coring is often in the 3–7 m range.[21,22] Another problem relates to possible channeling due to areal and vertical heterogeneities.

In fact, the oxidizing-gas flow rate program must be selected so as to take into account: (a) the minimum gas flux to sustain the combustion process in consideration of the vertical and horizontal sweep efficiencies, and (b) the injectivity of the wells.[1]

Assuming that a vertical combustion front advances in a piston-like manner away from the injection well, the oxidizing-gas flow rate into one injector of an inverted five-spot pattern is increased to a final value $\dot{v}_{gf}$ that can be evaluated as follows:

$$\dot{v}_{gf} = \alpha L E_V h_n V_{g\,min} \tag{10.16}$$

where αL is the final length of the combustion front in the horizontal plane, with L being the distance between adjacent producers in the pattern and α a numerical coefficient that may be evaluated as about 3·5. The parameter E_V is the vertical sweep efficiency by the oxidizing gas; h_n is the net thickness of the reservoir; $V_{g\,min}$ is the minimum gas flux through the combustion front required for sustaining the process. In consideration of the guidelines indicated above concerning the vertical coverage and the minimum air flux, it may be determined that the final air flow rate per well, generally ranges from 500 to 2000 sm³/h, for patterns with areas from 10 000 m² (2·5 acres) to 60 000 m² (15 acres).

The total volume of air (standard conditions) required for the combustion process to sweep the pattern, v_a, is:

$$v_a = E_V E_A h_n L^2 \frac{V_g}{u_b} \tag{10.17}$$

where E_V is the vertical sweep efficiency, and E_A is the areal sweep efficiency by the oxidizing gas. In homogeneous reservoirs, the areal sweep efficiency may reach about 65%. The gas requirement at injection, V_g/u_b, may be derived from laboratory tests.

The performance of the combustion process is characterized by the air/oil ratio (or the oxygen/oil ratio), i.e. the ratio of the amount of air (or oxygen) injected, determined under standard conditions, to the amount of oil produced. This parameter is generally expressed in sm^3/m^3 or in sft^3/bbl. The air/oil ratio ranges from 500 sm^3/m^3 to 3000 sm^3/m^3 in most air-fireflooding field projects.

Several investigators have proposed correlations to predict the air/oil ratio as a function of reservoir parameters.[1,20,23] It is generally assumed that the unburned oil from the zone swept by combustion is displaced towards the producers. Moreover, some oil is displaced from (a) the steam zone downstream from the burned zone, and (b) the underlying hot zone.[1,20,21] Therefore, the fraction of the initial oil that is displaced is significantly higher than the fractional volume burned.[20] Just as for steam-drive prediction, a capture coefficient can be used to convert the amount of oil displaced to the amount of oil actually produced. Oil recovery factors of 50% or more have been reported.

The basic principles for designing a water-fireflooding project are similar to those relevant to dry fireflooding, provided that the oxidizing-gas requirement and the extent of the steam-swept zone are estimated accordingly. The air/oil ratio should have a lower value in water-fireflooding than in dry combustion due to better thermal efficiency.

10.5 In-situ combustion operations

The first step in a combustion operation is ignition. If the oil in place is sufficiently oxidizable under bottomhole conditions, spontaneous ignition may occur in the formation in the vicinity of the injection well after several days of oxidizing-gas injection.[24–25] This phenomenon takes place more often in formations where the reservoir temperature is around 60°C or more. Otherwise, ignition is obtained by artificially heating the formation in the vicinity of the injection well bore with an electric heater, a burner, by hot-fluid injection or by injecting highly oxidizable chemicals.[1,26]

During the implementation of the combustion project, regular monitoring is recommended to control the process, i.e. gas analysis in the producers, temperature measurements, etc. Special safety precautions are required for enriched-air fireflooding.

Numerous in-situ combustion pilots as well as several large-scale operations have taken place and are described in the literature.[1,23,27] Typical examples are the South Belridge test in California,[20,22] the Bellevue project in Louisiana,[28] and the Suplacu de Barcau field operation in Romania.[21,29] Several pilot tests of oxygen-fireflooding have been reported.[30] The most successful operations have been performed in sand or sandstone formations containing heavy- to medium-gravity crude oil (13–35° API). Moderate thickness (2–20 m) and depth (100–1500 m), good permeability (more than 100 mD), and a minimum oil content of 0·1 m^3/m^3 are recommended.

10.6 References

1. Burger, J., Sourieau, P. & Combarnous, M., *Thermal Methods of Oil Recovery,* Editions Technip, Paris, and Gulf Publishing Company, Houston, 1985.
2. Alexander, J. D., Martin, W. L. & Dew, J. N., Factors limiting fuel availability and composition during in-situ combustion. *J. Petroleum Technol.,* **14** (1962) 1154–64.
3. Burger, J. & Sahuquet, B., Chemical aspects of in-situ combustion. Heat of combustion and kinetics. *Soc. Petroleum Engineers J.,* **12** (1972) 410–22.
4. Vossoughi, S., Willhite, G. P., Kritikos, W. P., Guvenir, I. M. & El Shoubary, Y., Automation of an in-situ combustion tube and study of the effect of clay on the in-situ combustion process. *Soc. Petroleum Engineers J.,* **22** (1982) 493–502.
5. Wilson, L. A., Reed, R. L., Reed, D. W., Clay, R. R. & Harrison, N. H., Some effects of pressure on forward and reverse combustion. *Soc. Petroleum Engineers J.,* **3** (1963) 127–37.
6. Burger, J. & Sahuquet, B., Laboratory research on wet combustion. *J. Petroleum Technol.,* **25** (1973) 1137–46.
7. Garon, A. M. & Wygal, Jr, R. J., A laboratory investigation of fire-water flooding. *Soc. Petroleum Engineers J.,* **14** (1974) 537–44.
8. Moore, R. G., Bennion, D. W., Belgrave, J. D. M., Gie, D. N. & Ursenbach, M. G., New insights into enriched air in-situ combustion. *J. Petroleum Technol.,* **42**(7) (1990) 915–23.
9. Petit, H., In-situ combustion with enriched air. SPE Paper No. 16741, 62nd Annual Techn. Conf. and Exhibition of Soc. Petroleum Engineers, Dallas, 27–30 September 1987.
10. Petit, H., Audibert, A. & Gadelle, C., In-situ combustion in consolidated cores. Fifth European Symp. on Improved Oil Recovery, Proc., Hungarian Hydrocarbon Inst., Budapest, 1989, pp. 505–14.

11. Hansel, J. G., Bennings, M. A. & Fernbacher, J. M., Oxygen in-situ combustion for oil recovery: combustion tube tests. *J. Petroleum Technol.*, **36** (1984) 1139–44.
12. Showalter, W. E., Combustion-drive tests. *Soc. Petroleum Engineers J.*, **3** (1963) 53–8.
13. Dietz, D. N. & Weijdema, J., Wet and partially quenched combustion. *J. Petroleum Technol.*, **20** (1968) 411–15.
14. Parrish, D. R. & Craig, Jr, F. F., Laboratory study of a combination of forward combustion and waterflooding—the COFCAW process. *J. Petroleum Technol.*, **21** (1969) 753–61.
15. Coats, K. H., In-situ combustion model. *Soc. Petroleum Engineers J.*, **20** (1980) 533–54.
16. Le Thiez, P. & Lemonnier, P., An in-situ combustion reservoir simulator with a new representation of chemical reactions. *SPE Reservoir Engineering,* **5**(3) (1990) 285–92.
17. Thomas, G. W., A study of forward combustion in a radial system bounded by permeable media. *J. Petroleum Technol.*, **15** (1963) 1145–9.
18. Chu, C., Two-dimensional analysis of a radial heat wave. *J. Petroleum Technol.*, **15** (1963) 1137–44.
19. Parrish, D. R., Rausch, R. W., Beaver, K. W. & Wood, H. W., Underground combustion in the Shannon Pool, Wyoming. *J. Petroleum Technol.*, **14** (1962) 197–205.
20. Gates, C. F. & Ramey, Jr, H. J., A method of engineering in-situ combustion oil recovery projects. *J. Petroleum Technol.*, **32** (1980) 285–92.
21. Carcoana, A., Turta, A., Burger, J. & Bardon, C., Balayage et récupération d'huile lourde par combustion in-situ. Int. Symp. on Hydrocarbon Exploration, Drilling and Production Techniques, Editions Technip, Paris, 1975, pp. 117–34.
22. Gates, C. F., Jung, K. D. & Surface, R. A., In-situ combustion in the Tulare Formation, South Belridge Field, Kern County. *J. Petroleum Technol.*, **30** (1978) 798–806.
23. Brigham, W. E., Satman, A. & Soliman, M. Y., Recovery correlations for in-situ combustion field projects and application to combustion pilots. *J. Petroleum Technol.*, **32** (1980) 2132–8.
24. Tadema, H. J. & Weijdema, J., Spontaneous ignition of oil sands. *Oil and Gas J.*, **68**(50) (14 December 1970) 77–80.
25. Burger, J., Spontaneous ignition of oil reservoirs. *Soc. Petroleum Engineers J.*, **16** (1976) 73–81.
26. Strange, L. K., Ignition: key phase in combustion recovery. *Petroleum Engineer,* **35**(12) (November 1964) 105–109; **36**(13) (December 1964) 97–106.
27. Chu, C., State-of-the-art review of fireflood field projects. *J. Petroleum Technol.*, **34** (1982) 19–36.
28. Joseph, C. & Pusch, W. H., Engineering and economics of the Bellevue In-Situ Combustion Project. Second Int. Conf., *The Future of Heavy Crude and Tar Sands,* McGraw Hill, New York, 1984, pp. 844–50.

29. Carcoana, A., Results and difficulties of the world's largest in-situ combustion process: Suplacu de Barcau Field, Romania. SPE Paper No. 20248, SPE/DOE Seventh Symp. on Enhanced Oil Recovery, Tulsa, 22–25 April 1990.
30. Garon, A. M., Kumar, M. & Cala, G. C., The state-of-the-art of oxygen fireflooding. *In situ*, **10**(1) (1986) 1–26.

UNCONVENTIONAL EOR TECHNOLOGIES

11 Unconventional EOR concepts

George J. Stosur
US Department of Energy, Washington, DC, USA

11.1	**Microbial enhanced oil recovery**	342
11.1.1	Background	342
11.1.2	Microorganisms and metabolic products	343
11.1.3	Microbes and their products useful in EOR	345
11.1.4	Potential field applications	349
11.1.5	Reservoir characteristics for MEOR projects	352
11.1.6	Design considerations for MEOR projects	352
11.1.7	Environmental considerations	354
11.1.8	Economic considerations	355
11.2	**Electrical and electromagnetic heating**	355
11.2.1	Conductive heating	356
11.2.2	Electromagnetic heating	357
11.2.3	Single-well heating concept	359
11.2.4	Field experience	359
11.3	**Oil mining**	360
11.3.1	Early attempts at mining for petroleum	361
11.3.2	Surface and underground mining	361
11.3.3	Oil mining field tests in Canada, USA and USSR	362
11.3.4	Mine safety and environmental aspects	363
11.4	**Nuclear explosive fracturing to recover oil and gas**	364
11.4.1	Phenomenology and effects	364
11.4.2	Chimney and fracturing effects	365
11.4.3	Current status of nuclear-stimulation projects	368
11.4.4	Soviet experience	370
11.4.5	Economic and environmental considerations	370
11.5	**References**	371

This chapter covers several methods of EOR not covered by the preceding chapters, and often referred to as unconventional or novel

EOR processes. In fact, they are not novel, since the concepts were recognized several decades earlier, however, they are not used widely. The application of these methods is generally not considered commercial, except under very specific, unusual circumstances. Four such methods are described in this chapter: microbial EOR, electrical and electromagnetic heating, mining-assisted oil recovery (with and without wellbores) and nuclear explosive fracturing to recover oil and gas.

11.1 Microbial enhanced oil recovery

11.1.1 *Background*

The first suggested use of microorganisms for the recovery of oil was made by Beckman in 1926, although little work was done until ZoBell started a series of systematic laboratory experiments in the 1940s.[1] The first field tests were conducted in Europe and the United States in the 1950s.

Microbial enhanced oil recovery (MEOR) processes generally consist of the injection of a microbial population with some form of nutrient (molasses, corn syrup, etc.). The microorganisms feed on the nutrient and produce a number of byproducts, such as carbon dioxide and other gases, certain acids, alcohols, surfactants and/or polymers. These byproducts are expected to mobilize additional oil and move it to producing wells through actions related to one or more of the following: supplying additional energy through reservoir repressurization, reduction of interfacial tension, reduction of oil viscosity and selective plugging of the most permeable zones. The use of microorganisms to enhance oil production has been thoroughly reviewed in a number of recent publications.[2–4]

The use of microorganisms is thought to be mostly applicable to depleted oil reservoirs and to stimulation of low productivity oil wells. There are two ways of using microbial or biological processes: (a) microbial production of the desirable product at the surface and the subsequent injection into a reservoir, and (b) direct injection of microorganisms into a reservoir and in-situ generation of the desirable product. The first approach includes the production in surface facilities of such products as biopolymers for mobility control and biosurfactants for reducing the interfacial tension between the oil and water. MEOR is usually restricted to the second approach. It involves growth

Table 11.1 Microbial products and their contribution to EOR

Microbial products	*Expected effect/mechanism*
Gases (CO_2, CH_4, H_2)	—Increased pressure —Viscosity reduction —Oil swelling —Decrease of interfacial tension —Permeability increase (dissolution of carbonates by CO_2)
Polymers	—Mobility control —Injectivity profile modification
Surfactants	—Decrease of interfacial tension —Emulsification —Viscosity reduction (by emulsification)
Solvents	—Viscosity reduction —Oil dissolution
Acids	—Permeability increase (in carbonates) —Emulsification —CO_2 production
Biomass	—Deep selective plugging —Wettability alteration —Emulsification —Viscosity reduction (by emulsification) —Oil desulfurization

and metabolism of the injected microbes *in situ.* The *in situ* generated byproducts such as carbon dioxide, methane, hydrogen, surfactants and polymers can either replenish reservoir energy and increase its capacity to produce oil, modify oil property so that higher oil recovery is possible, or/and modify reservoir property by selective plugging of the most permeable parts of the reservoir and thereby divert displacing fluids to those parts of the reservoir that still contain displaceable oil (Table 11.1). Only bacteria are considered promising for MEOR; other microorganisms such as protozoa, algae, fungi and viruses are not suitable for a variety of reasons, some of which are their size, tolerance to high water salinity, temperature, pressure and ability to grow *in situ*.

11.1.2 *Microorganisms and metabolic products*

Thousands of different species of microbes have been classified and examined to determine the products of their metabolism, their nutrient requirements and the limits of environmental conditions that they tolerate. Based on environmental tolerance, three very broad classes of microbes have emerged:

(a) *Aerobes*—capable of life processes only in the presence of oxygen, but they can exist dormant in the absence of oxygen (spore-forming species). When the microbes are supplied with nutrient, a wide variety of products can result, depending on the type of microbes, environmental conditions and the type of oil. The products can be carbon dioxide, low-molecular-weight acids, alcohols or biopolymers.
(b) *Anaerobes*—exist without oxygen and derive their energy from degradation of oxygenated molecules. Anaerobes use low-molecular-weight organic compounds such as sugars as a source of carbon and energy. In the process of metabolism, microbes release various products such as carbon dioxide, hydrogen, acids, alcohols, ketones and aldehydes. Some species of anaerobic genes have been found to produce all of these compounds.[5] These anaerobes may also react with sulfur occurring as inorganic sulfides or as part of the molecular structure of organic compounds to form hydrogen sulfide. Oil production has been known to become sour (i.e. produce H_2S with oil and natural gas) when infected by certain bacteria from injection of surface or sea water. Anaerobes can also produce biosurfactants useful for lowering the oil–water interfacial tension and for promoting oil emulsification. Biopolymers (primarily polysaccharides), useful as mobility control agents, can also be produced.
(c) *Facultative microbes*—able to exist and thrive with or without oxygen. Their products of metabolism are of the kind produced by aerobes and anaerobes, listed above.

These three broad categories can be further subdivided into mesophilic (capable of life below 45°C) and thermophilic (capable of life at constant temperature above 45°C) types.

Microbial cells always contain a large number of both organic and inorganic compounds in ratios that are characteristic of each type and of the environment. Nutrients are assimilated and chemically altered within the cells to products that are excreted as metabolic wastes. Heat may also be produced in the reaction. Although many distinct microbes have been isolated and characterized, many more are believed to exist at the surface and within oil reservoirs. Still more are capable of existence within oil reservoirs if the required nutrients are supplied for their metabolism.

To facilitate comparison with other EOR methods, biochemical products from microbial activities can be arranged so that they correspond to the better known EOR methods. These products are: organic solvents (e.g. alcohols, acetone), surfactants (e.g. emulsifiers), biopolymers (e.g. polysaccharides), and gases (e.g. carbon dioxide, methane, hydrogen). Most organisms produce several of these products.

The following paragraphs review the status of research on each of the four major applications of microbes or their products useful in EOR: solvent production, biosurfactant production, biopolymer production and gas production. Table 11.2 contains a summary of the findings of researchers over the past few decades and lists some of the most important microorganisms and their products useful in MEOR. For brevity, these organisms will not be discussed in detail, but Refs 6–20 in the last column of the table indicate where more detailed information can be obtained for each species. The list is by no means exhaustive.

11.1.3 *Microbes and their products useful in EOR*

Since microorganisms can produce a wide variety of chemicals that may assist in enhanced oil recovery, it is very important to match the product to the type of oil mobilization mechanism desired. For example, if there appears to be an obvious channeling problem in the reservoir, it would be desirable to use a microbial formulation that helps through the improvement of sweep efficiency, and therefore a polymer producer would be selected. If mobilization of residual oil saturation after waterflooding is contemplated, then a surfactant and/or alcohol producer might be selected to reduce oil–water interfacial tension. In a single well application, increased pressure would be most important, and gas-producing bacteria could be used. A combination of several effects is possible for best effects.

(a) *Organic solvent production.* Organic solvents such as ethanol, *n*-butanol and acetone are waste byproducts from the microbial fermentation of carbohydrates such as glucose, sucrose, lactose,

Table 11.2 Products and microbes useful in MEOR

Products	*Microorganisms*	*Oxygen requirements*	*Spore former*	*Reference*
Solvents (*n*-butanol, acetone, ethanol)	*Clostridium acetobutylicum*	oa	Yes	6
	Clostridium butylicum	oa	Yes	7
	Clostridium butyricum	oa	Yes	8
	Clostridium pasteurianum	oa	Yes	9
	Clostridium saccharoper butylacetonicum	oa	Yes	10
	Clostridium tetanomorphum	oa	Yes	11
	Zymomonas mobilis	fa	na	12
Surfactants	*Acinetobacter calcoaceticus*	a	na	12
	Arthrobacter paraffineus	oa	No	13
	Bacillus licheniformis	fa	Yes	14
	Bacillus subtilis	fa	Yes	12
	Clostridium pasteurianum	oa	Yes	15
	Corynebacterium fasciens	a or fa	No	16
	Pseudomonas rubescens	a or fa	No	12
	Rhodococcus erythropolis	a	na	17
Biopolymers (polysaccharides)	*Bacillus polymyxa*	fa	Yes	18
	Brevibacterium viscogenes	a	No	19
	Lactobacillus species	fa	No	9
	Leuconostoc mesenteroides	fa	na	18
	Xanthomonas campestris	a	No	20
Gases (carbon dioxide, methane, hydrogen)	*Clostridium spp.*	oa	Yes	9
	Enterobacter aerogenes	fa	na	9
	Methanobacterium spp.	oa	na	9

a = aerobe, oa = obligate (strict) aerobe, fa = facultative aerobe, na = not available.

Table 11.3 Weight per cent yield on glucose substrate

Chemical product	Clostridium acetobutylicum[6]	Clostridium *spp.*[7]	Clostridium *spp.*[9]	Zymomonas mobilis[12]
n-Butanol	12	16	20	—
Acetone	5	—	2	—
Ethanol	1	—	2	49
Acetate	3	4	—	—
Butyrate	1	4	—	—
Isopropanol	2	3	—	—
Carbon dioxide	45	50	55	49
Hydrogen	28	20	—	—
Total conversion	97	97	79[a]	98

[a] Remaining 21% assumed to be hydrogen gas.

mannose, etc. The production of these solvents by some species of bacteria may be quite high, depending on the carbohydrate substrate used and the conditions to which the culture is subjected. This observation has been used to attempt to develop anaerobic microbes that might be used in MEOR, because organic solvents, particularly *n*-butanol, are known to decrease the viscosity of oil if they come into contact with it, and because they mobilize residual oil from the rock surface. This results in greater recovery in a flooding operation.

Several microbial cultures have been characterized in the laboratory and tested in the field to estimate their potential for *in situ* production of solvents by anaerobic fermentation. Table 11.3 summarizes some recent findings pertaining to *Zymomonas mobilis* and several species of *Clostridium*.

Table 11.3 illustrates two important points: (i) that microbial cultures produce mixtures rather than single products, and (ii) that the amount of the desired product(s) can vary considerably from one type of culture to another. Indeed, in any culture medium, there is normally a distribution of cells that yield variable product slates and have different metabolic rates. This variability permits the selection of cells for optimum growth or for specific metabolic traits.

The initial feed cost of the carbohydrate (sugar) as a source of nutrient is a major obstacle to an economically viable process for *in situ* microbial solvent production. Because of this factor, a number of investigators have examined the use of alternative nutrient sources for *Clostridium* growth. The use of waste products from either the dairy or paper industries could significantly reduce the cost of nutrients for *Clostridium* growth. Whether these wastes could be used *in situ* to produce solvents is a question that must first be examined in core studies *in vitro*. Another unanswered question is whether the yields of solvents from the waste products would be sufficient for use in EOR.

(b) *Surfactant production.* The ability of some microorganisms to produce surfactants is useful in certain types of chemical flooding (see Chapter 3). They allow some reduction of the interfacial tension between water and oil. In addition to this, some biosurfactants can also act as emulsifiers; they cause the dispersion of very fine droplets of oil in water to form a suspension. These two characteristics make biosurfactants potentially very useful in EOR. Most biosurfactants are lipids of one type or another: glycolipids, phospholipids, trehalose lipids, etc.

Laboratory experiments have shown that many different bacteria produce biosurfactants (Table 11.2). Most of these bacteria are strict aerobes, and they could not be used *in situ,* even though they may grow on hydrocarbons. However, there are several species of anaerobic or facultatively anaerobic bacteria that do produce biosurfactants (e.g. *Clostridium pasteurianum* and *Bacillus licheniformis*). Unfortunately, these microbes cannot grow on hydrocarbons. To date, no anaerobic, surfactant-producing, hydrocarbon-utilizing species of bacteria has been isolated. Of those species capable of producing a biosurfactant under anaerobic conditions, none has been thoroughly tested *in situ.* Field experiments employing microbes with predominantly biosurfactant production capability have not been conducted, although solvent-producing microbes that may also produce biosurfactants have been tested. It should be noted that oil recovery due to biosurfactant activity cannot be separated from that due to solvent activity.

Microbes producing emulsifiers include *Acinetobacter calcoaceticus* and *Pseudomonas rubescens,* among others. Again, both of these species are aerobic. In fact, it should be mentioned that there is no clear evidence of obtaining very low interfacial tension ($\approx 10^{-3}$ mN/m) with biosurfactants. So the mechanism of displacement involved would rather be an emulsification process or possibly an effect on the rock wettability.

(c) *Production of biopolymers.* Another use of microbes for EOR is in the production of biopolymers, which are often used to increase the viscosity of the driving fluid in a waterflood operation. A number of different bacteria have been identified that are capable of producing useful biopolymers (Table 11.2). The best known of these is *Xanthomonas campestris,* which produces the biodegradable biopolymer xanthan gum (see Chapter 2). Xanthan gum has been used in numerous field tests, however, it may be degraded by indigenous bacteria.

(d) *Biological gas production.* Gas produced by microorganisms is often a byproduct of the breakdown of carbohydrates in a fermentation process, or the result of using alternative organic energy sources in methanogenesis. Through this process, carbon dioxide, hydrogen and methane may be produced, although carbon dioxide is usually the predominant gas formed by most microbes. This prolific gas production has been examined for potential use in EOR, primarily as a means of repressurizing a formation, but also for decreasing oil viscosity through solubility.

The microbes listed in Table 11.2 are some of the most promising ones for use in MEOR gas production, but there are many more that have not been examined for this specific trait. The primary advantage of these microbes is that they can all grow anaerobically. In addition, the *Clostridium* spp. are all spore formers and, as such, may be able to penetrate deeper into the formation than those species that do not form spores. Spores are small enough to propagate even in a very low permeability formation.

The *Clostridium* genus has many different species, all of which are capable of CO_2 and H_2 production. In one case, the conversion of glucose substrate to carbon dioxide was found to be as high as 55%. Such a production rate could have a significant impact on reservoir repressurization. The *Methanobacterium* spp. produce methane (CH_4) as a result of the reduction of various substrates such as acetate, methylated amines, carbon monoxide, methanol, formate and carbon dioxide. This activity could be exploited to repressurize a reservoir.

11.1.4 *Potential field applications*

It is estimated that more than 400 MEOR tests have been conducted in the United States, alone. Unfortunately, most of these treatments were done on poorly documented, low productivity wells, so that reliable data are sparse. Significant sustainable increases in oil recovery have been observed in some cases, but, due to initial very low rates of production, the increase in the production rate (when reported as a percentage) often appears extraordinary.

There are three major applications of microorganisms in enhanced oil recovery:

(a) *Single-well treatments*. The largest number of tests were conducted as single-well stimulation treatment in marginal oil-producing wells. A typical single-well stimulation treatment might increase the rate of production from 0·2 to 0·4 tons of oil per day (1·5 to 3·0 barrels per day) and sustain the increased rate for 2–6 months without additional treatments. This makes the technology attractive for application by independent producers, who mostly operate so called stripper wells. A stripper well is a well that produces less than 1·4 tons of oil per day (10 barrels per day). There are about 470 000 stripper wells in the United States (1990) and the average stripper well produces only 0·4 tons of oil per day (2·6 barrels per day).

The relatively low cost of a single-well treatment explains the popularity of the technique among independents. Actual microbial injection fluid is relatively inexpensive, and the whole cost of such a treatment can be very low if no cost of laboratory design work is included. In fact, most single-well treatments are done without proper laboratory tests, which probably accounts for the widely variable results. Laboratory compatibility studies have demonstrated the importance of determining the interactions of the injected microbes with both the indigenous microorganisms, nutrients, oil and rock and their activity as a function of salt concentration and temperature.

Typically, microbial culture is injected with molasses, and the well is shut-in for a few days to a few weeks to allow for fermentation. The well is then opened and the in-situ generated gas (normally carbon dioxide) expels additional residual oil and any particles that may have obstructed oil flow before the stimulation. The bacteria may also attack paraffins that often build up in the vicinity of the wellbore due to lowering of the temperature of the oil as it enters the well. Changes in capillary pressure phenomena at the wellbore may also be partially responsible for the additional oil production.

The results to date indicate that MEOR technology can be applied to single-well stimulation treatments, but the ultimate oil recovery will remain relatively small since the wells themselves are low-productivity wells and since the actual volume of the reservoir that is treated is also small. Exceptions may occur in cases where production rates are hindered by paraffin deposits that can be removed by microbial treatments, restoring the well productivity. A series of well engineered tests with good background data are needed to properly evaluate the method of helping establish the oil recovery efficiency of single-well treatments.

(b) *Microbial-enhanced waterflooding.* There have been relatively few well-documented field flooding tests using microorganisms. The results are equally mixed, but some appear promising. The US Department of Energy currently is beginning or has in progress additional field tests that will help to provide sorely needed information regarding the performance of MEOR under well-to-well displacement conditions.[3] Field testing will help to identify questions that are important in assessing the technical and economic feasibility of the technology and to focus the research program on answering the critical questions. The transfer of bacteria, coccus or bacillus type, is one of the most crucial points. In particular, retention of bacterial cells may occur, due to attractive colloidal forces that exist between cellular surface and rock.[21]

(c) *Profile modification treatments.* The use of microorganisms for profile modification treatments is largely in the laboratory stage, although several field tests have been reported in the United States and Canada. In this application, microbes are used to plug high-permeability zones in oil-bearing formations to improve sweep efficiency and displace bypassed oil. One technique, which recently underwent pilot testing, employs so-called microbacteria.[22] These are actually a selected strain of common *Pseudomonas* bacteria whose food supply has been deliberately cut off. Faced with a shortage of nourishment, the bacteria shrink in size from a few microns down to less than one-tenth of a micron. The shrinking bacteria loose their usual tendency to attach themselves to nearby surfaces and become more mobile. Because of their minuscule size and increased mobility, the microbacteria can be injected and moved through the formation to where plugging is required. A specially selected nutrient package, which follows, induces regrowth after a predetermined time interval, producing the desired plugging. By selecting different nutrients, the operator can vary the time interval between injection and the start of bacteria regrowth from hours to months. This could make it possible to control the placement of plugging within the formation.

11.1.5 *Reservoir characteristics for MEOR projects*

Since the reported field tests yielded widely variable results ranging from failures to successes and since very few were sufficiently well documented to draw conclusions, an attempt has been made to construct a database for MEOR field projects.[4] Such information could indicate limitations of MEOR processes and could help in constructing screening criteria for their application. The types of information that could constrain the application of microorganisms include reservoir permeability, salinity, temperature, and concentration of metals such as selenium, arsenic, mercury, nickel, etc. Unfortunately, very few MEOR projects to date have reported this kind of information, and therefore accurate assessment of field limitations cannot be made until more projects report these data.

The National Institute for Petroleum and Energy Research, under contract to the US Department of Energy, initiated the MEOR database with a view to future field application. The available information has been collected on only 39 projects, of which 30 projects have reported increased oil production. The reported maximum increase in oil production ranged from 1 to 730%, with the average maximum increase of oil production rate being 75%. Figure 11.1 gives the maximum increase in oil production reported for all the field projects in the database.

11.1.6 *Design considerations for MEOR projects*

Any MEOR field project should be uniquely tailored to the target reservoir. A microbial formulation that is able to grow and produce the desired metabolites under the reservoir conditions must be selected. Compatibility studies between reservoir fluids and rock must be conducted with the microorganisms and with the nutrients to be used. There are several reasons for compatibility testing: (a) to check whether microbes can grow and produce metabolites in the presence of crude oil; (b) tolerance of the bacteria to formation water salinity; (c) tolerance to any toxic metals or minerals present in oil, formation water or leachable from the rock; and (d) determination of whether microbes already present in the reservoir or in the nutrient could 'out-compete' or interfere with the injected microbial formulation.

Another important consideration is that the microorganisms must

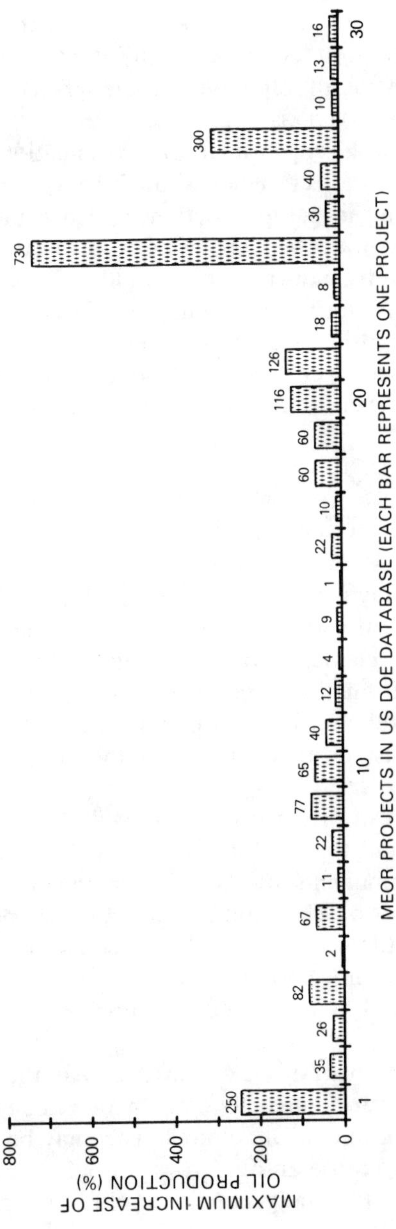

Fig. 11.1 Reported maximum percentage of oil production in US Department of Energy data base on MEOR projects. (From Ref. 4, reproduced by permission of the US Department of Energy.)

not only grow under the reservoir conditions, but that they must reproduce the correct types of metabolites that will improve oil mobilization. In turn, it is equally important that the project be designed for the type of oil mobilization mechanism desired. For example, if the reservoir is highly stratified with sharply varying permeabilities of the layers, a microbial formulation producing polymers to improve sweep efficiency should be considered and not a formulation resulting in gas production, which could exacerbate the problem. In a post-waterflooding MEOR project, microbial strains producing surfactants and alcohols could be used effectively to mobilize the oil left after waterflooding. Similarly, gas production and reservoir repressurization would be important for single-well injections, where oil is produced in a cyclic mode. Still another formulation would be required where removal of paraffin deposition (skin effect) is required.

11.1.7 *Environmental considerations*

MEOR processes may modify the immediate reservoir environment in a number of ways that could cause damage to production hardware or to the formation. Damage to the formation may be caused by indigenous slime formation (e.g. the *Bacillus* species), leading to plugging at the borehole. Some nonindigenous cultures also produce slime, and care must be taken to select the appropriate culture for in-situ use.

Certain bacteria can produce hydrogen sulfide (H_2S) as a result of sulfate reduction, and this acid can then corrode pipelines and other iron-containing components of the recovery equipment. H_2S is also toxic to animal life at concentrations above 10 ppm (parts per million). Another source of the sulfates that these bacteria use to produce H_2S could be the water injected into the reservoir. However, such a condition can easily be anticipated and preventive measures can be taken.

Another area that warrants close attention is the development of standard operating procedures to contain or eliminate unwanted or out-of-control *in situ* microbial cultures. This may be necessary when: (i) nutrient additions cause an overgrowth of microbes that then cause plugging problems, (ii) indigenous microbes overgrow the nonindigenous cultures because of the nutrient additions, (iii) bacterial byproducts cause the crude to become sour, or (iv) pathogenic or allergenic microbes become too prevalent.

Assessment of the potential environmental impact of injecting microorganisms into a reservoir should include the following: (a) environmental tests should be done to assure that undesirable products (such as H_2S) will not be produced with oil; (b) disposal of produced water, which may contain metabolites, should be considered; (c) the project area and nearby wells should be monitored to ensure that the microbes do not migrate out of the pre-established well pattern; and (d) drinking water from any nearby wells should be closely monitored.

11.1.8 *Economic considerations*

Despite numerous tests of MEOR (mostly single-well applications), the state-of-the-art of microbial oil production enhancement is one of uncertain promise. The technology has advanced from laboratory concepts to field applications, but the task of establishing how to apply the technology to give reliable and predictable improvement in oil recovery still remains.

Currently, the most promising applications of the microbial technology are for single-well treatments, primarily due to the low cost of treatments and the relatively quick response. However, since operators have reported widely varying degrees of success (or failure) in such projects, reliable data are scant and no firm conclusions as to the economics of the technology can be made at this time.

11.2 Electrical and electromagnetic heating

There are two basic methods for heating oil-bearing formations with electric energy; conductive heating and dielectric, or electromagnetic heating. Conductive heating results when a deposit contains water, which provides a conductive or ionic path. Dielectric heating results from periodic alteration of the shape of atoms; since water is a polar molecule, it can be heated dielectrically in much the same way that water is heated in a microwave oven. In both cases, the object is to heat certain parts of the reservoir, usually near the wellbore, to reduce oil viscosity and, in some cases, to provide additional driving energy through liberation of natural gas or generation of steam in the vicinity of the wellbore. Since viscosity reduction is the primary mechanism, these methods are mostly helpful in heavy, viscous oil reservoirs, where even a little heating around the wellbore results in a significant increase in pressure drop in the immediate vicinity of the wellbore and thus facilitates gravity drainage. The most efficient applications include removal of the skin effect—a thin area around the wellbore with reduced flow capacity caused by a variety of reasons, one of which is paraffin deposition.

11.2.1 *Conductive heating*

An *in situ* electric heating method is based on heat generation obtained by electric current flowing through the oil reservoir (Fig. 11.2). The electric circuit is formed by connate water in the reservoir between two or more wells acting as terminals. Either direct current or low frequency current is used. Due to the radial geometry of the system, the highest rate of heating occurs near the wellbore, where conductivity decreases as moisture evaporates and effective heating cannot continue. Thus, cooling of the well may be necessary, as well as resupply of moisture, to provide conductivity and to suppress this detrimental effect. Another disadvantage is nonuniform heating; high current densities at the electrode cause inefficient distribution of electrical energy.

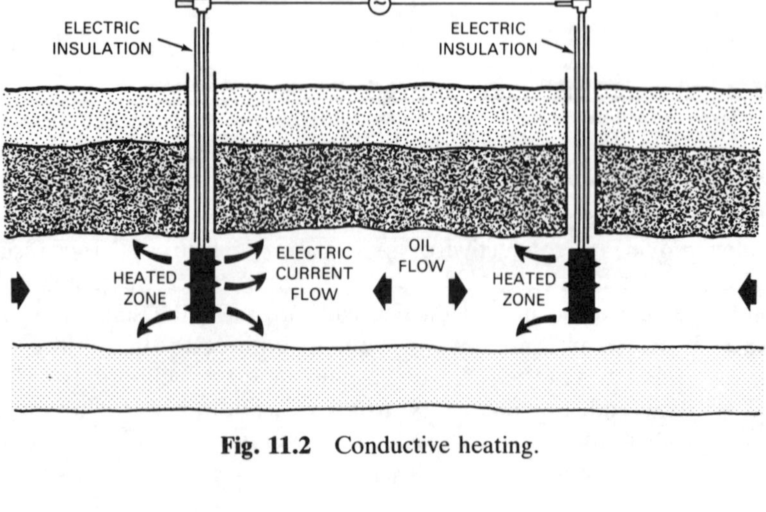

Fig. 11.2 Conductive heating.

The earliest attempts to use electrical energy for the direct heating of oil-bearing formations consisted of such simple techniques as burying a pair of electrodes to which 60-cycle electrical energy was applied.[23] In 1969, a field experiment in the Little Tom oil field in Texas was reported as successful. Production from four wells reportedly increased from about 0·1 to 3 tons of oil per day.[24] A few other field tests were reported within a few years.[25] In 1987, a pilot test was conducted in the Rio Panon oil field in Brazil. Results showed a significant increase in temperature around the wellbore accompanied by a notable increase in the oil production rate. This particular test is significant because it was used as the field test to validate a numerical simulation model of the process.[25] The results showed that most of the heating effects were concentrated in the region close to the wellbore and that a temperature increase of less than 2°C was observed at a distance of 30 m after five years of heating. In this sense, the method can be considered as a well stimulation process.

11.2.2 *Electromagnetic heating*

Electromagnetic heating, often referred to as radio-frequency heating, employs frequencies used in radio communications. The frequencies can be as low as 10 Hz or as high as 10 GHz. The key to reservoir heating is the ability of electromagnetic energy to penetrate the deposit away from the borehole, and of electromagnetic energy to be absorbed (Fig. 11.3). The ability of the reservoir to absorb electromagnetic energy depends strongly on the moisture content (connate water). The higher the water saturation, the greater the ability to absorb electrical energy.

An oil-bearing formation with a typical ≥10% water saturation enables a flow of current at 60 Hz from one electrode to any other electrode—even one several hundred meters distant. As the moisture content of the formation decreases, whether naturally or because it is evaporated and driven out, the electrical conductivity of the near-wellbore region decreases, since it is increasingly difficult to conduct current at 60 Hz.

However, at high frequencies, electromagnetic waves are almost completely absorbed by such a formation within only a short distance by a formation with 10% or higher water saturation. For example, at 10 MHz, an electromagnetic wave cannot effectively penetrate the

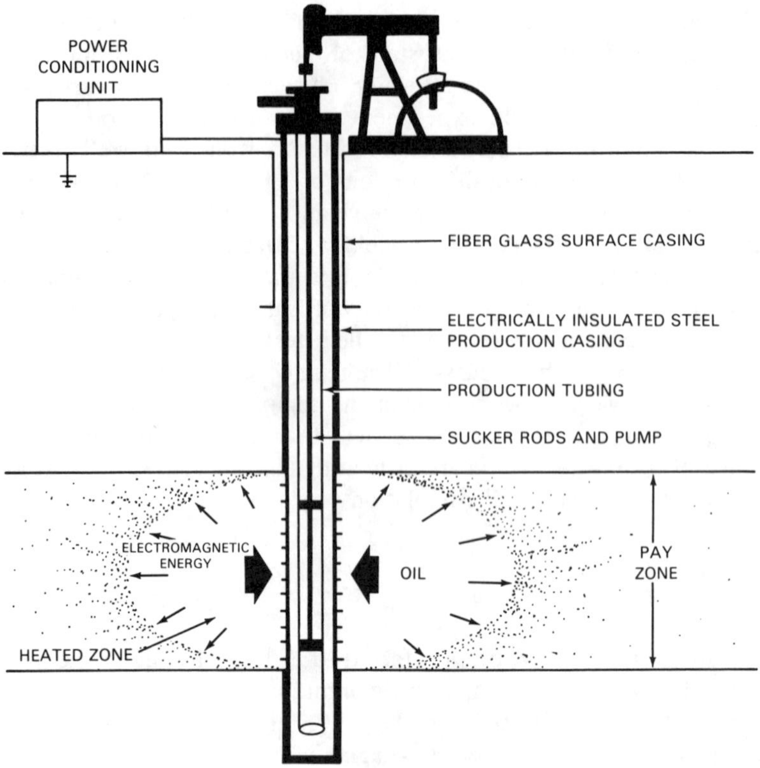

Fig. 11.3 Electromagnetic heating.

formation for more than 5–10 m, so that areas beyond that range are not effectively heated. Nevertheless, even if there is no more water, high frequency electromagnetic waves can propagate for greater distances, due to the sufficient electrical conductivity of the formation itself, so that deeper regions within the formation can also be heated.[26]

A region devoid of moisture near an overheated electrode effectively has a very large resistance at 60 Hz, but it readily enables propagation of a high frequency wave so that, even though heating at 60 Hz might stop, it could resume at higher frequencies. This phenomenon can be used to enlarge the downhole heating pattern by periodic increases of frequency so that energy is radiated through the nonabsorbing zone, farther out in the reservoir where connate water still exists. This makes possible periodic extensions of the heated zone, within the limits of the system.

11.2.3 *Single-well heating concept*

The electromagnetic heating process is most effective when removing impediments to oil flow in a well that may suffer from precipitation of paraffin, the presence of water-sensitive clays that prohibit other ways of the stimulation, or simply for reasons of high oil viscosity that limits well productivity. Heavy oil reservoirs with high *in situ* viscosity and light oil reservoirs with paraffin deposition problems are most suitable for electric or electromagnetic heating because even a relatively small amount of heat near the wellbore can dissolve the paraffin or lower oil viscosity enough to significantly improve well productivity. One of the system's advantages is that the energy can be applied continually to the reservoir through a single well, at the same time as the well is producing.

Deeper penetration of the electromagnetic energy radially into the reservoir tends to increase the stimulation effect of the process, but also reduces the efficiency of the process by increasing the surface area of the heated zone, which in turn increases the conductive heat losses in the vertical direction to the overburden and underburden. There is a definite economic trade-off between production stimulation and utilization efficiencies of the power transmitted.[27] Proper electrode or antenna design and placement, together with control of power input and pay zone temperature levels, can optimize this relationship.

In many wells, paraffin deposition takes place within tubing as oil flows upwards and is cooled, in perforations, or within the face of the formation and in the pore spaces within a fraction of a meter away from the wellbore. The deposition problem begins when the temperature of the oil at the wellbore falls below the cloud point of the crude, upon pressure reduction and gas evolution. Electromagnetic stimulation solves the problem by heating and dissolving the paraffin, all without interruption of oil production.

11.2.4 *Field experience*

Several successful field tests of both conductive heating and electromagnetic heating have been performed in the United States, Canada, Mexico and Brazil. For an example of conductive heating application, the reader is referred to the description of field tests in the US and Mexico.[24,28] Electromagnetic stimulation and field results from the US, Brazil and Canada can be found in Refs 26, 27 and 29.

Electrical preheating often precedes a conventional thermal recovery operation in very heavy oil reservoirs or tar sands, which may contain bitumen with an extremely high viscosity. Under such conditions, fluid injection schemes face a major difficulty related to the very low initial formation injectivity. A practical solution then is to raise the initial reservoir temperature sufficiently high, so as to decrease the viscosity of oil and make subsequent displacement possible with the steam or hot water.[29] Preheating can be accomplished with either conductive or electromagnetic heating.

Since electricity or electromagnetic energy is a form of energy that is efficiently convertible to mechanical energy, it would seem advisable to limit its use for stimulation or for preconditioning a reservoir, before applying a less expensive recovery process.[30]

While the process of electric and electromagnetic heating is well understood in principle and can be adequately simulated, a few practical problems remain to be solved. For low-frequency electrical heating, the areas that need most development are in well completion technology. Methods of insulating the tubing from the casing, creating extended electrodes in the formation, artificially enhancing formation conductivities near electrodes and reducing wellbore electrical losses, all at reasonable cost, must be developed further.

Shale interbedding may adversely affect electrical circuitry; conductive shale zones reduce the reservoir's total resistance and lessen conversion of electrical energy to heat. Above all, electric power is more costly than any other form of energy and therefore is applicable for heating only a relatively small part of the reservoir. It must also be said that some operation is possible so that areas normally not reachable by steam injection could be heated by electrical energy.

11.3 Oil mining

The term 'oil mining' as used by the petroleum industry has two different meanings. The first refers to the more traditional concept of mining and involves surface removal of oil-bearing sands and subsequent processing to remove the oil and dispose of the sand (e.g. Athabasca deposits in Canada). The second refers to deeply buried oil-bearing formations to which shafts and tunnels are dug, but oil is produced from oil wells that are drilled from underground chambers (e.g. the Soviet Thermal Mining Method at Yarega). Both methods can be considered as an alternative to enhanced oil recovery.

11.3.1 *Early attempts at mining for petroleum*

Oil mining is actually one of the oldest methods of oil recovery. Primitive methods of mining shallow oil-bearing formations have been used for ages by merely digging pits or shallow wells and driving short shafts into oil sand outcrops. The oil seepage was then bailed out. More recently, mining methods were practiced several centuries ago in Germany and France, some of which continued through the early 20th century.[31] In the United States, underground methods were practiced prior to the development of conventional well drilling. Following that, rapid development of conventional drilling operations and oil recovery techniques took place, and little attention was given to oil mining.[32]

Oil mining regained interest in the early 1970s due to the sharply higher oil prices and the realization that huge amounts of oil remained in known but depleted oil fields. Higher recovery efficiency was postulated and expected from selected oil fields that had experienced poor recovery using conventional techniques.

11.3.2 *Surface and underground mining*

Open-pit and strip mining are the two fundamental methods of surface mining. The choice of mining method and the mine configuration is dependent upon the size, shape and resource concentration of the oil-bearing formation and on the topographic features of the ground surface. Open-pit mining is generally used in formations that have limited lateral extent but are relatively thick vertically. Problems basic to open-pit mining include waste disposal, slope stability and dust control.

The extensive Canadian projects of mining the Athabasca tar sands are good examples of strip mining. The Athabasca tar sands are water wet, which makes it easier to separate the hydrocarbon from the host rock. In the United States, much of the reservoir material containing similar hydrocarbons is oil wet, which makes the separation process much more difficult.

When oil deposits are too deep for effective surface mining, the reservoir can be accessed through shafts and tunnels. Underground galleries can then be excavated, and either horizontal or inclined wells can be drilled for effective oil drainage (Fig. 11.4).

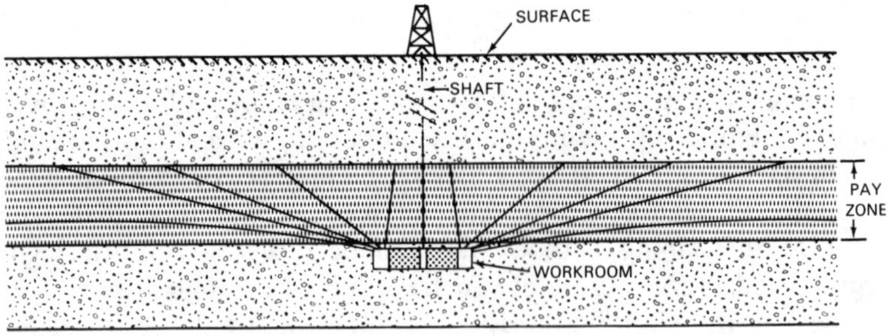

Fig. 11.4 Mine and directional well arrangement in improved oil mining.

11.3.3 *Oil mining field tests in Canada, USA and USSR*

Since the early 1970s, several large-scale field tests and commercial projects have been conducted in several countries. Canada's oil mining industry is perhaps the most advanced, thanks to the very large, shallow and rich Athabasca tar sand deposits in the province of Alberta. A good example of an underground mining project is a Canadian project known as the Underground Tunneling Facility (UTF). As a research facility, UTF was constructed to test several promising ideas involving oil production through inclined wells drilled from chambers dug underneath the Athabasca formation. Three sets of two horizontal well-pairs were drilled; one directly above the other (Fig. 11.5). The top well is used for steam injection and the bottom well for producing oil, formation water and condensed steam. The process involves countercurrent gravity segregation and therefore depends primarily upon the density difference between the steam and the liquid phases and upon effective vertical permeability. The steam rises, gives off heat to the formation as it condenses, and the hot mobilized bitumen and condensed water flow downward to the bottom well, countercurrent to the upward steam movement.[33]

In the past, oil mining in the United States did not attract sufficient attention, primarily due to the availability of other options with higher rates of return. Nonetheless, more than 10 small oil mining tests and commercial projects were completed, although their results were mixed.[34] One of the better known underground oil mining projects was conducted by Hopco in the Kern River field, California, started in 1981. The sometimes-called Hotplate project used cyclic steam injection into a heavy oil formation in a hub and spoke pattern

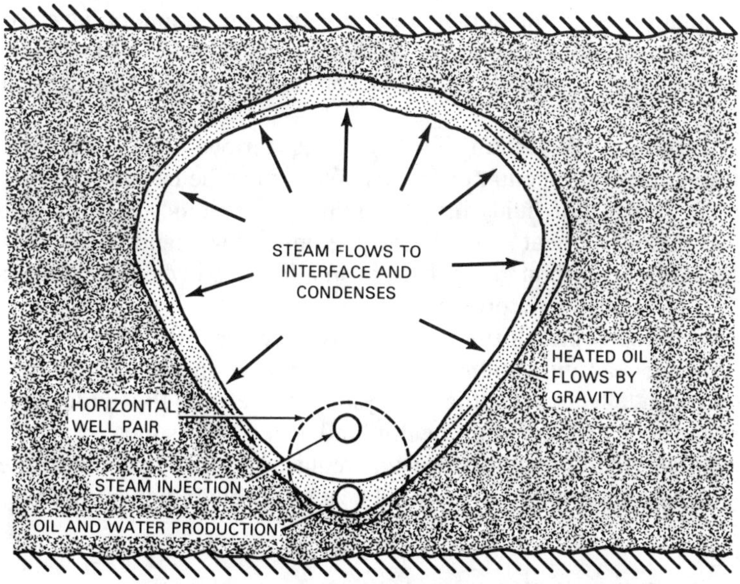

Fig. 11.5 Steam chamber process for the recovery of bitumen from tar sands. (From Ref. 33, reproduced by permission of AOSTRA.)

from lateral wells. Initially, a feasibility study estimated that the Hopco method could achieve 77% recovery of oil in place in about 6 years; however, the project was terminated sooner, with less than satisfactory results.

The Soviet Union's thermal-mining method, in operation since 1982, has received the most publicity. Located in the Yarega oil field near Ukhta, it was the first operation involving steam injection from a mine to assist heavy oil recovery. The process is expected to recover up to 60% of the oil in place, compared to only 2% recovery by conventional methods.

11.3.4 *Mine safety and environmental aspects*

Aside from economic considerations, environmental requirements are probably the most serious obstacle to a wider application of oil mining. Open-pit mines require a lengthy permit process and, at the minimum, filling, recontouring and revegetating are required. Large pits have to take cognizance of the surrounding geology, and aquifers have to be sure that they do not interfere with the groundwater table. Aquifer protection may be required and at some sites could be prohibitively expensive.

Underground oil mining/drilling faces serious safety hazards in addition to environmental problems. Wells must be carefully sealed to prevent seepage of fluids into the mine. These fluids may be under pressure due to heat injected at the mine level and may contain dissolved hydrocarbon gases that become liberated when the fluids are flushed to atmospheric pressure.

A continuous monitoring program is necessary to detect toxic and explosive gas accumulations in the mining environment. Miner safety requires an effective ventilation system to refresh the underground working area. In general, health and safety measures, as well as preventive environmental steps are required by law and constitute a very significant amount of the total cost of oil mining operations.

11.4 Nuclear explosive fracturing to recover oil and gas

In the past two decades, the United States and the Soviet Union have investigated the use of nuclear explosives to stimulate production from very low permeability reservoirs that do not produce at commercially attractive rates under natural flow, or even after stimulation by conventional techniques such as hydraulic fracturing. It has been postulated that the very high instantaneous energy available from nuclear explosives could be used to enlarge the wellbore and to fracture the adjacent formation, thus improving both deliverability and ultimate oil and gas recovery.

11.4.1 *Phenomenology and effects*

The nuclear explosive is emplaced at a depth selected after consideration of yield, geological environment and the desired results. After the emplacement hole has been properly plugged up with sand or concrete, the explosive is detonated. In a few microseconds after detonation, temperatures and pressures at the shotpoint reach millions of degrees and atmospheres. Within milliseconds, the shock wave generated by these extremely high pressures vaporizes, melts and fractures the surrounding rock and creates a cavity that expands to maximum size within a fraction of a second. Cavity growth stops when the pressure within has declined to approximately the overburden pressure. Within seconds to a few minutes, temperature decreases sufficiently to permit the vaporized rock to condense. The molten rock then collects at the cavity base in a pool, trapping most of the solid radioactive products (Fig. 11.6).[35]

In most formations the overlying rock usually begins to collapse into the cavity after the temperature declines sufficiently for the water vapor to condense. This may require several minutes or even hours. Overlying rock continues to fall, and the upward progression of this collapse forms a cylindrical chimney that can reach a height of up to 15 times the cavity radius (Fig. 11.7). The ultimate height of the chimney depends on the yield of the explosive and the rock properties. The porosity in the rubble zone (void space between the fallen blocks) ranges from 20 to 30% and the permeability is assumed to be infinite.

The distance to which fractures will radiate outward from the shotpoint and beyond the chimney wall depends on the characteristics of the subsurface environment. Fractures have been observed at 3–10 cavity radii from the shot point. In a fully contained explosion no radioactivity is released to the atmosphere.

11.4.2 *Chimney and fracturing effects*

Three nuclear explosive fracturing field tests were conducted in the United States from 1967 to 1973.[36] All were designed to fracture extremely low permeability, natural gas bearing formations in the southwestern United States.

Reservoir response to nuclear-explosive fracturing in terms of producing rate and ultimate recovery depends on the contribution of three elements: the chimney, the fractured zone surrounding the chimney and the unaltered formation beyond the fractured zone. The significance of the radial dimensions can be illustrated by using one of the tests, such as the most recent US project, Rio Blanco, and the attendant reservoir characteristics. Project Rio Blanco consisted of the simultaneous detonation of three 30-kiloton nuclear explosives. The upper cavity had a radius of 22 m, and the fractured zone extended to a distance of about three cavity radii from the wellbore, according to the results from pressure buildup and drawdown data.

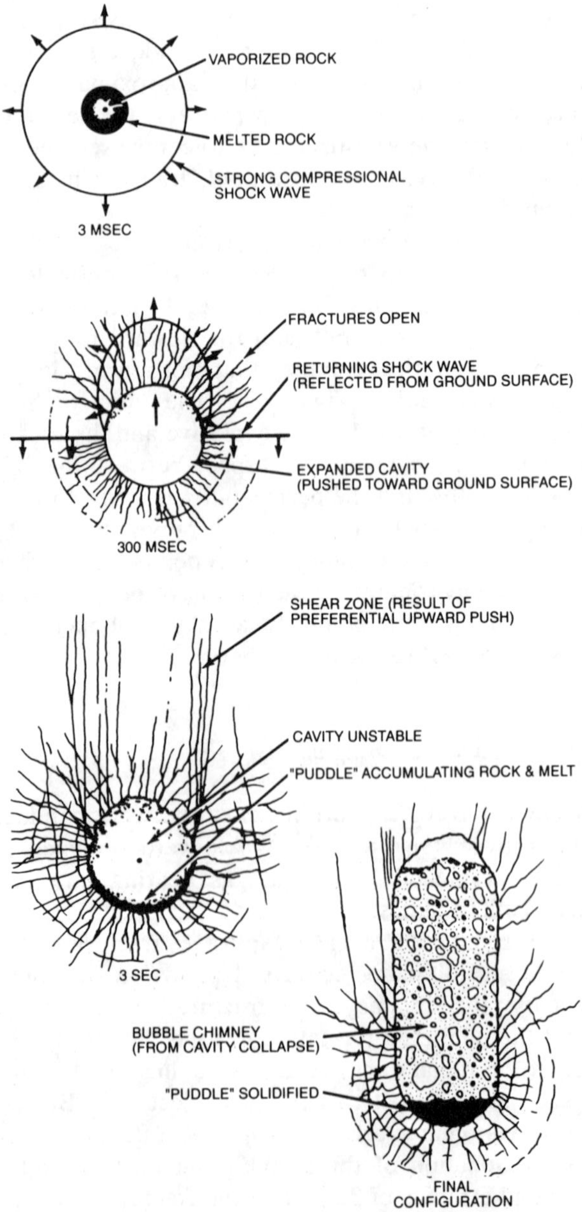

Fig. 11.6 Cavity (chimney) formation history for contained nuclear explosive. (From Ref. 35.)

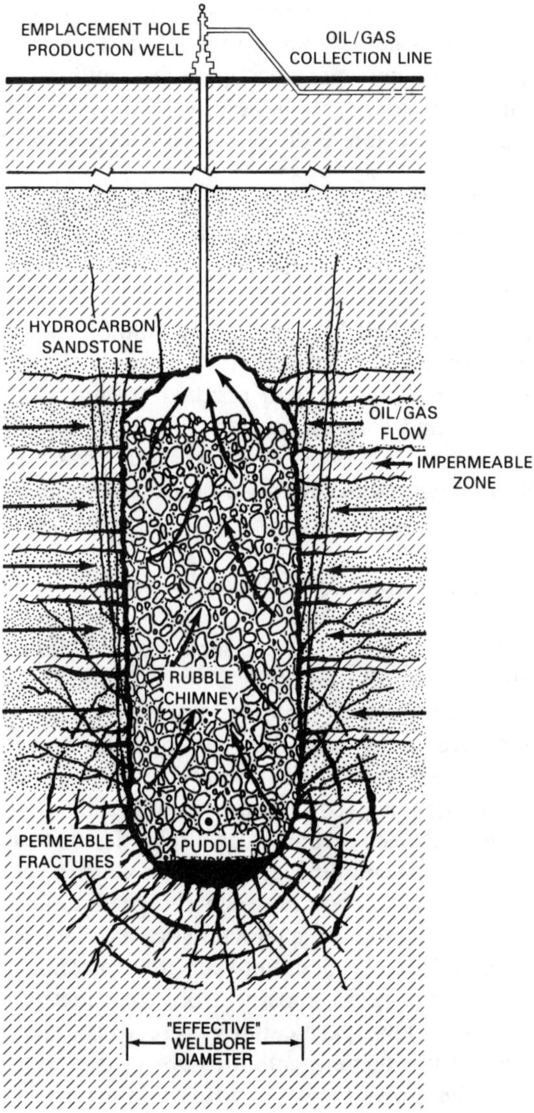

Fig. 11.7 Geometry of a well fractured with a nuclear explosive.

The distinct advantage of nuclear-explosive well fracturing lies in its capability to fracture very large intervals. This capability is particularly useful in areas where productive formations are numerous and separated by impermeable shales with thicknesses of several meters. This advantage may, however, be a liability under less than ideal conditions, such as proximity to an aquifer, because of the necessity of the disposal of large volumes of water contaminated by radioactive elements.

A new, so-called Diamond series of nuclear devices was designed by the Atomic Energy Commission (now US Department of Energy) specifically for project Rio Blanco. The small diameter of the explosive used in the test helped cut drilling and well-completion costs, since standard off-the-shelf bits and casing could be used. The design of the device is different from the standard one; it is only 19 cm in diameter and produces much less residual tritium than the devices used for Gasbuggy and Rulison projects (Table 11.4). Only about 5% of the total tritium is contained in the gaseous phase; it is estimated that about 40% is trapped in the melted rock and about 55% in water.

11.4.3 *Current status of nuclear-stimulation projects*

The three nuclear-explosive tests conducted in the United States were projects Gasbuggy, Rulison and Rio Blanco (Table 11.4). If generalizations on the three nuclear-explosive fracturing tests were possible, then it would have to be said that: (a) predictions with respect to physical effects and cavity size were within, or close to, the predetonation predictions, (b) the amount of residual tritium from the Gasbuggy to Rio Blanco events was dramatically reduced, largely owing to the new Diamond device, and (c) in all three cases there was an apparent large discrepancy between the predetonation evaluation of the reservoir and reservoir characteristics as determined after the events. The large discrepancy between the preshot and postshot formation evaluation stems mostly from the fact that formation evaluation of very low permeability reservoirs has always been difficult, but when in-situ permeabilities are in the range of $10^{-6}\ \mu m^2$ (1 microdarcy), the current state of formation evaluation stretches to the limits of available technology.

The reason for the lower than expected production performance of the three nuclear well fracturing tests were never clearly established,

Table 11.4 Comparison of relevant data for projects Gasbuggy, Rulison, and Rio Blanco

	Gasbuggy	*Rulison*	*Rio Blanco*
Date	10 December 1967	10 September 1969	17 May 1973
Industrial sponsor	El paso Natural Gas Co.	Austral Oil Co.	CER Geonuclear Corp.
Location	San Juan Basin, New Mexico	Piceance Basin, W. Colorado	Piceance Basin, W. Colorado
Device size (kt)	29 ± 3	43 ± 8	(simultaneous detonation of three 30 kt explosives)
Diameter of device (m)	0·49	0·25	0·21
Depth of emplacement hole (m)	1 450	2 860	2 590
Total tritium produced (Ci)	40 000	10 000	<1 000
Depth of explosion (m)	1 388	2 713	1 921, 2 049, 2 200
Weight of device and cannister (t)	9·1	0·7	—
Length of device (m)	9·8	4·9	—
Device run on	string of 7 in casing	3/4 in wire line	string of 7 in casing
Size of hole drilled (m)	0·76 (special bit)	0·41 (stock bit)	0·41 (stock bit)
Size of casing set (m)	0·55 (special item)	0·3 (stock item)	0·3 (stock item)
Sealing material used	cement	sand and gravel	cement and gelled water
Cavity size, diameter × height (m)	59 × 131	49 × 89	43 × (sphere)
Re-entry after detonation	30 days	6 months	4 months

but several hypotheses were mentioned:[36]

(a) Overestimated formation permeability prior to the detonations appears to be by far the most important reason behind the lower than expected stimulation effects. In some cases, the formation porosity and net sand count were also overestimated. After the initial test data became available, it was apparent that the matrix permeability of the unfractured formation was very much lower. It appears that the overestimate ranged from about 2–3 orders of magnitude for project Gasbuggy, about 1–2 orders of magnitude for project Rulison and a factor of 6–10 for the upper chimney region of project Rio Blanco. The bulk of reservoir rock beyond the unfractured region is still dominated by the ultralow matrix permeability that results in a lower than expected gas influx into the chimney from regions immediately adjacent to fractures emanating from the chimney. Lower deliverability and lower ultimate recovery than predicted would then be expected.
(b) Closing of newly created, unsupported fractures: increasing the water influx into the cavity would decrease the relative permeability to gas.
(c) Glazing of initial cavity walls by the melted rock, most of which collects in the bottom of the cavity in the form of a glassy slug, but some may solidify on cavity walls and create a skin effect. This would apply mainly to a cavity that did not collapse or a partially collapsed chimney.

11.4.4 *Soviet experience*

About ten nuclear fracturing tests were done in the Soviet Union but information is very sparse.[37] Tests involving three nuclear explosions in the Yakut Autonomous Republic were directed towards 'increasing oil and gas flows from formations with poor permeability lying among thick beds of rock salt.'[38]

11.4.5 *Economic and environmental considerations*

Economic aspects are closely tied to the successful solution of technical problems. Many economic feasibiltiy studies have been performed, and, even though there are sharp disagreements, the consensus is that nuclear-explosive fracturing can be economically viable. Apart from the technical and economic aspect of nuclear-explosive fracturing, public acceptance of the technology is equally important. It has never been favorable, and, in fact, publicity at the time seems to have created apprehension.

The magnitude of the problems that the projects' sponsors must contend with is exemplified by environmental statements prepared for project Wagon Wheel,[39] which was designed as an ultimate test involving five nuclear explosives to be set off sequentially. The project was finally terminated. The environmental and political climate seems to pose the single largest obstacle at this time, and it appears that no improvement can be expected in the foreseeable future. Until public acceptance is reversed, there seems little hope for the future of the technology. It is also apparent that the use of nuclear explosives in the future would require legal and political changes. Political expediency dictates that the current efforts in the development of very tight gas reservoirs be placed within the context of conventional technology, such as massive hydraulic fracturing.

11.5 References

1. ZoBell, C. E., Bacterial release of oil bearing materials. *World Oil,* **126** (1947) 36, 47 and **127** (1947) 35–41.
2. Smith, R. J. & Collins, A. G., State-of-the-art of microbial enhanced oil recovery: A literature survey. US Department of Energy, NIPER 12, Report No. DE8401743 (1984).
3. Bryant, R. S., Donaldson, E. C., Yen, T. F. & Chilingarian, G. V., Microbial enhanced oil recovery. In *Enhanced Oil Recovery, II—Processes and Operations,* ed. E. C. Donaldson, G. V. Chilingarian & T. F. Yen, Elsevier, Amsterdam, 1989, pp. 423–50.
4. Bryant, R. S., MEOR data base and evaluation of reservoir characteristics for MEOR projects. US Department of Energy, Report No. DE 89000764 (1989).
5. Grula, M. M. & Sewell, G. W., Microbial interactions with polyacrylamide polymers. Int. Conf. on MEOR, Afton. NTIS, Springfield, 1982, pp. 129–34.
6. Spivey, M. J., The acetone/butanol/ethanol fermentation. *Process Biochemistry,* **13** (1978) 25.
7. Jones, D. T. & Woods, D. R., Acetone-butanol fermentation revisited.

Microbial Rev., **50** (1986) 484–520.
8. Ulmer, D. C., Moreira, A. R. & Linden, J. C., The acetone-butanol fermentation of *Clostridium acetobutylicum*. In *Microbial Enhanced Oil Recovery*, ed. J. E. Zafic, D. C. Cooper, T. R. Jack & N. Kosaric, Pennwell Books, Tulsa, 1983, pp. 141–9.
9. Smith, R. J. & Collins, A. G., State-of-the-art of microbially enhanced oil recovery. NIPER (US Department of Energy), Energy Research Abstracts, **9**(22) (1984) Abstract No. 45429.
10. Soni, B. K., Das, K. & Ghose, T. K., Bioconversion of agrowastes into acetone-butanol. *Biotechnol. Lett.*, **4** (1982) 19–22.
11. Gottwald, M., Hippe, H. & Gottschalk, G., Formation of n-butanol from D-glucose by strains of the *Clostridium tetanomorphum* group. *Appl. Environ. Microbiol.*, **48** (1984) 573–6.
12. Zajic, J. E. & Seffens, W., Biosurfactants. *CRC Crit. Rev. Biotechnol.*, **1**(2) (1984) 87–107.
13. Cooper, D. G. & Zajic, J. E., Surface-active compounds from microorganisms. *Adv. Appl. Microbiol.*, **26** (1980) 229–53.
14. Cooper, D. G., Biosurfactants. *Microbiol. Sci.*, **3** (1986) 145–8.
15. Cooper, D. G., Zajic, J. E., Gerson, D. F. & Manninen, K. I., Isolation and identification of biosurfactants produced during anaerobic growth of *Clostridium pasteurianum*. *J. Fermentation Technol.*, **58** (1980) 83–6.
16. Zajic, J. E., Hostack, J. & Seffens, W., The effect of HLB on the surface activity and bitumen extraction capability of *Corynebacterium fasciens*. Int. Conf. on MEOR, Afton. NTIS, Springfield 1982 pp. 15–19.
17. Kretschmer, A., Bock, H. & Wagner, F., Lipids from *Rhodococcus erythropolis* grown on n-alkanes. *Appl. Environ. Microbiol.*, **44** (1982) 864–70.
18. von Heningen, J., DeHaan, A. J. & Jansen, J. D., Process for the recovery of petroleum from rocks. The Netherlands Patent No. 89 580 (1958).
19. Xiuyuan, W., A microbial polysaccharide produced from crude oil or liquid paraffin and its application in the petroleum industry. Int. Conf. on MEOR, Afton. NTIS, Springfield, 1982, pp. 29–37.
20. Sandvik, E. I. & Maerker, J. M., Application of xanthan gum for enhanced oil recovery. In *Extracellular Microbial Polysaccharides*, ed. P. A. Sandford & A. Laskin, ACS Symposium Series, **45** (1976) 242–64.
21. Jang, L.-K., Chang, P. W., Findley, J. E. & Yen, T. F., Selection of bacteria with favorable transport properties through porous rock for the application of microbial-enhanced oil recovery. *Appl. Environ. Microbiol.*, **46**(5) (1983) 1066–72.
22. Anon., Bacteria to assist in oil recovery. *The Tar Paper*, Alberta Oil Sands Technology and Research Authority (AOSTRA), Edmonton, **12**(2) (1989) 4.
23. Flock, D. L. & Tharin, J., Unconventional methods of recovery of bitumen and related research areas particular to the oil shale of Alberta. *J. Canadian Petroleum Technol.*, **14**(3) (1975) 17–27.
24. Anon., AC current heats heavy oil for extra recovery. *World Oil*, **170**(6) (1970) 83–6.

25. Pizarro, J. O. S. & Trevisan, O. V., Electrical heating of oil reservoirs: numerical simulation and field test results. *J. Petroleum Technol.*, **42**(10) (1990) 1320–6.
26. Spencer, H. L., Bennett, K. A. & Bridges, J. E., Application of the IITRI/UENTECH electromagnetic stimulation process to Canadian heavy oil reservoirs. Paper No. 42, Fourth UNITAR/UNDP Conf. on Heavy Crudes and Tar Sands, Edmonton, 7–12 August 1988.
27. Anon., Electromagnetic reservoir stimulation shows promise in Brazil. *Oil and Gas J.*, **86**(46) (14 November 1988) 106–8.
28. Hiebert, A. D., Vermeulen, A. D., Chute, F. S. & Capjack, C. E., Numerical simulation results for the electrical heating of Athabasca oil sand formations. *SPE Reservoir Engineering*, **1**(1) (1986) 76–84.
29. Chute, F. S. & Vermeulen, F. E., Present and potential applications of electromagnetic heating in the in-situ recovery of oil. *AOSTRA J. of Research*, **4** (1988) 19–33.
30. Burger, J., Sourieau, P. & Combarnous, M., Thermal methods of oil recovery, Editions Technip, Paris and Gulf Publishing Company, Houston, 1985, pp. 415–6.
31. Rice, G. S., USBM Bulletin 351 (1932).
32. Energy Development Consultants, Mining for petroleum: feasibility study. A report to the US Department of Energy, 1978.
33. Haston, J. A., Executive Summary, UTF Project Evaluation, AOSTRA, 1985.
34. Protopapas, T. E. *et al.*, Detailed technical and economic feasibility of light oil mining in the United States. A report to the US Department of Energy, 1983.
35. Weaver, L. E., Peaceful uses of nuclear explosives. University of Arizona Press, Tucson, 1970, pp. 23–61.
36. Stosur, G. J., Status of project Gasbuggy, Rulison and Rio Blanco and an appraisal of nuclear-explosive fracturing potential for the future. Proc., UNITAR Conf. on Energy and Future, 5–16 July 1976, p. 697.
37. Anon., Soviets report wider use of nuclear stimulation. *Oil and Gas J.*, **87**(39) (25 September 1989) 30.
38. Anon., Soviets claim oil boost with nuclear explosions. *Oil and Gas J.*, **85**(25) (22 June 1987) 22.
39. Anon., Wagon Wheel gas stimulation project. US AEC, Wash-1524 (1972).

CONCLUSIONS ON THE MAIN EOR PROCESSES

12 Methodology and screening criteria—implementation of EOR processes

Pierre Chaumet
Institut Français du Pétrole, Rueil-Malmaison, France

and

René H. Cottin
Société Nationale Elf Aquitaine, Pau, France

12.1	**Preliminary remarks**	378
12.1.1	Economic and local constraints	378
12.1.2	Project management	380
12.2	**Flow chart of an EOR project**	380
12.3	**Analysis of the reservoir performances**	382
12.3.1	Reservoir characterization	382
12.3.2	Evaluation of oil saturation distribution	385
12.3.3	Reservoir modeling and performance forecasting	385
12.3.4	Influence of infill drilling on recovery	386
12.4	**Preselection of EOR methods**	387
12.5	**Studies of preselected methods**	390
12.5.1	Laboratory studies	390
12.5.2	Phenomenological simulation	391
12.6	**Pilot testing**	391
12.6.1	The need for pilot tests	391
12.6.2	Design of pilot tests	392
12.6.3	Monitoring pilot operations and special measurements	393
12.6.4	Interpretation of pilot results	394
12.7	**Design of large-scale EOR operations**	394
12.7.1	Starting time	394
12.7.2	Spacing and pattern	395
12.7.3	Type of well and completion	395
12.7.4	Amount of active chemicals	397
12.7.5	Decision for large-scale application of EOR methods	397
12.8	**Status of EOR development: conclusion**	398
12.9	**References**	400

12.1 Preliminary remarks

12.1.1 *Economic and local constraints*

Before describing the methodology of an EOR project, it may be wise to stress the important role played by economic and environmental constraints in its development.

The ultimate recovery rates obtained in the laboratory, or calculated by numerical simulation with a reservoir model, are theoretical values never reached in practice. As a matter of fact, production is stopped as soon as the value of the oil produced no longer pays the expenditures. Figure 12.1, which illustrates the production of a field as a function of time, shows that this break-even point is such that:

$$P(\mathrm{d}c/\mathrm{d}t) = \mathrm{d}(F + T)/\mathrm{d}t$$

where P is ex-field crude selling price, c is cumulative production, t is time, F is cumulative operating expenditures (investments + operating costs), and T is cumulative fees and taxes.

Thus, economic recovery is lower and far from ultimate theoretical recovery if:

- the crude market price P is low,
- operating expenditures F are high,
- taxes T are high.

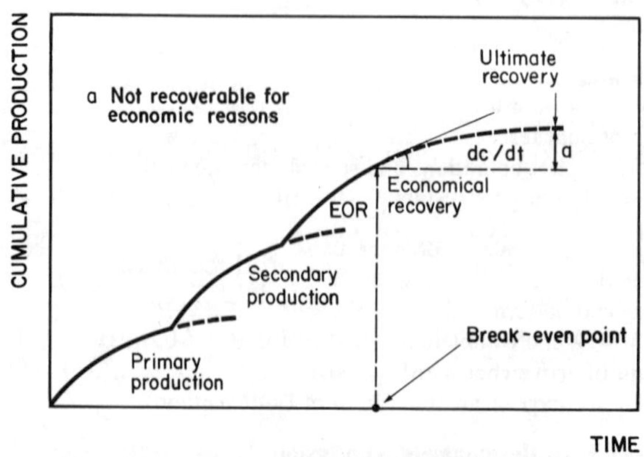

Fig. 12.1 Recovery and production history.

In the fluctuating economic situation prevailing since the first oil crisis in 1973, crude price forecasting has become extremely difficult. This is the factor of the greatest uncertainty in designing projects. Work is generally done with several price development hypotheses to evaluate the risk incurred. In statistical studies, EOR reserves are nearly always given as a function of the crude price.[1]

Production expenses F depend not only on the type of recovery process but also on the geographic conditions. For example, they are much higher if the reservoir is located in northern Canada, far from any refining center, with temperatures that may need operations to interrupt, than if it is located in California in favorable industrial and climatic environment. Similarly, offshore conditions are especially penalizing. As an example, an economic analysis concerning a steam flood pilot located offshore in the Gulf of Guinea (water depth 65 m, favorable climatic conditions), showed that the same pilot, located onshore, all other things remaining equal, would cost 40% less.[2]

Another case which demonstrates the importance of the industrial environment in EOR is that of CO_2 injection. The development of this process, technically highly efficient, is economically feasible only if the CO_2 sources are abundant and not too far from the operating site. This largely explains why the process has grown considerably in the American Mid-Continent, whereas its success is limited in other parts of the world.

The environmental laws may have a non-negligible influence on the economics of an EOR project.[3] In the USA, for example, the Clean Air Act and the Clean Water Act of 1977 have created difficulties for many steam injection projects in the Bakersfield area, for which the purchase of scrubbers to purify the flue gases from steam generators and the purchase of efficient oil separators for waste water treatment have severely increased the operating expenses.

Taxes T depend on the policy set by the local authorities. Large tax incentives are often granted to encourage the operators to develop difficult projects. For reasons of national policy, the producing States are reluctant to abandon the primary production of reservoirs located on their territory, without trying to enhance recovery, even at the cost of a smaller tax benefit. Yet, as pointed out by Chierici,[4] the system of tax reductions based on the oil quality can cause regrettable distortions in the choice of EOR methods.

In conclusion, for a given EOR method and a given type of reservoir, a project may prove to be profitable in a certain place at a certain time, or negative in another area and in a different economic and political situation. *Hence, no general rules are universally applicable for an a priori estimate of the value of an EOR project for a given reservoir.* Each case is a specific one, and requires a separate technico-economic analysis. Argument by analogy must be discarded.

12.1.2 *Project management*

An EOR project is a multidisciplinary task, requiring the services of many specialists: research and development engineers, geologists, reservoir engineers, production engineers, marine specialists (if offshore), and economists.

Management of such a multidisciplinary project is not easy due to the number of specialists coming from very different departments in a company. The fact that EOR requires advanced techniques, generally developed in research laboratories and unfamiliar to reservoir or production engineers, makes problem of coordination even more difficult.

For maximum effectiveness, it is recommended to entrust the design of the EOR project to an in-house multidisciplinary team, set up exclusively for the duration of the project, and led by an engineer with a wide general background (Fig. 12.2). However, the implementation of the project will be entrusted to the local operational departments, not specialized in EOR, but thoroughly familiar with the field. For this system to work smoothly, it is necessary for the team to define precise operational programs, and for each engineer to have clearly defined responsibilities.

12.2 Flow chart of an EOR project

The flow chart of an EOR project is shown in Fig. 12.3. The successive phases of a project are as follows:

(i) *Detailed analysis of the reservoir and of its performance*: gathering of all the available information, characterization of the reservoir, analysis of drainage mechanisms, construction of the reservoir model, fitting of the model to the production record, extrapolation of performance to the end of primary (or secondary) period of production.

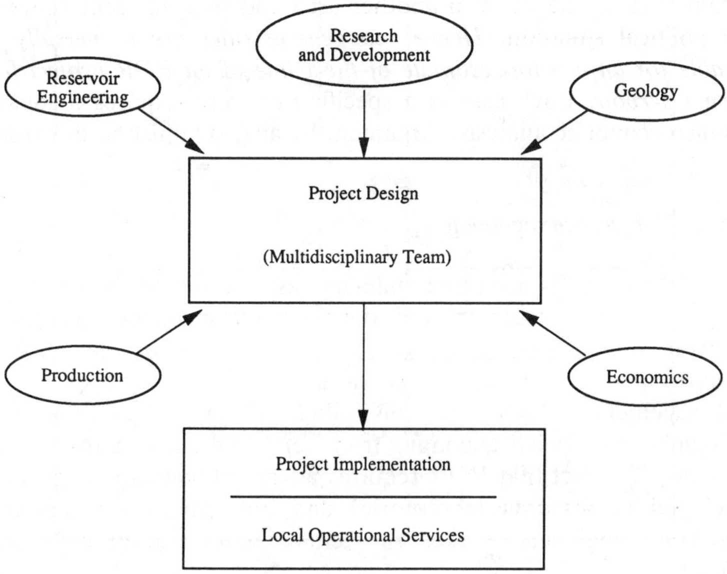

Fig. 12.2 Organization chart for an EOR project.

(ii) *Preselection of recovery methods*: comparison of the screening criteria for each feasible EOR process with the characteristics of the reservoir and with the local constraints (discarding of incompatible processes).

(iii) *Study of preselected processes*: laboratory experiments on cores under reservoir conditions to determine microscopic efficiency, analyses of macroscopic sweep on phenomenological models.

(iv) *Pilot tests*: definition of pilot tests, choice of size, injection systems, selection of products to be injected, forecast simulations, operation follow-up, interpretation of results.

(v) *Optimization of operational parameters*: starting date of development, spacing, patterns, well completion, amount of active products, size of slugs, surface treatment.

(vi) *Final project*: resumption of the initial reservoir model by introducing the information provided by the pilot tests, comparative simulations of different scenarios, economic analyses with various crude prices and tax assumptions, final decision, implementation schedules.

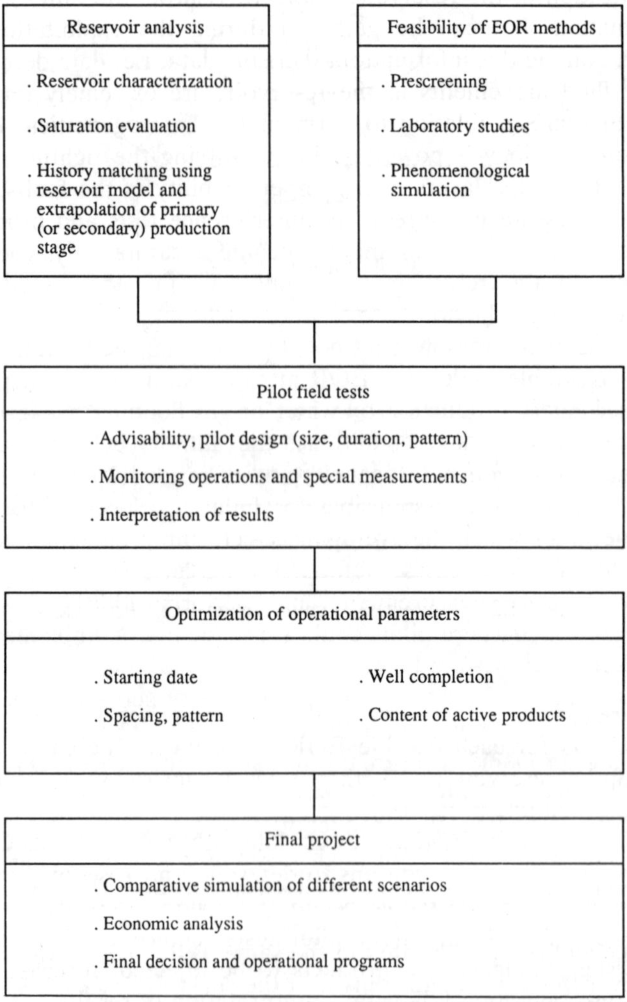

Fig. 12.3 Flow chart of an EOR project.

12.3 Analysis of the reservoir performances

12.3.1 *Reservoir characterization*

Experience has proved that many failures encountered in EOR are due to a poor understanding of the reservoir, particularly insufficient characterization of the reservoir. Table 12.1 shows the different types of information that can be gathered during the exploitation of a reservoir. Among this information, dynamic data, i.e. data determined from the fluid movements in the reservoir, are extremely important because they are in relation to permeability heterogeneities. In fact, conventional geology is powerless in quantifying the tightness of the barriers between the layers or the impermeability of the faults, in the absence of pressure measurements, interference tests and production logs. These dynamic measurements become accurate only when fluid movements in the reservoir are significant, in other words after several years of production.

This is one of the reasons (but not the only one, see Section 12.7.1) why it is preferable to develop EOR methods after primary depletion, or even secondary production by water or gas flooding. Nevertheless, even under these tertiary conditions, the authors believe that poor knowledge of reservoir heterogeneity and of residual oil saturations at the time of flooding is responsible for failures, especially for small-sized projects for which the averaging effect cannot be exploited. This means that, in many cases, before starting an EOR project, it is necessary to take measurements and perform additional studies, especially in zones where pilots could subsequently be implemented.

Considerable efforts are being made by various companies to obtain a good description of the reservoirs by the use of geostatistical models constrained by the lithological data at the well, taking into account the type and the probable scale of the heterogeneities by observation in similar deposits and information provided by new seismic surveys. The purpose is to improve production forecasts taking into account reservoir heterogeneities.[5] The high resolution seismic methods used to characterize reservoirs include well seismic techniques:

- source and receiver located in the same well,
- receiver in the well and source on the surface,
- cross-hole seismic surveys (detectors in one well, source in a neighboring well): this method, which is emerging from the experimental domain, is highly promising for the future.[6]

Conventional well pressure tests have also made considerable progress, thanks to new interpretation methods. Some of these new methods use the derivative of pressure with respect to time, providing a more precise investigation, and, *inter alia,* an evaluation of the

Table 12.1 Data available in each phase of field development

STAGE	YEAR	NUMBER OF WELLS	SEISMIC		WELL LOGGING			CORE ANALYSIS	PRESSURE MEASUREMENT	PVT	WELL TESTS	INTERFERENCE TESTS	PROD. HISTORY	OUTCROP STUDIES
			2D	3D	STAND.	PROD.	S_{OR} EVAL.							
DISCOV.	1	1	×		×			×	▨	×	▨			
EVALUATION	2	3		×	×	▨		×	▨	×	▨	▨		
	3	6			×	▨		×	▨	×	▨	▨		×
DEVELOPMENT	4	12			×	▨		×	▨			▨		×
	5	18			×	▨		×	▨			▨		
PRODUCTION	6	–				▨			▨				▨	
	7	–				▨			▨				▨	
	8	–							▨				▨	
	9	–					▨		▨				▨	
	10	24			×			×	▨			▨	▨	
	11	–				▨			▨				▨	
	12	–					▨		▨				▨	
	13	–							▨				▨	

× STATIC DATA   ▨ DYNAMIC DATA

degree of fracturing of the reservoir rock or the degree of communication between two layers.[7]

Inter-well hydraulic diffusivity can also be measured by interference tests and especially by pulse tests.[8]

12.3.2 *Evaluation of oil saturation distribution*

The initial saturations in the reservoir are determined by three main methods: electrical logs, direct measurements on cores, and capillary correlations with respect to water levels. In the case of discrepancy between the results of these three types of evaluation (which frequently occurs), the values derived from resistivity logs are generally selected, although the formation factor and the resistivity index used in the calculations are often incorrect.[9] Hence the initial saturations are not always reliable.

Furthermore, these saturations change during production in a non-uniform manner, especially for reservoirs subjected to a water drive, a gas cap drive, or enhanced recovery. The problem is to determine what the residual saturations are when EOR begins. This problem, difficult to solve, is very important for the assessment of an EOR project. Several approaches are available:

—material balance: the average saturation in the reservoir is obtained;

—simulation with a numerical model: the spatial distribution of the oil saturation is obtained if the reservoir heterogeneity is effectively described;

—analysis of cores sampled with special core barrels (pressure core barrels or sponge core barrels) in newly drilled wells;

—by special logs in cased producing wells[10] or by special measurements such as log-inject-log[11] or single well tracer[12] performed on available wells or, better, on newly drilled wells.

Unfortunately, the results obtained from these methods are rarely in good agreement. As a consequence, their interpretation is difficult.[13]

12.3.3 *Reservoir modeling and performance forecasting*

To assess an EOR project, it is necessary to calculate the incremental oil that the enhanced recovery process would provide in comparison with the production method under way, as a function of time. Hence the primary (and possibly secondary) production of the reservoir must be extrapolated.

Usually, a numerical model is built in the very first years of the exploitation of the reservoir, and subsequently undergoes several adjustments. An EOR project needs a complete updating of this model, incorporating the latest information gathered in terms of geology, petrophysics, thermodynamics and production data. After a careful fitting of the production data, a forecast simulation is made of the primary and secondary recovery of the reservoir, up to the economical limit.

In these forecast simulations, it is important to change the operating conditions as the reservoir declines: shut-in of wells flooded by water or invaded by gas, recompletion of wells in uninvaded zones, infill drilling, application of pumping or gas-lift, lowering of the pressures at the separators by the use of boosters, etc.

12.3.4 *Influence of infill drilling on recovery*

The primary production of a reservoir begins with wide spacing to limit the initial investment. Oversized surface equipment and producing the wells far below their potential are avoided. For these economic reasons, fields are generally developed in two or three drilling campaigns separated by several years.

Infill drilling during the primary (or secondary) production stage can increase recovery, which competes closely with the EOR project. This is especially true of some types of very heterogeneous reservoirs, where the inter-well connectivity is poor (and specifically where EOR methods are inefficient). Fluvio-deltaic deposits, in which the sand beds separated by clay beds are stacked in a random configuration, or complex dolomite and anhydrite reservoirs with cyclic sequences and poor lateral connections, are examples of very heterogeneous reservoirs.

Figure 12.4 shows the impact of 10 acre ($40{\cdot}5 \times 10^3$ m^2) infill wells in production in two waterflooding pilot areas in the Means field, West Texas, initially drilled to 40 acre (162×10^3 m^2) spacing.[14] Increased recovery resulting from infill drilling is the production represented by the area between the total production and extrapolated decline curve without infill drilling. Additional recovery from infill wells is estimated to vary from 2 to 5% of the original oil in place.

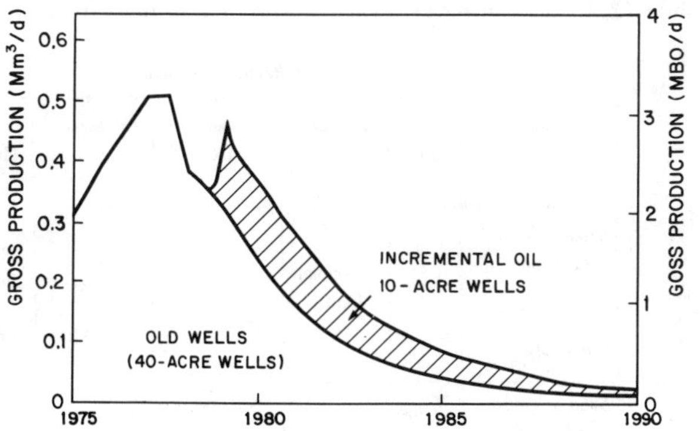

Fig. 12.4 Comparative production of 40-acre and 10-acre spacings in Mean San Andres Unit. (From Ref. 14, reproduced by permission of The Society of Petroleum Engineers.)

12.4 Preselection of EOR methods

For a given reservoir, two aspects have to be taken into account:

- How can additional oil be produced apart from conventional production methods?
- At what cost will the additional oil be produced?

There is no single answer to these two simultaneous questions. Additional production depends on the configuration and heterogeneity of the reservoir, on oil saturation (as stated in Section 12.3.2), and also on the mineralogical properties of the rock and the oil quality. The cost of production by one of the EOR methods depends on all these reservoir parameters, on the products injected, on the well and surface equipment required for injection and production, and also on local conditions. Hence it is not possible to establish criteria for the application of EOR methods, but merely points of reference.

Table 12.2, taken essentially from Ref. 15, gives, for each process, the limit values required for a number of parameters: reservoir, rock and fluids. It is understood that apart from any local particularity, each of the criteria is limitative either for technical reasons (e.g. thermal stability of the polymers available today) or for obvious economic reasons (e.g. with steam flooding, a quantity of oil that is too small per unit volume of rock). However, the application of one of the EOR methods would probably be inappropriate in a reservoir for which all

Table 12.2 Screening criteria for enhanced oil recovery methods[15]

<table>
<tr><th colspan="3" rowspan="2"></th><th colspan="7">Reservoir properties</th></tr>
<tr><th>Nature</th><th>Depth (m)</th><th>Pressure (MPa)</th><th>Temperature (°C)</th><th>Thickness, h (m)</th><th>Permeability, k (10^{-3} μm^2 or mD)</th><th>ΦS_{o_i} (vol/vol)</th></tr>
<tr><td colspan="3">In situ combustion</td><td>Sand, sandstone
Carbonate</td><td>>100</td><td><economic limit</td><td rowspan="4">Not critical</td><td>2–30</td><td>>100 if high oil viscosity
>25 if low oil viscosity</td><td>>0·07</td></tr>
<tr><td colspan="3">Steam stimulation</td><td rowspan="2">Sand or sandstone with low content in swelling clays
Carbonate</td><td><1 500</td><td></td><td>>5</td><td rowspan="3">>100</td><td rowspan="3">>0.1</td></tr>
<tr><td colspan="3">Steam drive</td><td><1 000</td><td></td><td rowspan="2">10–100</td></tr>
<tr><td colspan="3">Hot waterflooding</td><td>Sand
Sandstone
Carbonate</td><td><1 500</td><td></td></tr>
<tr><td rowspan="3">Miscible flooding</td><td rowspan="2">Hydrocarbon gas</td><td>Lean gas</td><td rowspan="3">Sand, sandstone and homogeneous carbonate</td><td>>2 500</td><td>>30</td><td rowspan="3"></td><td rowspan="2"><10</td><td rowspan="3">>10</td><td rowspan="2"></td></tr>
<tr><td>Rich gas</td><td>>1 500</td><td>20–30</td></tr>
<tr><td colspan="2">CO_2</td><td><3 000</td><td>>7</td><td></td><td></td></tr>
<tr><td colspan="3">Partially miscible CO_2 flooding (heavy oil)</td><td>Sandstone and carbonate</td><td></td><td>>6</td><td></td><td><15</td><td>>50</td><td></td></tr>
<tr><td colspan="3">Polymer flooding</td><td rowspan="2">Sand
Sandstone
Carbonate</td><td rowspan="2"></td><td rowspan="2">injection pressure < lithostatic pressure</td><td rowspan="2"><70–90°C depending on salinity</td><td rowspan="2"><20–30</td><td rowspan="2">>20</td><td rowspan="2"></td></tr>
<tr><td colspan="3">Micellar/polymer flooding</td></tr>
<tr><td colspan="3" rowspan="2">Alkaline flooding</td><td>Sandstone</td><td rowspan="2"></td><td rowspan="2"></td><td>>100°</td><td rowspan="2"><20–30</td><td>>50–100</td><td rowspan="2"></td></tr>
<tr><td>Carbonate</td><td>Not critical</td><td>>20</td></tr>
</table>

Crude oil			*Notes*	
Specific gravity 15°C	*In situ viscosity (mPa s or cP)*	*Nature*	*Favorable features*	*Detrimental features*
0·82–0·99	2–10 000	Sufficient content in heavy components Readily oxidizable	Low vertical permeability Oil with low sulfur content $\frac{kh}{\mu} > 5 \frac{mD.m}{cP}$[a]	Gas cap High permeability streaks High content in swelling clays (air/water injection) Thin overburden
0·82–1·00	>100	Not critical		Gas cap High permeability streaks High water content
	<10 000		$\frac{kh}{\mu} > 30 \frac{mD.m}{cP}$[a]	
	50–10 000			
0·78–0·85	<1	High content in light components	Primary production High dip angle	High permeability streaks Important gas cap
<0·90	<4	High content in intermediate components	Low gas saturation Natural CO_2 available	
<0·99	<1 000	Low content in dissolved gas		
<0·90	<60		Low clay content	Rock with large-scale fissuration
		Influence not well known		Highly heterogeneous formation Formation water with high content in divalent ions
<0·95	<40	Acid index ≥0·5 mg KOH/g crude oil	Oil with high content in polar compounds (resins, asphaltenes)	Formation water with high content in divalent ions High content in clays or anhydrite High heterogeneity

[a] Or 10^{-3} $\mu m^2.m/(mPa\ s)$.

the parameters would be equal to the limit values given in this table. For example, from considerations given in Chapter 2, it appears that polymer flooding is particularly well suited for reservoirs that have poor sweep efficiency by water flooding due to high oil viscosity or large-scale heterogeneity. However, if the oil viscosity is high (i.e. higher than 50 mPa/s), the likelihood of success is greater in more homogeneous reservoirs.

The screening criteria selected here have been demonstrated to be valid by field experiments. They may change in the years to come. For example, new chemicals (polymers or surfactants) already tested in the laboratory could prove their better resistance to the ionic environment and to the temperature in the reservoirs, and would hence help to broaden the field of application of chemical methods. Scleroglucans or sulfonated polymers, recently proposed, could make polymer flooding possible at temperatures higher than 100°C. Moreover, alkaline solutions are indicated in a restricted sense in Table 12.2. In 1990, specialists believe that better performances would be obtained by injecting alkaline products containing surfactants at low concentration, possibly followed by a polymer solution (see Chapter 4). In other respects, the prior injection of an alkaline solution decreases surfactant adsorption and could help to widen the field of application of surfactants to rocks whose adsorption capacity is considered prohibitive today.

More generally, any technical improvement in the exploitation of oil reservoirs can strongly modify the criteria for selecting an EOR method. For example, the technology of horizontal wells has been mastered during the 1980s. The use of this new technique has resulted in resuming the exploitation of Austin Chalk.[16] and of heavy oil reservoirs in Canada.[17] Similarly, since these horizontal wells produce high flow rates with low gradients, it is felt that they increase the recovery of oil rings.[18] It is certain that combining horizontal wells and EOR processes will significantly widen the area of application of these processes.

12.5 Studies of preselected methods

12.5.1 *Laboratory studies*

Once some EOR methods have been preselected, and residual oil saturations have been evaluated, experimental studies are initiated in the laboratory. The experiments are preferably conducted on large cores, and under the temperature and pressure conditions prevailing in the reservoir at the time when the EOR project is scheduled to begin. Since experiments on large cores under reservoir conditions are expensive, work can be done only on a small number of samples; hence it is important to select these samples with the greatest care, using the competence of production geologists.

The laboratory experiments are mainly designed to determine microscopic sweep characteristics: residual oil saturation behind the front, dispersion, adsorption, degradation of injected products, reactions with formation water, etc. The laboratory procedures are described in the previous chapters.

12.5.2 *Phenomenological simulation*

The volumetric sweep efficiency of the EOR method, which depends on the macroscopic heterogeneities of the reservoir, cannot be investigated on cores. Moreover, it is hardly possible to use the conventional correlations giving the sweep efficiencies as a function of the pattern, the mobility ratio, and a heterogeneity coefficient. Hence, in most cases, it is necessary to use specialized numerical models specific to each method:

- chemical models taking into account phase behavior, adsorption and degradation phenomena for polymer and micellar flooding,
- compositional models taking into account mass transfers between phases for miscible flooding,
- thermal models taking into account the various modes of heat transfer for thermal methods.

These models are constantly evolving, and it is difficult to provide definitive references. Their performances now approach those of the classic models used to simulate primary and secondary production, but their costs are much higher. This is why an attempt is made in preliminary simulations to represent only limited and simplified portions of the reservoir.

12.6 Pilot testing

12.6.1 *The need for pilot tests*

The uncertainties concerning the characteristics of a reservoir and the effectiveness of EOR methods are such that analyses on cores and by phenomenological models are rarely sufficient to go on directly to field development. Hence the need arises for intermediate field tests, called pilot tests. Two main types of pilot tests can be distinguished:

- the method pilot, of limited size, designed to test the technical feasibility of the EOR process, without any concern for profitability;
- the industrial pilot, more developed, designed to optimize operations and to prove their profitability; generally it follows the method pilot.

It must be pointed out that, between the start of its design and the end of its interpretation, a pilot test needs some five years of work. Hence it is an operation which, if not started sufficiently early, can seriously delay the final implementation of EOR. Furthermore, it may be difficult to extrapolate to the whole field, the results obtained locally on a pilot. The advisability of a pilot test is hence debatable. For a very well-known reservoir (long primary and secondary production data, small well spacing) and a fully-controlled EOR method (such as classic steam drive), it may be unnecessary to conduct pilot field tests, and merely go on to commercial application without incurring excessive risk, but this is rare. Pilot tests are necessary in all other situations.

12.6.2 *Design of pilot tests*

The purpose of a pilot field test is to identify the following parameters:

- microscopic displacement efficiency of the process
- volumetric efficiency
- reactions between injected fluids and fluids in place
- ratio of injected and recovered volumes
- injectivity index
- recovery rate

To obtain results within a reasonable time, the size of the pilot test must be small. The spacings are calculated so that field operations do not last more than two or three years. The pilot pattern must be as simple as possible to minimize costs. A two-well or a three-spot pattern can be sufficient for a method pilot; an industrial pilot requires more complex patterns (see Section 12.6.4).

Pilots must be established in the accurately characterized zones of the reservoir, e.g. zones in which there has already been a natural or artificial water drive, so that the fluid paths are already well known.

If two or more EOR methods are comparatively tested on a field, successive pilots must not be conducted in the same reservoir zone, using the same wells, because interpretation then becomes too difficult. Thus, at the Slaughter Field, successive floods were performed at the same point: one water flood, followed by a flood of $H_2S + CO_2$ slugs alternating with water, followed by a nitrogen flood.[19] These pilots yielded very interesting information from the technological standpoint, but their excessive interactions prevented any clear conclusion from being drawn as to the sweep efficiencies of each method.

12.6.3 *Monitoring pilot operations and special measurements*

If the reservoir is not too deep, before starting the pilot and after it is terminated, it is recommended to perform a series of core-drills designed to measure the variation in oil saturations (see Section 12.3.2). A maximum of measurements must be made during the operations. Observation wells must therefore be carefully located. Two types of observation wells can be distinguished:

—Unperforated cased wells, for regular well log recordings (pulsed neutron, carbon/oxygen log) which provide information about the change in saturation.[20] In some cases, Fiberglas casings are placed directly at the layer, for the use of an induction log. For steam pilots, the most widely-used tool in observation wells is the temperature log.[21]

—Open wells, completed conventionally and perforated in the reservoir for the sampling of the fluids.

It is often useful to add slugs of chemical or radioactive tracers to the injected fluids to determine the fluid paths and breakthrough times more accurately. The most commonly-used tracers are tritium, specific alcohols, and water-soluble ions contained in salts (ammonium thiocyanate and nitrate, potassium and sodium bromides and iodides). The choice of the tracer is made in the laboratory, based principally on solubility in water, low adsorption in the porous medium, absence of any reaction with formation water, and ease of handling.[22]

12.6.4 *Interpretation of pilot results*

The pilot test must be simulated on a numerical EOR model as it proceeds. The main problem is in setting the correct boundary conditions for the model. In fact, the edges of the pilot are not zero flow boundaries. Part of the liquid injected leaves the flooded pattern. Part of the produced oil comes from areas outside the pattern. This is why the reservoir zone represented in the model overlaps the pilot area and, preferably, includes a line of adjacent wells. If a pressure gradient exists in the pilot zone before the start of flooding, it is necessary to represent this gradient initially in the model. The simulation period must also start before the startup of the pilot, so as to adjust the model to primary (or secondary) production.

All the parameters for adjusting the model to the pilot (residual saturation, relative permeability, capillary pressure, dispersion and adsorption coefficients, etc.) are then introduced into the general reservoir model for simulation of the industrial development of the EOR method.

12.7 Design of large-scale EOR operations

12.7.1 *Starting time*

The optimal date at which EOR development should begin is a factor that has a direct influence on the efficiency and the profitability of the process. Apart from exceptional cases of reservoirs in which the oil is too viscous to allow an initial flow under reservoir conditions (such as tar sands), an EOR operation is never begun before a sufficient quantity of oil has been produced by primary or secondary exploitation methods. This is due to technical as well as economic reasons.

Technically, the operator, as we have already mentioned, must be thoroughly familiar with the dynamic characteristics of a reservoir before becoming engaged in the difficult operation that EOR represents. Economically, it is important to amortize the initial investments rapidly, by producing cheap primary oil before going on to the EOR phase, in which the operating costs are always high. However, if too much time elapses before enhanced recovery begins, it is possible that the production rate of the remaining oil in place in the reservoir may be insufficient to pay the costs of the EOR process. As a consequence, the starting date of an EOR application is the subject of a rather careful economic optimization.

12.7.2 *Spacing and pattern*

In primary or secondary production, recovery is influenced by the spacing only in some cases of strong heterogeneities (see Section 12.3.4). By contrast, in EOR most of the processes are sensitive to the spacing, irrespective of the type of reservoir.

This is especially true of steam flooding, in which heat losses to the surrounding formations lower the efficiency as the spacing rises. Besides, whichever process is adopted, it is clear that the reservoir is better known and the risk of failure is lower if the spacing is narrower. On the whole, from a technical standpoint, it is always better to adopt small spacings, smaller than those used in the primary or secondary phases. Here also, economic factors will impose their rule, because the drilling of additional wells is expensive. Hence the spacing must also be the subject of a technico-economic optimization.

The ratio of the number of producing wells to injection wells depends on the ratio of the fluid mobilities and on the reservoir pressure. For example, for steam flooding in a heavy oil reservoir, many more producers than injectors are needed, whereas for polymer flooding, the number of injectors and producers is equal. This ratio of the number of wells directly governs the choice of the flooding pattern. Yet another factor is involved, namely the horizontal sweep efficiency, which generally rises with the total number of wells in the pattern.

To minimize investments, it is also essential to make maximum use of existing wells, the arrangement of which is not always regular. Phenomenological simulations by numerical models may be needed to decide on the choice between different patterns.

12.7.3 *Type of well and completion*

The well equipment depends on the EOR process considered. Each method has its own constraints. For example, for steam injection the well equipment will have to be designed to reduce heat losses and to obtain sufficient downhole steam quality. For CO_2 injection, the well equipment will have to be chosen so as to resist to corrosion. For polymer injection, special precautions must be taken to avoid the presence of oxygen. These problems have been discussed in the previous chapters.

The use of horizontal wells has spread significantly in recent years.[23] In primary depletion, horizontal wells are ideal for:

- reservoirs with vertical heterogeneities (such as fissured rocks or karsts)
- reservoirs subject to coning
- compact reservoirs
- thin and slightly stratified layers
- reservoirs with very viscous oils

Thanks to progress achieved in drilling techniques, a well penetrating horizontally into a reservoir over a distance of 500–600 m only costs 1·5 to 2 times the price of a conventional vertical well, but with 4–5 times higher productivity or injectivity.

The weak points of these new wells are:

- the difficulty of guiding them with respect to the reservoir cap rock,
- the difficulty of successful cementing, selective completion, hydraulic fracturing,
- sometimes, the difficulty of finding adequate production equipment (e.g. in the case of pumping in very depleted wells).

If the drainage mechanism is forced sweep (generally the case in EOR), the conditions governing the application of horizontal wells are more severe. In particular, these wells are totally unsuitable for stratified or relatively thick reservoirs, in which they would cause very poor vertical sweep.

Designing an effective pattern by using very long horizontal wells in a reservoir with complicated geometric forms may also prove to be a difficult exercise. It is fortunately possible to set up mixed patterns combining horizontal wells with vertical wells. Thus, on the Chateaurenard polymer flood pilot in the Paris Basin, a horizontal producing well yielded results far better than those of the neighboring vertical wells, both in total productivity and in the change in the water/oil ratio. This is a thin and homogeneous sand reservoir. The well, the horizontal part of which measures 340 m, has been curved to match the shape of the oil bank displaced by polymers.[24]

Another interesting example is that of a horizontal steam flooding well drilled in the Lacq Supérieur reservoir, in the midst of vertical wells. This horizontal well, 350 m long, was drilled in a zone of compact limestone, and it serves to inject 650 tonnes of steam per day, although the permeability of the limestone is not more than 3 to $4 \times 10^{-3}\ \mu m^2$ (3 to 4 mD).[23]

The cementing of EOR wells has to be performed even more carefully than that of primary wells. Uncontrolled communication between layers could have disastrous effects on sweep.

12.7.4 *Amount of active chemicals*

The active product concentration of the injected fluids and the size of the slugs must also be optimized. Thus, in chemical EOR, for the same amount of surfactants or polymers, small slugs can be used with a high product concentration, or longer and more dilute slugs. In general, the total quantity of chemicals or solvent gases used must be optimized in accordance with the incremental oil recovered. Economically, it may be wise to recover less oil if the corresponding gain in terms of injected products is higher.

These technico-economic optimizations are cumbersome and costly. They generally involve the selection of several scenarios followed by forecast simulations for each of them, before comparing them for their economics.

12.7.5 *Decision for large-scale application of EOR methods*

After completion of preliminary studies, a number of cases will be discarded because they are clearly uncompetitive, and some of the parameters will have to be optimized. Yet doubts may subsist concerning:

- the process itself (for example, the choice between CO_2 flooding and improved water flooding),
- the development starting date,
- the scale of development (the entire reservoir or only part of it),
- the spacing (i.e. the number of additional wells to be drilled).

Several scenarios have to be compared by simulations that provide production profiles as a function of time. For each of these scenarios, forecasts of investments and operating costs are then compiled by production engineers and economists, also as a function of time. Simultaneously, assumptions as to the trend in the cost of crude and the applicable tax arrangements will be made by specialists. The economic analysis serves to help in the selection of a development scheme in accordance with the economic criteria selected (rate of return, discounted profit, payout, etc.).

12.8 Status of EOR development: conclusion

Many projects are cost effective today (steam flooding, carbon dioxide flooding, polymer flooding). The various EOR techniques are widely used around the world, covering a large range of applications from light to heavy crude oils. In 1989, world EOR production reached 75 million tonnes, with the USA ensuring half of this total production. The continued importance of enhanced oil recovery is reflected in a worldwide survey,[25] dealing with most EOR projects, but with the noticeable exception of the Hassi-Messaoud (Algeria) miscible gas large-scale operation. This survey provides extensive information: reservoir characteristics, size of pattern, production, profitability, etc.

Table 12.3 gives the EOR production by country for the three groups of methods.[26] It emphasizes the high development of steam flooding (USA, Venezuela, Indonesia), and also of miscible gas injection (carbon dioxide, low-molecular weight hydrocarbon) which leads to EOR production increase (USA, Canada, Algeria, Libya). The lesser development of the other techniques can be explained both by high risk and technical cost, which could be appreciably reduced by improving the technology through appropriate Research and Development.

The technological progress expected in the coming years will help increase the performances of EOR processes and, at the same time, will bring about a decrease in the cost of the oil produced. The incidence of such progress was already observed in the USA from 1940 to 1973; a period during which the price of oil remained stable, whereas the recovery factor after primary and secondary production stages was regularly increasing, according to the American Petroleum Institute statistics.

Improvement in the economics could reach 25–30% of the technical cost of the processes,[26] and therefore make them economically

Table 12.3 EOR production worldwide[26]

	Thermal		*Gas*		*Chemical*		*Total*	
	10^6 *m*^{3}/*year*	10^3 *bbl/d*	10^6 *m*^{3}/*year*	10^3 *bbl/d*	10^6 *m*^{3}/*year*	10^3 *bbl/d*	10^6 *m*^{3}/*year*	10^3 *bbl/d*
USA	26·35	454	11·08	191	0·69	11·9	38·12	656·9
Canada	0·46	8	7·37	127	1·0	17·2	8·83	152·2
Venezuela	6·24	108	0·64	11	0	0	6·91	119
USSR	1·16	20	5·22	90	2·90	50	9·29	160
Other (estimated)	9·92	171[a]	16·25	280[b]	0·09	1·5	26·26	452·5
Total	44·13	761	40·56	699	4·68	80·6	89·41	1 540·6

[a] Mainly Duri field (Indonesia).
[b] Mainly Hassi-Messaoud (Algeria) and Intisar (Libya) fields.

attractive. One major point to be emphasized is the necessary improvement of chemicals used in EOR. This concerns chemical flooding by surfactants and by polymers: optimization of molecular structures to improve the behavior regarding interfacial efficiency, adsorption, stability and mobility control. Thermal and miscible processes are also concerned in so far as foaming agents (surfactants) will be widely used to improve the sweep efficiency. For this challenge, a close association of petroleum and chemical industries appears to be highly desirable.

As mentioned above, each of the EOR processes has a specific range of application depending on the field characteristics (Table 12.2), but also on local considerations such as the availability of injection fluids (carbon dioxide, natural gas), market conditions, and economic incentives.[27] Consequently, each of the main techniques offers an important potential for oil recovery.

In the short term, the main effort to increase oil recovery will be devoted to improving the characterization of reservoirs by focusing attention on developing tools and techniques to define their internal architecture more accurately.[28,29] Also, much is expected from a synergistic association of EOR methods and production techniques, such as the use of horizontal wells, and from the contribution of new technologies, for instance those based on seismic methods.[30,31]

Production by natural depletion and by waterflooding is declining for many big oilfields and new discoveries are smaller in size. This is especially true in the USA.[30] Hence, the contribution of EOR to oil production should grow. In the next ten years, excluding any major crisis, the price of oil is expected to increase steadily. There is no doubt that the petroleum industry will take advantage of this favorable long-range economic situation to speed up the development of enhanced oil recovery processes.

12.9 References

1. Stosur, J. J. G., Frye, K. N. & Bohannon, J. M., The potential of EOR in USA for the next thirty years. Third European Meeting on Improved Oil Recovery, Rome, **2**, 16–18 April 1985. Agip SpA, Rome, 1985, pp. 393–402.
2. Cottin, R. H. & Sahuquet, B., Economics of heavy crude production by EOR methods in Europe and Africa. Third Int. Conf. on Heavy Crude

and Tar Sands, Long Beach. Proc., Unitar/UNDP Information Centre, New York, 22–31 July 1985, pp. 2192–203.
3. Donaldson, E. C., Environmental aspects of EOR. *Energy Sources,* **4**(3) (1979) 213–29.
4. Chierici, G. L., EOR. An overview of projects in different technical and economic environments. *Petrole Informations,* **1581** (1982) 23–26.
5. Guerillot, D. G., Lemouzy, P., Galli, A. & Ravenne, C., 3D fluid flow in porous media characterized by geostatistical method. SPE Paper No. 21081, Latin American Petroleum Engineering Conference, Rio de Janeiro, 14–19 October 1990.
6. Mari, J. L. & Coppens, F., La sismique de puits. *Revue de l'Institut Français du Pétrole,* **44** (1989) 135–60, 291–316, 419–42.
7. Daviau, F., *Interprétation des Essais de Puits par les Méthodes Nouvelles,* Editions Technip, Paris, 1986.
8. Kamal, M. M., Interference and pulse testing—A review. *J. Petroleum Technol.,* **35**(13) (1983) 2257–70.
9. Longeron, D. G., Argaud, M. J. & Bouvier, L., Resistivity index and capillary pressure measurements in reservoir conditions using crude oil. SPE Paper No. 19589, 64th Annual Techn. Conf., and Exhibition of Soc. Petroleum Engineers, San Antonio, 8–11 October 1989.
10. Desbrandes, R., *Encyclopedia of Well Logging,* Editions Technip, Paris, 1985.
11. Murphy, R. P., Foster, G. T. & Owens, W. W., Evaluation of waterflood residual saturation using log-inject-log procedures. SPE Paper No. 5804, Symp. on Improved Oil Recovery of Soc. Petroleum Engineers, Tulsa, 22–24 March 1976.
12. Sheely, C. Q., Description of fields tests to determine residual oil saturation by single well tracer. *J. Petroleum Technol.,* **30** (1978) 194–202.
13. Roest, J. A., Runia, D. J. & Trebolle, R., Reconciliation of contradictory ROS data from a Venezuelian light oil sandstone reservoir. 4th European Symp. on Enhanced Oil Recovery, Hambourg, Proc. DGMK, 27–29 October 1987, pp. 675–85.
14. Barber, A. H., George, C. J., Stiles, L. H. & Thompson, B. B., Infill drilling to increase reserves—actual experience in nine fields in Texas, Oklahoma and Illinois. *J. Petroleum Technol.,* **35**(9) (1983) 1530–8.
15. Burger, J., Sourieau, P. & Combarnous, M., *Thermal Methods of Oil Recovery,* Editions Technip, Paris, and Gulf Publishing Company, Houston, 1985.
16. Sheikholeslami, B. A., Schlottman, B. W., Siedel, F. A. & Button, D. M., Drilling and production aspects of horizontal wells in the Austin Chalk. SPE Paper No. 19825, 64th Annual Techn. Conf. and Exhibition of Soc. Petroleum Engineers, San Antonio, 8–11 October 1989.
17. Sharp, D. A., Winter reservoir—horizontal well project. Paper No. 29, Third Techn. Meeting of the Saskatchewan Section of CIM, Regina, 25–27 September 1989.
18. Anon., Norsk Hydro reports good flow troll west horizontal well test. *North Sea Letter* 733/5, 20 December 1989.
19. Ader, J. C. & Stein, M. H., Slaughter estate unit tertiary miscible gas pilot reservoir description. *J. Petroleum Technol.,* **36**(5) (1984) 837–45.

20. Roux, C., Fertl, W. H., Frost, E., Sterdman, D. & Elliot, D., Well logs evaluate and monitor heavy oil steam flood in Kern County, California. Paper No. 11, 22nd Annual SPWLA Logging Symp., Mexico City, 23–26 June 1981.
21. Oglesby, K. D., Blevin, T. R., Rogers, E. E. & Johnson, W. M., Status of the 10-pattern steamflood, Kern River Field, California. *J. Petroleum Technol.,* **34**(10) (1982) 2251–7.
22. Allison, S. B., Analysis and design of field tracers for reservoir description. Master Sciences Thesis, University of Texas at Austin, Austin, 1988.
23. Sahuquet, B., Spreux, A., Corre, B. & Guittard, M., Steam injection in a low permeability reservoir through a horizontal well in Lacq Superieur Field. SPE Paper No. 20526, 65th Annual Techn. Conf., and Exhibition of Soc. Petroleum Engineers, New Orleans, 23–26 September 1990.
24. Foxonet, F. & Bruckert, L., Enhanced oil recovery by a horizontal well located inside a polymer flood pilot. Fifth European Symp. on Improved Oil Recovery, Hungarian Hydrocarbon Institute, Budapest, Proc., 25–27 April 1989, pp. 329–38.
25. Anon., Annual production report, OGJ Special. *Oil and Gas Journal,* **88**(17) (23 April 1990) 49–82.
26. Simandoux, P. & Valentin, E., Improved recovery—A strategy option or not? Ninth Int. Offshore Northern Seas Conf. and Exhibition, Stavanger, 28–31 August 1990.
27. Taber, J. J., Environmental improvements and better economics in EOR operations. Annual Workshop of the IEA Collaborative Project on Enhanced Oil Recovery, Palo Alto, California, 4–6 October 1989.
28. Haldorsen, H. H. & Damsleth, E., Stochastic modeling. *J. Petroleum Technol.,* **42**(4) (1990) 404–12.
29. Déquirez, P. Y. & Richard, V. Application of inversion to integrated stratigraphic interpretation. *Revue de l'Institut Français du Pétrole,* **45**(3) (1990) 397–416.
30. Stosur, G. J., Singer, M. I., Luhning, R. W. & Yurko, W. J., Enhanced oil recovery in North America: Status and prospects. Fifth European Symp. on Improved Oil Recovery, Budapest. The Hungarian Hydrocarbon Institute, Budapest, 1989, pp. 43–52.
31. Mungan, N., Enhanced recovery under constrained conditions. *J. Petroleum Technol.,* **42**(8) (1990) 962–5.

Index

Absolute permeability
effect of mechanical stress, 281
effect of temperature, 282
Acentric factor, 165, 178
Acid number, 127
Activation energy, 253, 256, 264, 267
Additives (in combination with steam), 305
Adhesion, 140
Adsorption
polymers, 57–60
surfactants, 102, 109
Aerobes, 344
Air flux
effect on in-situ combustion, 328, 329
minimum value for viable fireflood, 334
Air injection flow rate, 334
Air/oil ratio, 335
Air requirement, 321, 329, 330, 331
minimum value for viable fireflood, 321, 330
Alkali, 123, 127
concentration (effect on pH), 126
consumption, 138, 141
costs, 127
Alkali-polymer process, 150
Alkali-surfactant-polymer process, 151
Alkaline flooding (field applications), 149
Alkalinity
definition, 127
useful, 142
Alteration
oil physical properties, in, 244
rock composition, in, 252
Alternate gas and water flooding (*see* Water alternate gas injection)
Amphiphilic molecules, 91
Anaerobes, 344
Analcime, 142
Analytical models
continuous injection of hot fluid, 302
cyclic steam injection, 299
Anhydrite, 135, 151
AOR (*see* Air/oil ratio)
Aquathermolysis, 251
Aquifer, 4
Asphaltenes, 128, 248, 255
effect on first contact miscibility, 207
precipitation by CO_2, 237
Atomic hydrogen/carbon ratio, 258, 318, 325, 326

Bacteria, 47, 65, 343
spore forming, 344, 349
types, 344
Balance (*see* Mass balance)
Bicontinuous structure, 95
Binary interaction parameter, 168, 179
Binary screening guides, 231
Biocides, 66
Biological gas, 349
Biopolymers (*see also* Polymers), 45, 342, 348

Biosurfactants, 342, 348
Break-even point, 378
Bubble point curve, 189
Buffered alkali systems, 150
Build up of gas behind the combustion front, 320, 330
Byproducts (*see* also Metabolites and Metabolic products), 342–345

Capillary forces, 11, 24
  heterogeneous reservoirs, in, 29
Capillary number, 18, 90, 108
  relative permeabilities, and, 105
Capillary pressure, 16, 280, 395
Carbonates, 60, 127, 343
  CO_2 effect, 223
Carbon dioxide (CO_2), 215, 342
  breakthrough, 236
  effect on clays, 223
  effect on oil viscosity, 220
  effect on water viscosity, 222
  phase behavior, physical properties, 216–220, 225, 233
  solubility in crude oils, 220
  solubility in water, 221
  viscosity, 218
Carbon dioxide flooding
  field applications, 216
  performance, 236
Carbon number distribution, 249
Carboxylic acids, 129
Casing temperature (hot fluid injection), 292
Catalysts, 326
Catalytic effect, 252
  metals, of, 63, 253, 260
Cation exchange (*see also* Ion exchange), 110
Cation exchange capacity (CEC), 110
Caustic coefficient, 128
Chemical flooding, 37, 41
Chromatography
  polymer, 68
  surfactant, 109
Clays, 253, 259, 260
  effect of CO_2, 223
  effect on in-situ combustion, 330
Clays—*contd.*
  swelling, 127
Coalescence, 101, 132
Coefficient of heat transfer
  surface line, for a, 289
  well, for a, 290
Coke, 253, 254, 262, 314, 325
Combustion, 244, 246, 258
  gas composition, 331
  reaction, 318
Combustion front temperature (peak temperature), 314, 328, 331
  minimum, 321, 330
Combustion front velocity, 328, 329, 332
Combustion tube tests, 327–333
Compaction, 281
Compressibility factor (CO_2), 218
Condensation, 245
Condensing gas drive, 196
  combined vaporizing-condensing gas drive, 203
Conduction (heat transfer by), 278
Conductive heating (*see* Heating)
Conductivity (thermal), 278
Contact angle, 12
Contact forces, 10
Contact lines, 11
Continuous multiple contact experiment, 227
Continuous thermodynamics, 180
Coreflood
  polymers, with, 75
  surfactants, with, 114
Corresponding states (law of), 163
Corrosion (with CO_2), 211, 237
Cosolvent, 95
COSTALD correlation, 173
Cosurfactant, 92, 95
Cracking, 245, 248, 250, 253, 256
Cricondenbar, 206
Critical macromolecule concentration (C^*), 51, 58
Critical flow rate, 33
Critical micelle concentration (CMC), 93, 109
Critical point, 170, 187
Critical tie line, 97, 187, 191, 198

Critical wavelength (in viscous fingering), 33
Crossflow, 78
Crude oils
 heat capacity, 278
 specific gravity, 127, 220, 330
 thermal conductivity, 280
 thermal expansion, 277
 vaporization, 283
 viscosity, 127, 275
Cyclic steam injection (*see* Steam stimulation)

Darcy's equation, 15
Dehydrogenation, 245
Demixtion curve, 96
Density
 gas, of, 276
 liquids, of, 276
Depletion, 69
Design of projects
 alkaline flooding, 149–152
 carbon dioxide flooding, 236–238
 hot fluid injection, 295–309
 in-situ combustion, 333–335
 miscible gas flooding, 211–212
 polymer flooding, 80–82
 surfactant flooding, 114–117
Dew point curve, 189
Differential Thermal Analysis (DTA), 257, 325
Diffusion, 160, 234
Diffusion coefficient, 51, 74, 160, 208
Diffusion-dispersion
 heterogeneous reservoirs, in, 30
 miscible flooding, in, 208
Displacement
 immiscible (*see* Immiscible fluid displacement)
 miscible (*see* Miscible fluid displacement)
Displacement efficiency, 9
Displacement mechanisms
 hot water injection, for, 296
 steam drive, for, 302
 steam stimulation, for, 298

Dissolution
 carbonates by CO_2, of, 225
 minerals by alkalis, of, 141
Dry combustion
 laboratory tests, 328–331
 prediction of performance, 335
 principle, 314–316
Dry forward combustion (*see* Dry combustion)
DTA (*See* Differential Thermal Analysis)

Efficiency
 end effect, 29
 microscopic, 76, 89
Effluent Gas Analysis (EGA), 257, 263, 265
Electrical heating (*see* Heating)
Electromagnetic heating (*see* Heating)
Electromagnetic wave, 357
Emulsification, 107, 348
 entrainment, and, 131, 132
 entrapment, and, 132
Emulsion screening test, 133
Enhanced recovery (*see* Recovery)
Enriched-air fireflooding, 313, 330, 331, 336
Enthalpy, free enthalpy, 57, 103, 279
 water and steam, of, 279
Entrapment (*see* Retention)
EOR (Price and tax effect on), 378, 390
EOR projects
 economic and local constraints, 378
 flow chart, 380–382
 large-scale operations, 394
 management, 380–381
 phenomenological simulation, 391
 screening criteria, 387–390
 well equipment and completion, 395–397
Equations of state, 162
 mixing rules, 166
Equilibrium quotient, 143
Excess phases, 93, 95, 98
Exclusion, 68, 74, 75

Extent of reaction, 250
Extinction
  combustion front, of, 333
  wet combustion regime, of the, 318, 322, 333

Facultative aerobes, 344
Fermentation, 345, 350
Fick's law, 208
Field applications
  alkaline injection, 149
  carbon dioxide flooding, 216
  cyclic steam injection, 300–301
  electrical and electromagnetic heating, 359–360
  hot water injection, 295
  in-situ combustion, 336
  MEOR, 351–352
  miscible flooding, 211–212
  nuclear explosive fracturing, 365–370
  oil mining, 362–363
  polymer flooding, 82–84
  steam drive, 303
  surfactant flooding, 117–119
Filtration velocity, 15
Fingering (viscous), 31, 235
Fireflooding (*see* In-situ combustion)
Flocculation, 132
Flow tube, 21
Flue gas, 186, 189
Foam, 235–236
  characterization, 306
  drive, 35
  mobility control, 308
  modeling flow, 308
  stability, 307
Forward combustion, 313
Fractionation
  oil (*see also* Partitioning and Pseudocomponents), 229
  polymers, 69
Free radicals, 63
Fuel availability, Fuel content
  experimental determination (dry combustion), 324–326, 330, 331
  gas requirement, and, 318–320
Fuel availability—*contd.*
  minimum value for viable dry combustion, 320–322
  oil properties, correlations with, 326
  wet combustion, in, 333
Fuel consumption (*see* Fuel availability)
Fuel content (*see* Fuel availability)
Fuel requirement (*see* Fuel availability)
Fugacity, 161, 173, 180

Gas
  gas cap, 4
  density, 276
  heat capacity, 277
  injection in combination with steam, 305
Gas/oil ratio (GOR), 220
Gas requirement (*see* Oxidizing-gas)
Gas storage behind the combustion front, 320, 329
Gravity
  acceleration, 15
  forces, 10
  override, tonguing, 235, 303–305, 334
Group contribution models, 180
Gypsum, 135, 151

H_2S (*see* Hydrogen sulfide)
Hardness, 107
Heat balance
  dry combustion, for, 320–321
  wet combustion, for, 324
Heat capacity
  per unit mass (specific heat), 277
  per unit volume, 278
Heat condition, 278
Heating using electric energy
  dielectric (or electromagnetic), 355, 357
  electric (or conductive), 355, 356
  field applications, 359–360
  radio-frequency, 357

Heat losses
injection wells, from, 289
surface lines, from, 288, 289
surrounding formations, to, 293
Heat of combustion, 321
Heat transfer
coefficient, 288
conduction, by, 278
Heating value of fuel (*see* Heat of combustion)
Heterogeneity, 27, 44, 387
Horizontal drilling, 305
Horizontal wells, 36, 390, 395–397
Hot water flooding–Hot water drive, 295
field applications, 295
Huff and Puff (*see also* Steam stimulation), 234
Huggins constant, 50, 52
Hydrodynamic chromatography (*see* Chromatography)
Hydrogen/carbon ratio (see Atomic hydrogen/carbon ratio)
Hydrogen sulfide (H_2S), 237, 251, 253, 344, 354
Hydrolysis (polymer degradation by), 64
Hydrolysis degree, 47
Hysteresis, 18
on relative permeability (by CO_2 flooding), 235

Ignition, 335
Immiscible fluid displacement, 10
Inaccessible pore volume, 74
Infill drilling, 386
In-situ combustion, 313
field applications, 335–336
monitoring, 336
performance, 335
with oxygen, 320, 331, 336
Instability (viscous), 74
Interfacial tension, 12, 90, 280, 348
low (lowering) IFT, 98, 114, 127, 129, 233
measurement, 99
transient, 130

Inverted micelles, 93
Ion exchange, 138
cation exchange capacity (CEC), 110, 135, 139
hydrogen exchange capacity (HEC), 139
total exchange capacity, 139
Irreducible water saturation (effect of temperature), 282

Kinetic models
mineral dissolution by alkalis, 143
chemical reaction (in in-situ combustion), 263–265
Kinetics of retention (polymer flooding), 61–62
Krafft point, 93

Laplace's formula, 12
Latent heat of vaporization, 278
Lines (*see* Contact lines, Surface lines)
Liquids
heat capacity, 277
thermal conductivity, 280
thermal expansion, 276
viscosity, 275
Low temperature oxidation (LTO), 259, 263, 265

Macromolecules
conformation, 48, 49
diffusion, 51, 74
flexibility, rigidity, 48
modeling, 48
persistence length, 49
radius of gyration, 48, 74
relaxation time, 51, 52
stretching, 53
Marx and Langenheim model, 293
Mass balance, 16
Mass transfer, 106, 186, 225
MEOR (*see* Microbial enhanced oil recovery)
Mesophase, 95

Metabolic products, 343
Metabolites (*see also* Byproducts), 352
Metals (*see* Catalytic effect)
Micellar solution, 93
  viscosity, 100
Micelles, micellization, 93
Microbes (*see also* Bacteria), 342
Microbial cells, 344
Microbial enhanced oil recovery, 341
  economics, 355
  enhanced waterflooding, 351
  field applications, 351–352
  profile modification treatment, 351
  screening criteria and design considerations, 352, 354
  single-well treatment, 350
Microemulsions, 95
  stability, 103
  viscosity, 100
Microgels, 45
Microorganisms (*see also* Bacteria), 342
Middle phase, 95
Migration (primary and secondary), 4
Mineral matrix, 244, 252
Minimum miscibility composition (MMC), 200
Minimum miscibility presure (MMP), 190, 228
  determination, 195, 230
Mining (*see* Oil mining)
Miscibility, 205, 225
  dynamic, 186
  first contact, 186, 205, 228
  minimum miscibility composition (MMC), 200
  minimum miscibility pressure (MMP), 190, 228
  multiple contact, 186, 190, 222
  partial, 20, 233
  thermodynamic, 24
  total, 19
Miscible flooding (field applications), 211–212
Miscible fluid displacement, 20, 157
Miscible transition zone, 186
Mobility, relative mobility, 25, 105
Mobility buffer, 117
Mobility control, 78, 105, 344
  WAG process (*see also* Water alternate gas injection), 236
  effect in miscible gas flooding, 207
Mobility ratio, 25, 30, 106, 235, 282
Mobility reduction, 69
Modeling
  in-situ combustion, 333
  hot fluid injection (*see* Analytical models)
  polymer flooding, 79–82
  reservoir, of, 385–386
  surfactant flooding, 112–113
  miscible gas flooding (*see* Equations of state)
Models (*see* Analytical models)
Molasses, 342, 350

Nitrogen, 186, 189
Non-Newtonian behavior (micellar solution), 100
Non-wetting fluid phase, 12
Nuclear explosives, 364, 365
  cavity and chimney, 365
  field applications, 365–370
Numerical models (*see* Modeling)
Nutrient, 342, 344

Oil
  chemical changes, 244, 249, 280
  saturation evaluation, 385
  thermal reactivity, 254
  upgrading, 251
Oil bank, 104
Oil fraction, 249
Oil mining
  field applications, 362–363
  safety and environmental aspects, 363
  surface mining (or open-pit mining), 361
  underground mining, 361
Oil/steam ratio
  definition, 298
  for steam drive operations, 301, 303

Oil zone, 4
Open-pit mining (*see* Oil mining)
Optimum salinity, 97, 115
Organosulfur compounds, 251
Overburden and underburden (heat loss), 293, 294
Oxidation, 244, 246, 258, 262
Oxidizing-gas
  flow rate program, 334
  flux, 320, 322, 334
  requirement, 318–320, 321
  storage, 320, 330
Oxygen
  requirement, 329
  utilization, 319, 332

Paraffins (precipitation by CO_2), 237
Paraffin deposits, 351, 356
Partial miscibility (*see* Miscibility)
Partitioning (*see also* Pseudocomponents), 96, 101
Partition coefficient, 177
Pattern (*see also* Pilot test), 391, 395
Peclet number, 211
Peng–Robinson equation of state, 168
Permeability (*see also* Absolute permeability, Relative permeability), 15, 79, 225, 253
Permeability reduction, 71, 78
Peroxides, 63
Phase behavior, 96, 159, 216
  representation by equations of state, 162
  surfactant-containing system modeling, 102
pH (*see also* Alkali)
  polymer adsorption, and, 59
  polymer stability, and, 64
  surfactant adsorption, and, 109
  surfactant stability, and, 93
Pilot test (*see also* Field applications)
  design, pattern, 392
  methodology (*see also* EOR projects-flow chart), 391, 394
  monitoring and interpretation, 393–394
Plait point, 97
Plugging (*see* Selective plugging)
Polyacrylamide (PAM, HPAM), 45
Polydispersity index, 47
Polymers, water soluble polymers, 44, 342
  adsorption, 57, 60
  compatibility with alkaline reagents, 137
  compatibility with surfactants, 102, 110
  precipitation, 48
  retention (*see* Retention)
  stability (*see* Polymer stability)
  structures and molecular weights, 46–48
Polymer floodings, 35, 44
  field applications, 82–84
  modeling, 79–82
Polymer solution
  rheology, 51
  Newtonian regime, 52, 70–73
  shear flows, shear rates, 52
  shear-thickening regime, 53, 67
  shear-thinning regime, 52, 68
Polymer stability, 63–67
  biological, 66
  chemical, 63
  mechanical, 65
Porphyrins, 129
Precipitates
  carbonate, 135
  silicate, 135
Precipitation (asphaltenes), 237
Pressure (capillary), *see* Capillary
Pressure
  critical pressure, 164, 217
  minimum miscibility pressure (*see* Miscibility)
Primary recovery (*see* Recovery)
Profile modification treatments, 351
Pseudocomponents (*see also* Partitioning), 97, 101, 176–179, 187
Pseudophase model, 103
Pyrobitumen, 249
Pyrolysis, 244, 245, 246, 255

Radial diffusion (*see* Diffusion)
Radio-frequency heating (*see* Heating)
Radius of gyration (*see* Macromolecules)
Random walk model, 48
Rate
  dissolution, 145
  equation, 146
  reaction, 142
Ratio of heat loss per unit length
  surface lines, on, 286, 289
  wells, in, 289
Reaction order, 256, 267
Recovery (enhanced), 9
Recovery (primary), 6
Recovery (secondary), 6
Recovery (tertiary), 9
Redlich–Kwong equation of state, 168
Relative permeability, 16, 395
  effect of temperature, 282
  equations, 15, 16
Relaxation time, 51
Repressurization, repressurizing, 342, 349
Reservoir, reservoir rock, 3
  architecture, 400
  characterization, 382–385, 400
  modeling, 385–386
Residual oil saturation, 395
  effect of temperature, 282
Resins, 129, 248, 255
Retention (*see also* Adsorption)
  hydrodynamic retention of polymers, 56, 61
  mechanical entrapment of polymers, 56, 60
  surfactants, 107
Rock matrix
  catalytic effect, 326, 331
  surface area, 326, 330
Rubble zone, 365

Salinity (optimum), 97, 115
Salinity gradient, 115–117
Saturation (fluid), 16
Scales, 135
Scleroglucan, 48
Screening criteria for EOR projects (*see* EOR projects)
Secondary recovery (*see* Recovery)
Segregation (*see* Gravity override)
Selective plugging, 342
Self avoiding walk model, 49
Shafts, 360, 361
Shear flows, shear rates (*see* Polymer solutions)
Shock wave, 364
Shotpoint, 364
Silicate (ortho and meta), 127
Simonet–Behar–Rauzy equation of state, 169
Single-well heating process, 359
Slip (no slip condition), 10
Soave–Redlich–Kwong equation of state, 168
Sodium hydroxide, 127
Solubilization, 93
  interfacial tensions, and, 98
  parameter, 97, 114
  R-theory, 95
Spacing, 395
Specific gravity (crude oils), 127, 220, 330
Spinning drop method, 99
Spore formers (*see* Bacteria)
Steam drive, steamflooding, 301
  field applications, 303
  modeling, 302
  performance, 303
Steam injection
  additives, with, 305
  continuous (*see* Steam drive)
  cyclic (*see* Steam stimulation)
  gas, with, 305
  slug size in cyclic, 298, 300
  underlying water zone, into an, 305
Steam properties, 279
Steam quality, 298
Steam soak (*see* Steam stimulation)
Steam stimulation, 298
  field applications, 300
  modeling, 299
  performance, 299
  soaking period, 298, 300

Stimulation, 350
Stokes' law, 150
Strains, 354
Stripper wells, 350
Sulfonates, petroleum sulfonates, 91
Sulfonation degree, 91
Superwet combustion, 318, 333
Surface activity, 93
Surface lines (heat losses from), 289
Surfactants, 90, 235, 305, 342
  adsorption, 102, 109
  compatibility with polymers, 102, 110
  foams, for, 235, 305
  partitioning, 96
  precipitation, 102, 107
  retention, 107, 108
  stability, 93, 136
  structures and molecular weights, 91–93
Surfactant flooding
  field applications, 117–119
  modeling, 112–113
Surfactant solution
  modeling, 102–103
  phase behavior, 96–98
Sweep efficiency, 9, 35
  areal (or horizontal), 25
  CO_2 flooding, in, 234–236
  hot fluid injection, in, 303–304
  in-situ combustion, in, 334–335
  miscible gas flooding, in, 212
  polymer flooding, in, 76, 77–79
  vertical, 26
Swelling (*see also* swollen phase), 198
  oil, 216, 220, 223, 233
  clay, 127
Swelling factor, 233
Swollen phase, 107

Temperature
  and oil viscosity (ASTM diagram), 275
  critical temperature, 164, 217
  effect on permeability, 282
  LTO (*see* Low temperature oxydation)
Temperature—*contd.*
  of combustion front (*see* Combustion front)
Tension (*see* Interfacial tension)
Tertiary recovery (*see* Recovery)
TGA (*see* Thermogravimetric analysis)
Thermal conductivity, 278, 280
Thermal expansion
  crude oils, of, 276, 280
  solids, of, 276, 281
  water, of, 276, 280
Thermal methods, 36, 241
Thermogravimetric analysis, 257, 265, 325
Tracers, 393
Transition zone
  flow tube, in a, 35
  immiscible fluid displacement, in, 22
  miscible fluid displacement, in, 23
Translation factor, 173, 175
Tunnels, 360, 361

Underground combustion (*see* In-situ combustion)

Van der Waals equation, 162
Vaporization (crude oil), 197, 216, 225, 283
Vaporizing gas drive (*see also* Condensing gas drive), 186
Visbreaking, 248
Viscosity
  definitions, 50–54
  effective, 69
  interfacial, 11
  gas, of, 218, 272
  liquids, of, 127, 275
  microemulsions, of, 100
Viscosity ratio, 281
Viscosity reduction, 216, 220, 233, 342, 355
Viscous fingering, 235
Viscous instability, 74
Volume translation, 172–175

WAG *see* Water alternate gas injection
WAR (*see* Water/air ratio)
Water
  density, 276
  enthalpy, 279
  heat capacity, 278
  latent heat of vaporization, 278
Water alternate gas injection (WAG process), 35, 212, 222, 235
Water gas shift reaction, 251
Water-fireflooding, 316–318, 333, 335
Water/air ratio (for wet combustion), 322, 324, 333
Water/gas ratio, 235
Water/oil ratio (effect on surfactant phase behavior), 101
Water soluble polymers (*see* Polymers)
Well
  horizontal (*see* Horizontal wells)
  injection or production
Wet (preferentially-, totally-), 12
Wet combustion
  compared with dry combustion, 332, 333, 335
  laboratory tests, 331–333
  principle, 316–318
  water/oxidizer ratio in, 322–324
Wettability, 11, 90, 280, 282, 348
  alteration, change, 107, 123, 140
  dependency on pH, 141
  reversing of, 124, 140
Wetting fluid phase, non-wetting fluid phase, 12
Winsor (*see also* Solubilization (R-theory))
Winsor, types I, II, III and IV, 93–95

Xanthan (gum) 45, 349

Young's equation, 12